Pioneers of the U.S. Automobile Industry

Michael J. Kollins

Other SAE books on this topic:

Ford: The Dust and The Glory, Volumes 1 and 2
by Leo Levine
Order No. R-292.SET

The E-M-F Company
by Anthony J. Yanik
Order No. R-286

The Franklin Automobile Company
by Sinclair Powell
Order No. R-208

For more information or to order this book, contact SAE at
400 Commonwealth Drive, Warrendale, PA 15096-0001
(724) 776-4970; fax (724) 776-0790
e-mail: publications@sae.org
web site: www.sae.org/BOOKSTORE

Pioneers of the U.S. Automobile Industry

Volume 2: The Small Independents

Michael J. Kollins

Society of Automotive Engineers, Inc.
Warrendale, Pa.

Library of Congress Cataloging-in-Publication Data

Kollins, Michael J.
    Pioneers of the U.S. Automobile Industry / Michael J. Kollins
      p. cm.
    Includes bibliographical references and index.
    Contents: v. 1. Big three -- v. 2. The small independents -- v. 3. The financial wizards -- v. 4. Design innovators.
      ISBN 0-7680-0904-9 (set) -- ISBN 0-7680-0900-6 (v.1) -- ISBN 0-7680-0901-4 (v.2) -- ISBN 0-7680-0902-2 (v.3) -- ISBN 0-7680-0903-0 (v.4)
    1. Automobile engineers--Biography. 2. Automobile industry and trade--History. 3. Automobiles--History. I. Title

TL139 .K65 2001
338.7'6292'0922--dc21
[B]
                                                              2001049584

Editing, design, and production by Peggy Holleran Greb
301 Brookside Road
Baden, PA 15005

Copyright © 2002 Society of Automotive Engineers, Inc.
400 Commonwealth Drive
Warrendale, PA 15096-0001 U.S.A.
Phone: (724) 776-4841
Fax: (724) 776-5760
E-mail: publications@sae.org
http://www.sae.org

ISBN 0-7680-0901-4 (Volume 2)
ISBN 0-7680-0904-9 (set)

All rights reserved. Printed in Canada.

Permission to photocopy for internal or personal use, or the internal or personal use of specific clients, is granted by SAE for libraries and other users registered with the Copyright Clearance Center (CCC), provided that the base fee of $.50 per page is paid directly to CCC, 222 Rosewood Dr., Danvers, MA 01923. Special requests should be addressed to the SAE Publications Group. 0-7680-0901-4/ $.50.

SAE Order No. R-251

Preface

A hundred years ago a trip by automobile was as much a test of manpower as of horsepower; "Men had to be men." In those days, "Get out and get under" (the song was not composed then) had a direct meaning to the adventuresome, soiled, and grease-stained motorist chauffeurs. Happily, these crude, cumbersome, horseless carriages are no more. Here and there a restored one may be found, hidden among the array of glistening new vehicles of modern achievement. Those pioneer vehicles were in fact as naive as the ancient chariots of Egypt and Rome.

The early horseless carriages were big, heavy, uncomfortable, noisy, and smelly wagons or carriages, powered by engines having huge cylinders that were gluttons for fuel. Or they were small, fragile, uncomfortable, noisy, and smelly buggies, powered by small engines, hardly big enough to propel the buggy. Clanking chains rotated the rear wheels, while noisy engines dripped oil like a sieve. They emitted billowing clouds of smoke, as the drab vehicles trembled on wobbly wheels that seemed ready to collapse.

Although the average asking price for one of these "headaches on wheels" was $1,000 or more, the greatest expense came later for maintenance and repairs. These vehicles were plagued with engine, clutch, transmission, steering, brake, wheel, and fuel troubles, let alone problems from the weather. The cost of broken and worn-out parts greatly exceeded the cost of operation. Axle shafts fractured, universal joints failed, crankshafts broke or scored, pistons cracked, cylinders scored or wore rapidly, connecting rods broke, bearings burned out, clutches slipped, transmission gears stripped and chattered. Adjustments and overhaul procedures were common operating procedure. The cost of replacement parts was high because of the lack of standardization and volume.

These strange and crude-looking vehicles spit, coughed, belched, groaned, backfired, and stalled unexpectedly. They were the source of distrust, despair, doubt, ridicule, and embarrassment to their owners. It was soon quite obvious that the smelly, noisy, imperfect, and expensive automobile needed much refinement and performance proof to convince a skeptical public that it was a viable alternative to travel by horse. Fortunately, the novelty, rarity, or scarcity attracted enough buyers to keep some manufacturers in business, while the brilliant minds of these stalwart men worked to solve the problems, and to regain the public's confidence, despite the negatives.

These vehicles were the direct ancestors of our modern-day cars, and from their trials, failures, and successes, came knowledge and improvements. It is the purpose of this publication to acknowledge the accomplishments, give credit to, and honor those various, selfless individuals who risked all their possessions and toiled to acquire a better means of transportation, which has led to a better and fuller life for all Americans.

Specifications

The specifications and other pertinent information in this publication may differ from other contemporary automotive publications. The data and prices shown were taken from official records and/or automobile journals of the time.

The prices as shown are based on the lowest-priced runabout as listed in the ALAM (Association of Automobile Manufacturers) through 1911. From 1912 through 1924 the price shown is for the five-passenger touring car as listed by the NACC (National Chamber of Commerce). From 1925 on, the prices listed were the advertised price for the five-passenger sedan model.

Prior to 1911 the horsepower ratings listed were the ALAM rating, calculated by squaring the engine bore size in inches, multiplying by the number of cylinders, and dividing by 2.5. (The stroke was not considered.) From 1912 through 1924 this responsibility was taken over by the NACC, and the same calculation was used. Beginning in 1925, the horsepower rating was taken over by SAE. However, most manufacturers preferred to use the dynamometer brake horsepower in their publications, referred to in this book as the advertised (adv) horsepower.

The sales volume amounts were taken from the official R.L. Polk Company published records.

Acknowledgments

The author acknowledges the help, cooperation, friendship, encouragement, inspiration, and patience of those who helped to acquire and accumulate the research documents, photographs, and all related information. First of all, I want to thank my wife, Julia A. Kollins, for the patience and encouragement of such an enormous achievement. Also, I thank our children: Michael L. Kollins, Richard P. Kollins, and Laura G. Kollins, for their patience and doing without during this project. Thanks to our grandchildren, namely Monika Kollins, Chloe Beardman, and Fiona Beardman, who gave the inspiration to push ahead.

The author expresses his gratitude for the patience, friendship, teaching, and guidance of his mentors. Among these are his friends, classmates, teachers, professors, journeymen leaders, and the chief engineers, presidents, department directors, and other executives of the automobile manufacturers. These individuals made it possible for the author to obtain the hands-on experience and knowledge of automobile design, engineering, manufacture, and merchandising of automotive products.

Next, the author expresses his thanks to the many libraries, archives, museums, historical societies, and the personal collectors who were willing to share their historical material, data, and related documents. A special thanks to the directors of historical organizations, their curators and librarians, who took a sincere personal interest in this publication and unselfishly gave their assistance.

Due to the enormous size of this publication, it is humanly impossible to include all persons engaged in the automotive industry. However, a sincere effort is made to include the information pertinent to the history of its founding, development, and growth. The author sincerely apologizes to those who may have been omitted.

Individuals

Roy Abernathy
Paul Ackerman
Rodney F. Ackerman
Fred W. Adams
Alex Adastik
George Adastik
Terrence Adderle
Chester Advent
Freddie Agabashian
Mary Alexander
Robert H. Aller
David Andrea
Charlotte Andres
Emil Andres
Mary Atkinson

Lawrence Ball
Jeremy Ball
John Ball

Thomas Ball
Clay Ballinger
John Barnes
Paul Barrett
Oliver E. Barthel
Chloe Beardman
Fiona Beardman
Harold Beardsley
Maurice Beardsley
Mary Benda
John Benedict
Mary Lou Bennethum
William Bennethum
J. Gordon Betz
Albert Bizer
William Bizer
Albert Bloemker
Dushon Boich
Henry Scripps Booth

Max Booth
William Boyd
Albert Brandt
Eric Braun
Helen Braun
Melbourne Brindle
Frank Brisko
Dorothy Broomhall
Gregg Buttermore

Eugene (Jep) Cadou
Dorothy Caldwell
Eula Caldwell
Russell Caldwell
Thomas Cahill
Edna Callahan
Thomas Carnegie
Edna Carpenter
Jack Carson

Russell Catlin
Eugene Cattabiani
James Chaskel
George T. Christopher
Harold Churchill
James Circosta
Ralph Circosta
Rose Circosta
Harley Citron
Joseph J. Cole
LeRoi Cole
George Collick
John A. Conde
William C. Conley
Goerge Connor
Elizabeth Cook
Donald Cummins
Lyle Cummins
Edward P.J. Cunningham
Michael Cunningham
George T. Cutler
Ronald Cutler
James Cypher

Arthur Dau
Donald Davidson
Michael W.R. Davis
Roy Dean
Jack Decker
Raymond Denges
Peter DePaolo
Thomas P. DePaolo
Robert Derleth
Frank DeRoy
John Dovaras
Denny Duesenberg
Fritz Duesenberg
Gertrude Duesenberg
Harlan Duesenberg
Len Duncan
Zora Arkus Duntov
Reeves Dutton
Ralph Dunwoodie
Nicholas Dyken

Helen J. Early
Christopher Economaki
John Eichorn
Donald Eller

Charlene Ellis
Frances Emmons
Erik Erikson
Harold Estelle
William Evans Sr.
Donald Everett

Matthew Fairlie
Harlan Fengler
J. Robert Ferguson
Hugh Ferry
William Finley
Milton Forester
Arthur W. Fowler
Anton J. Foyt
Roy H. Frailing
Kay Frazier
Howard P. Freers
Norman Froelich
Barbara Fronczak
John A. Fugate

George Galster
James (Radio) Gardner
Jerry E. Gebby
Joseph P. Geschelin
Howard Gilbert
Jack Gilmore
Jeffery Godshell
Reneé (Chevrolet) Goeke
David R. Graham
Mitzi Graham
Ronald Grantz
William H. Graves
Karl M. Greiner
Mrs. C. Grimes
Louis (Jerry) Grobe

Roscoe Hadley
Theodore Halibrand
Ira Hall
Samuel Hanks
Robert S. Harper
John J. Harris
Jack Harrison
Harry Hartz
Leslie R. Henry
Franklin Hershey
Jack Hinkle

Hugh W. Hitchcock
Wilma (Ramsey) Hoffman
Gertrude M. Hope
William Hope
Lindsey Hopkins
Raymond House
Ronald Householder
Herbert Howarth
John Hranko
Howard Hruska
Wallece S. Huffman
Frederick Hunty

Helen Cole Imbs
Ralph H. Isbrandt

J.A. Jacks
Frank Jenkins
Marvin Jenkins
Louis Jilbert
Mary Johnson
Myrtle Johnson
Raymond Johnson
Patricia Jones
Charles Jordan
Roam Jordan
Steve Jordan

August Kapecky
Joseph Karschner
John Kay
Ralph Kegler
Kaufman T. Keller
Nancy Kennedy
Mildred Kernen
Russell S. Kettlewell
Lester Kimbrell
Richard Kirchner
Ward Kirkman
James Kirth
Manuel Klasky
Nathan Klasky
Albert Kline
Julia A. Kollins
Laura G. Kollins
Michael L. Kollins
Monika Kollins
Richard P. Kollins
Lee Kollins
Ada Knight

Acknowledgments

Theodore Lake
Robert Lakin
James Lamb
H.G. Langston
Ben Lawrence
Myron Lawyer
Robert Laycock
P.R. Leatherwood
Harry LeDuc
Matthew Lee
Donald Leeming
Shirley Leeming
Robert H. Lees
Lugjie Lesovsky
David Lewis
Robert A. Lindau
Jim Locke
Joseph D. Lovely
George Lucas
Helen Lucas
Steve Lyman

Claude T. McClure
J. Kelsey McClure
Alvan Macauley
Edward R. Macauly
Raymond Macht
Casmere Maday
Joseph Madek
Freddie Mangold
John Mangold
Jack L. Martin
William Marvel
John Matera
James Melton
Anthony Merchell
Lawrence Merchell
John C. Merchanthouse
Marshall Merkes
William Meyer
Louis Meyer Sr.
Louis Meyer Jr.
Arthur Meyers
Walter Meyers
Jack Miller
Milford Miller
John W. Mills
Thomas Milton Sr.
Thomas W. Milton III

Herbert L. Misch
Guy Monahan
George Moore
Wilma (Ramsey) Musselwhite

Dennis (Duke) Nalon
Adelia Newhouse
Frank Newman
Byron Nichols
John F. Nichols
John Notte III
Robert G. Nowicke

William O'Brien
Rebecca Obsniuk
George J. Oeftger
Mary Lou Oles
William Opdyke
Patrick O'Riley
Cletus O'Rourke
Edward O'Rourke
Carol Osborne
Bruce Owens

Andrew Pardovich
Johnny Parsons
Clyde R. Paton
Eleanor Paton
Mark Patrick
U.E. (Pat) Patrick
Manfred Paul
Johnny Pawl
Curtis Payntor
Howard Pemberton
Marcel Periat
Charles Phaneuf
Frank Plovick
Thomas Poirer
Carl Preuhs
Vera Prindle

Earl Ramsey
William Ramsey
Louis Rassey
Philip Renick
John J. Riccardo
Maurice Rice
John Rinehart
Rita Rittenhouse

Frank Rose
Ross Roy
Frank Russell
August Russo
Robert Russo
Johnny Rutherford

Harry St. Bernard
Aaron Sawyers
Nell Sawyers
Richard Scharchburg
Edward Schipper
Harold Schram
Carmen Schroeder
Gordon Schroeder
Paul Scupholm
Johnny Shannon
William (Wilber) Shaw
Lawrence Shinoda
John Shipman
Igor Sikorsky
Albert Silver
Powell Sinclair
Michael Sizemore
Joseph Slomski
Ruth Smiley
Earl H. Smith
Jack Smith
John Martin Smith
Mike Smith
Peggy Smith
William J. Smyth
William Snively
John Snowberger
Russell Snowberger
Arthur Sparks
William D. Sparks
Tony Spina
Thomas Spragle
Stanley Steiner
Myron Stevens
Peggy Stevens
Raymond Stevens
Steven Stevens
Russell C. Stone
Frank E. Storey
Robert A. Stranahan
Gary M. Stroh
John W. Stroh

Russell Sutton
Peggy Swalls
Robert P. Swanson
Theodore Swiontek

William F. Taylor
Richard Teague
Robert Tebelman
Theodore P. Thomas
David Thompson
Richard Thompson
Thomas Thompson
Donald Thornber
Milton Tibbetts
Michele Tinson
Alexander Tremulis
Floyd Trevis
Robert Trobek
Roy Trowbridge

Bobby Unser
Roy Utley

Charles Van Acker
Eleanor Van Dyne
Leland Van Dyne
Leonard Van Dyne
Andrew Vesely
Joseph Vichich
Patrick Vidan
Robert Vieth
Jesse G. Vincent
Anthony Viviano
Donald Vnasdale
Rolla Volstedt
Axel Von Bergen
Eric A. Von Braun
Magnus Von Braun

Charles R. Wade
Richard Wallen
Roger Ward
Marsden Ware
A.J. Watson
Ronald Watson

I. Milton Watzman
A.W. Webb
Reuben Webb
Erwin A. Weiss
Emerson R. White
Roger H. White
R.L. Williams
Lester Wood
James Wren
Lester Wright

Anthony Yanick
Wilson A. York
Fred M. Young
Louis Young
Bruce Youngs
Smokey Yunick

Steven Zdanski
Susan Zemmens
Richard J. Zimmer
John S. Zink
Raymond Zink

Publications Used in Research

Automobiles of America by Automobile Manufacturers Incorporated, 1962
American Cars, 1805-1942
American Machinist, November 7, 1895
Automobile, Volume I, September 1899, through Volume II, September 1901
The Automobile, Volume VIII, January 1903, through Volume XXXVII, July 1917
Automobile & Automotive Industries, Volume XXXVII, July 1917, through Volume XXXVII, October 25, 1917
Automotive Engineering, SAE, 1952 through 1999
Automotive Industries, Volume XXXVII, November 1, 1917, through Volume LXXVII, December 1942
Automobile & Motor Review, June 1902
Automobile Topics, Volume I #1 through Volume CCLIV, December 1942
Automobile Trade Journal
Automotive News, 1926 through 1999
Cars With Personalities, by John A. Conde, October 1, 1982
Carriage Monthly, 1890 through 1902
Chilton Catalogue and Directory
Chronicle of the Automobile Industry, Automobile Manufacturers Association, 1893 through 1949
Commercial Car Journal, 1948
The Complete Motorist, by Elwood Haynes, 1913-1914
Cycle & Car, 1895 through 1905
Detroit Free Press, 1928 through 1970
Detroit News, 1928 through 1999
Detroit Saturday Night, 1946-1947

Detroit Times, 1928 through 1950
Duesenberg—Mightiest American Motor Car, by J.L. Elbert, 1950
Horseless Age, Volume I #1, January 4, 1902, through Volume XLIV #3, May 1, 1918
Motor Magazine, 1903 through 1926
Motor Age, Volume I #1, October 4, 1900, through Volume CI, July 1940
Motor World Wholesale
R.L. Polk Company, 1921 through 1972
SAE Journal, 1952 through 1999
SAE Transactions, 1952 through 1999
SAE Roster, 1952 through 1999
Scientific American, 1893 through 1999
Smithsonian Institution Bulletin #198
The Story of Speedway, Indiana, Speedway Civic Committee, 1976
History of the Studebaker Corporation, by A.R. Erskine, 1852-1923
Ward's Automotive Quarterly, 1965
Ward's Auto World, 1965 through 1999

Archives, Libraries, and Museums Visited by Personal Contact or Correspondence

Ann Arbor Public Library, Ann Arbor, Michigan
Auburn-Cord-Duesenberg Museum, Auburn, Indiana
Bendix Historical Archives, South Bend, Indiana
Beloit Public Library, Beloit, Wisconsin
Bloomfield Hills Library, Bloomfield Hills, Michigan
Bridgeport Public Library, Bridgeport, Connecticut
Buick Motor Division Archives, Flint, Michigan
Cadillac Motor Car Division Museum, Detroit, Michigan
Carnegie Library of Pittsburgh, Pittsburgh, Pennsylvania
Case Western Reserve Institute Archives, Cleveland, Ohio
Chrysler Corporation Museum, Auburn Hills, Michigan
Children's Museum, Indianapolis, Indiana
Cincinnati Public Library, Cincinnati, Ohio
Cleveland Public Library, Cleveland, Ohio
Cornell University Library, Ithaca, New York
Crawford Automotive and Aircraft Museum, Cleveland, Ohio
Detroit Historical Museum Archives, Detroit, Michigan
Detroit Public Library, Detroit, Michigan, including Burton Biographical Collection, Legal and Patents Collection, and National Automotive Historical Collection
Elkhart Public Library, Elkhart, Indiana
Flint Public Library, Flint, Michigan
General Motors Institute Foundation Archives, including W.C. Durant Papers, Kettering University Archives, C.S. Mott Papers
Henry Ford Museum Archives, Dearborn, Michigan
Freeport Public Library, Freeport, Illinois
Goshen Public Library, Goshen, Indiana
Grosse Pointe Historical Society, Grosse Pointe, Michigan
Hartford Public Library, Hartford, Connecticut
Elwood Haynes Historical Museum, Kokomo, Indiana

Indianapolis Motor Speedway Museum, Indianapolis, Indiana
Kalamazoo Public Library, Kalamazoo, Michigan
Kenosha Public Library, Kenosha, Wisconsin
Milwaukee Public Library, Milwaukee, Wisconsin
New Haven Public Library, New Haven, Connecticut
New Jersey Historical Society, Newark, New Jersey
Newman and Altman Museum, South Bend, Indiana
Novi Motorsports Hall of Fame Museum, Novi, Michigan
Ohio State Library, Columbus, Ohio
R.E. Olds Museum, Lansing, Michigan
Packard Motor Car Company Engineering Records, Detroit, Michigan
Packard Museum, Warren, Ohio
Racine Public Library, Racine, Wisconsin
Rockford Public Library, Rockford, Illinois
Rose-Hulman Institute, Terre Haute, Indiana
South Bend Public Library, South Bend, Indiana
Studebaker Historical Museum, South Bend, Indiana
Syracuse Public Library, Syracuse, New York
University of Toledo Library, Toledo, Ohio
Vigo County Library, Terre Haute, Indiana
Warren-Trumbull County Library, Warren, Ohio
Willard Library, Battle Creek, Michigan
Youngstown Packard Museum, Youngstown, Ohio

Contents

Introduction 1

Charles and Frank Duryea 3

Studebaker 17

The Pratt Family and the Elcar Motor Car Company 53

Joseph Moon 67

Russell Gardner 85

Louis Clarke 93

George Pierce and Charles Clifton 107

Packard/Joy/Macauley and the Packard Motor Car Company 129

Edwin Thomas 197

Ransom Olds 211

Peerless 249

Fred and August Duesenberg 273

Kissel Brothers 297

Hupp/Drake/Hastings/Young and the Hupp Motor Car Corporation 307

Walter Flanders 333

Chapin/Coffin/Bezner/Jackson/Hudson/McAneeny and The Hudson Motor Car Company 347

Harry Stutz 375

Harry Ford 401

Graham Brothers 407

Charles Nash 427

Index 461

About the Author 473

Contents of Other Three Volumes

Volume 1 – The Big Three

Louis Chevrolet
Walter Chrysler
Dodge Brothers
William Durant
Henry Ford
General Motors Corporation

Volume 3 – The Financial Wizards

Allison/Fisher/Newby/Wheeler and the Indianapolis Motor Speedway
Benjamin Briscoe
Hugh Chalmers
Frederick Chandler
E.L. Cord
Harry Jewett
Henry Leland
Charles Matheson
David Parry
Albert Pope
Edward Rickenbacker
Thomas White
John Willys

Volume 4 – The Design Innovators

Elmer and Edgar Apperson
Vincent Bendix
James Scripps Booth
Alanson Brush
David Buick
Joseph Cole
Clyde Coleman
Claude Cox
Herbert Franklin and John Wilkinson
Elwood Haynes
Frederick Haynes
Thomas Jeffery
Edward Jordan
Charles King
Howard Marmon
Jonathan Maxwell
Percy Owen
Raymond and Ralph Owen
Andrew Riker
Frank Stearns
Thomas J. and Thomas L. Sturtevant
C. Harold Wills
Alexander Winton

Introduction

The purpose of this publication is not so much to furnish statistics and technical information on automobiles, as these can be found in libraries and many technical publications, but to expose the warm, human, compassionate, and romantic relationship of the persons involved and the products of their labors.

The normal, restless nature of man is such that he is never satisfied with things as they are. Therefore, in his quest for perfection, he has brought about evolutionary changes in transportation methods, not by need alone, but by the compassionate relationship between man and machine.

The historical (chronological) data is furnished primarily for the purpose of establishing the background, trend of the times, and the environment in which these stout-hearted men developed their marvelous machines. The evolution of mobility can be traced to the invention of the wheel, several millenia before the days of Caesar, to make the movement of large masses of material possible.

The first recorded appearance of a self-propelled land vehicle occurred in France in 1769, when Nicholas Cugnot, a French army captain, built a self-propelled, steam-powered tractor to move artillery field caissons.

In Redruth, England, in 1784, an assistant to James Watt (and against his objections) built a three-wheeled land vehicle powered by a high-pressure steam engine. In 1801 Richard Trevithick, with the assistance of André Vivian, built a rear-engine steam-powered carriage, capable of carrying twelve passengers at speeds up to 9 mph. By the late 1820s, with improved economy and road conditions, many steam carriages appeared on England's roads, many large enough to carry 20 passengers, capable of speeds up to 15-20 mph.

Steam propulsion started to create interest in the United States in 1805 when Oliver Evans moved his massive dredge "Orukter Amphibolus" through the streets of Philadelphia and down the banks of the Schuylkill River. The steam-powered driving wheels gave way to paddle wheels that propelled the dredge upstream to the dredging location.

Steam tractors had a great fascination for the author, because even at the age of four, he developed an insatiable love for these mechanical marvels. The most exciting days of his life were the days when the steam tractors, belching smoke from their smoke stacks, would pull the thresher and baler down the lane to the grain stacks and the barn on the farm. Steam tractors performed a remarkable service to man, in the agricultural fields as well as the railroads. Stationary steam power plants provided a source of power for lumbering, pumping, electrical generation, and marine propulsion uses.

The extensive use of self-propelled personal vehicles did not occur until after the discovery and refinement of gasoline in 1859, and the invention of the hydrocarbon-fueled internal-combustion engine. Ironically, the internal-combustion engine principle was invented almost a century before James Watt developed his steam engine.

In 1680 Christian Huygens, a Dutch scientist/astronomer, working on a proposal by Father Jean deHautefeville of Versailles, France, for a device to pump water, used gunpowder for the explosive driving force, acting directly to drive a piston. Steam was used only as a cleaning agent to clean the cylinder after the explosion. Gasoline was unknown at that time. Work was continued by Dionysius Papin and Thomas Newcomen. In 1804 Isaac DeRivaz, a Swiss engineer, was able to obtain a driving force using hydrogen, and built an engine accordingly. It had been said that DeRivaz adapted this engine to a vehicle.

Pioneers of the U.S. Automobile Industry

In 1860 Etienne Lenoir, a Frenchman, developed the first practical gas-fueled internal-combustion engine. Lenoir used coal lighting gas for fuel, and ignited it with a spark from an induction coil. In 1862 Lenoir built a vehicle using this engine, and drove it on the roads in France.

While many had made attempts to discredit Lenoir, even a German publication, *Zur Frage der Freien Concurrenz im Gasmotorenbaue*, published in 1883, acknowledged the existence of the Lenoir engine and vehicle. In the famous Selden Patent case against Henry Ford, a working model of Lenoir's vehicle was built to prove antecedence, and the case was won by Henry Ford.

During this period in 1865 Siegfried Marcus, an Austrian, invented and constructed a practical benzene-fueled internal-combustion engine using the two-stroke-cycle principle, and installed the engine in his self-propelled vehicle. He wasn't interested in getting the engine or vehicle patented, he just enjoyed inventing and creating. Chronologically his engine and vehicle might have been the first. The development was continued further by DeLamarre DeBoutteville in France, who in 1883 invented what is known as a carburetor.

Europe had a head start on the United States because many of their hard-surface roads were built centuries before for military purposes. However, the extensive use of the personal self-propelled vehicle was not the result of any single invention or individual, but rather the result of the combined efforts of many inventors, in many countries of the world, who were not aware of what other inventors were doing at the time. The author does not take exception to the method of record of the accomplishments of the automotive inventors; however, for the benefit of the reader we offer the facts as they pertain to the inventor.

The first recorded successful American gasoline-engine-powered vehicle was operated on September 21, 1893, in Springfield, Massachusetts, using the design of Charles and Frank Duryea. The breakthrough for personal self-propelled vehicles did not happen until after the "Spindletop" gusher oil well erupted in 1901, and gasoline prices dropped to such a low level that automobile travel was made affordable.

The development and use of the liquid-fuel-engine-powered vehicle has virtually changed every aspect of American life. The family car has become the American way of life, and it has literally remade rural America. This individual and flexible form of transportation has provided Americans with a new-found freedom, a mobility ending isolation in the city as well as on the farm. This movement gave stimulus for the development of the land and resources, and contributed to the vigorous growth of the American economy. Automotive transportation has certainly enriched the lives of Americans, and has provided jobs for one-seventh of the total United States work force.

The most important aspect of this story is that, while its beginning is not too fully described, and each chapter is concluded, it has no ending. Its fullest meaning lies in the promise of greater accomplishments to come, in the unending story of progress that is the epic of America.

Charles Duryea. (Source: M.J. Duryea)

J. Frank Duryea.

Charles and Frank Duryea

When Edgar Apperson asked J. Frank Duryea what he knew of the facts in the early controversy "as to the builder of America's first automobile," Duryea smiled and asked, "What difference does it really make?" Apperson smiled, too, and said, "In a great land like America, someone else would have done it anyhow." They sought no credit, plaudits, nor accolades. To them the reward was in the labor itself.

* * *

Charles E. Duryea was born on a farm near Peoria, Illinois, on December 15, 1861, and J. Frank Duryea was born on October 8, 1869. Charles attended the Giddings Seminary in LeHarpe, Illinois. His graduation thesis was on the subject of rapid transportation in which he emphasized the need for personal transportation. He also predicted great strides in air transportation, whereby the travel time would be reduced to a bare minimum, and the social changes that would accompany such rapid mass transportation. Charles was a visionary, his personal charm and character attracted affluent people, and through his gift of great salesmanship, he could promote his ideas and obtain financial support for his business ventures.

In 1888 Charles moved to Washington, D.C., and was employed by the Owens Manufacturing Company, in charge of bicycle manufacturing and service shops. Shortly afterward, Frank joined Charles in Washington, D.C., and, upon Charles' recommendation, was hired by the Owens Manufacturing Company, working under Charles' supervision. Frank was grateful for this move, since it gave him a chance to gain a practical knowledge of the bicycle and its manufacture, through hands-on experience. He looked up to his older brother (as such Charles was Frank's mentor). Frank learned the machinist's trade thoroughly and qualified to become a skilled machinist journeyman.

Meanwhile, Charles spent much of his time at the Patent Office in Washington, D.C., studying the patent applications and the patents that were granted. At that time Charles and Frank shared their dreams and aspirations to create a self-propelled vehicle, which was destined to come.

The Early Partnership

After a couple of years at Owens Manufacturing Company, Charles and Frank moved to Chicopee Falls, Massachusetts. There they were employed by the Ames Manufacturing Company, where they both learned about the bicycle manufacturing business. Shortly afterward they moved to Springfield, Massachusetts, since at that time Springfield was considered the center of the tool manufacturing industry. By March of 1892, Charles convinced Frank to leave his job and engage full-time in the building of an automobile prototype, to which Frank agreed. Meanwhile, Charles purchased a used (horse-drawn) buggy, on which he intended to mount a gasoline engine for propulsion by means of belts, pulleys, and chains.

However, because of Charles' wanderlust nature, he left Springfield on September 22, 1892, and returned to Peoria, Illinois, to engage in bicycle manufacture. This might have left Frank in a quandary, except for

1892 Duryea Buggyaut.
(Source: Horseless Age)

1894 Duryea phaeton.
(Source: M.J. Duryea)

Frank's tenacious nature which pushed him to work on the proposed vehicle toward completion. Frank encountered many obstacles, including the failure of the engine to run. He worked patiently and diligently, and one by one he overcame the obstacles. The failure of Charles' enterprises in Peoria, Illinois, caused him to return to Springfield, Massachusetts, to rejoin Frank and to work with him on the project. Frank designed, engineered, and built all the components of the propulsion mechanism.

On February 10, 1893, they succeeded in getting the engine to operate properly to propel the vehicle, and put it through a "pulling test," according to an article in a Springfield newspaper dated September 10, 1893. After the successful operation on the streets of Springfield on September 21, 1893, Frank built a second vehicle of improved design, and continued the testing.

The *Times-Herald* Race

In April 1895, a successful businessman, H.H. Kohlsaat, purchased the *Chicago Times-Herald* newspaper, and by July 1895, had publicly announced that the *Times-Herald* would sponsor and hold a motocycle (automobile) contest. Kohlsaat witnessed the Paris to Rouen race in France in 1892. He became very interested because he knew a good thing when he saw it. Thus, the *Times-Herald* race was planned for November 2, 1895, to promote, encourage and stimulate the invention, development, and general adoption of the use of motor vehicles (motocycles).

The prizes to be awarded amounted to a total of $5,000. The first prize was $2,000 in cash and a gold medal to the vehicle which, in the opinion of the judges, was the most practical from the standpoint of general utility, ease of operation, and adaptability to the various travel loads the vehicle would be expected to carry.

Second prize was $1,500 to the speediest vehicle. Third prize was $1,000 to the vehicle of economy to purchase and maintain. Fourth prize was $500 for the lowest cost of operation and maintenance. Fifth prize was $200 for the best design and appearance.

By mid-October 1895 there were 100 entries from 84 firms and individuals from 15 states. However, as the scheduled date of November 2, 1895, drew nearer, most of the entrants felt they could not be ready, and were asking for a postponement of the race. As the result of the pleas, Kohlsaat rescheduled the event for Thanksgiving Day, November 28, 1895. However, he insisted on holding a contest, "an unofficial trial run," on November 2nd. This proved to be a wise decision, as a trial run whetted the appetite and increased the interest in the "Big Race" scheduled for November 28th. On November 2nd the weather was ideal and the trial run was very successful in promoting the big event.

But, when the day of the race approached, so did "old man winter" in all his fury. On Tuesday, November 26, 1895, a blizzard of bitter cold and 18 in. of heavy snow was dumped in Chicago and its environs. Thursday morning, November 28, 1895, found the planned course concealed by deep and drifted snow. A four-horse-drawn snow plow cleared a small area at the starting line. But only six contestants showed up for the start: the Duryea Motor Wagon, the De La Vergne Machine, the Mueller Manufacturing Company entry of Decatur, Illinois, the Morris and Shalom (electric) of Philadelphia, the Sturgis Electric Motocycle of Chicago, and the R.H. Macy entry of New York. The Macy, De La Vergne, and Mueller entries were powered by a German Benz engine, while only the Duryea was powered by an American engine. The first machine to get started and going, at 8:55 a.m., was the Duryea, with Frank driving and Charles as the passenger. The other contestants followed a few minutes apart; the Mueller Benz was the last to get started.

About a mile out on the course, the De La Vergne stopped because it could not get traction. The Sturgis Electric dropped out after about 12 miles, for lack of electrical power. The R.H. Macy Benz was eliminated because of a series of mishaps, dropping out at 6:15 p.m. That left only the Duryea, the Morris and Shalom,

J. Frank and Charles Duryea, 1st-prize winners of the 1895 Times-Herald *race. (Source: Horseless Age)*

and the Mueller-Benz in competition. While Charles B. King had intended only to be an umpire, he had to complete the driving of the Mueller-Benz when Mueller became sick. The Duryea crossed the finish line at 7:18 p.m., winning the race in 10 hours and 28 minutes elapsed time, averaging 7.50 mph for the distance of 54 miles. The Duryea accepted the $2,000 first prize money and the gold medal.

Duryea Motor Wagon Company

The early successes of the Duryea brothers in competition aroused considerable interest in vehicles and provided the Duryeas with additional capital, with which Charles and Frank organized the Duryea Motor Wagon Company of Springfield, Massachusetts, for the manufacture of gasoline-engine-powered vehicles. By 1898 the Duryeas had built 13 three- and four-wheeled vehicles. While Duryea vehicles gleamed brightly in competition, manufacturing operations at the Springfield plant did not progress satisfactorily. The partnership of Charles and Frank Duryea was dissolved in 1898.

Duryea Motor Wagon Company, 1896. (Source: Horseless Age)

Charles' Path

Charles E. Duryea moved to Reading, Pennsylvania, in 1900 and organized the Duryea Power Company to manufacture an improved lighter vehicle.

Charles' vehicle was powered by a three-cylinder, four-stroke-cycle, 6-hp engine, with the cylinders inclined horizontally so that the engine could be mounted under the seat. Lifting the seat, the engine was exposed for inspection or servicing. The vehicle was a tricycle-type, supported in the rear by two 36-in. wheels with 3-in. pneumatic tires driven by sprocket and chain from the two forward speed and reverse transmission. The front wheel was a 30-in. wheel mounted with a 3-in. pneumatic tire. Steering was accomplished by a single lever (tiller) pivoted at the front edge of the seat. The engine throttle and transmission and clutch were controlled by the same lever. This single-lever control enabled the driver to steer the vehicle by moving the lever transversely, change gear speeds by moving the lever fore and aft, and open or close the throttle by twisting the handle, providing one-hand control similar to motorcycle control. The list price was $1,500 f.o.b. Reading, Pa. The following year a surrey body type was available: a four-passenger vehicle of four-wheel construction, supported on three, fully elliptical springs. The Duryea central one-hand control was used, and the passengers were seated facing forward. The price of the surrey was $1,600 f.o.b. Reading, Pa. For one reason or another, the Duryea Power Company cars did not sell well. It was possible that the three-wheel construction, one-hand control, and peculiar body form did not appeal to buyers.

1899 Duryea delivery wagon. (Source: M.J. Duryea)

At that time Charles felt that his problem was not one of product acceptance, but rather one of geographical location. Thus, in 1902 he set up shop in Waterloo, Iowa. After several noble attempts with no success, Charles sold his interests to a group of New York bankers, and returned to Reading, Pennsylvania. There he re-established a business and started to build the "Buggyaut" again, building three-wheelers, four-wheelers, and an occasional four-wheeled motor truck (delivery wagon). By 1904, he had added phaetons, surreys, and tonneau models to his line-up, priced at $1,250 to $2,000. All models were built on the same chassis of 72-in. wheelbase, powered by the same type of three-cylinder engine, rated at 15 hp.

By 1908 Charles Duryea had regressed to building the original, high-wheeler "buggyaut." It seems that everything was going in reverse for Charles. Still blaming geographical location, he took his next venture to Saginaw, Michigan, in 1911, where he formed the Duryea Automobile Company. There he tried building the "Electa," some buggyauts, and a few delivery wagons. His business failed in January 1914, so he again returned to Reading, Pennsylvania. His next attempt was a cyclecar with side-by-side seating. By 1917, he built another tricycle-type motor vehicle called the "GEM." It had one wheel up front, and two driving wheels in the rear, powered by a two-cylinder engine. With that venture failing, Charles Duryea moved to Philadelphia where he spent the remainder of his life.

Frank's Path

J. Frank Duryea stayed in Springfield, Massachusetts, and in 1900 built a prototype car and organized the Hampden Automobile and Launch Company.

Stevens-Duryea Automobile Company

Frank and his prototype vehicle attracted the attention of Jay Stevens, owner of the J. Stevens Arms and Tool Company. Stevens offered Frank a deal he could not refuse, whereby the J. Stevens Arms and Tool Company

1902 Stevens-Duryea runabout.

*1903 Duryea three-wheeler.
(Source: M.J. Duryea)*

purchased the Hampden Automobile and Launch Company. The equipment, tooling, and the entire Hampden operation (lock, stock, and barrel) were transferred to Chicopee Falls, Massachusetts, and set up in the vacant plant of the former Overman Automobile Company factory.

J. Frank Duryea became the vice president of the Stevens-Duryea organization in charge of engineering and production, and the Hampden prototype became the 1901 Stevens-Duryea automobile. It was a handsome runabout powered by an opposed two-cylinder, four-stroke-cycle engine of 7 hp, propelled through a three-speed planetary transmission, chains and sprockets, to power the 69-in. wheelbase chassis on wire-spoke wheels and rubber tires. The vehicle was light, weighing only 850 lb, providing economy of operation of 30 mpg. It was placed on the market in March 1902 priced at $1,200. Because of the splendid reputation of J. Frank Duryea and the J. Stevens Arms and Tool Company, the vehicle was able to practically sell on its merits. Stevens-Duryea was able to franchise many reputable firms in the major cities as distributors.

The Stevens-Duryea remained the same through 1902 and 1903, except for a device that was added to make starting from the driver's seat possible, and the price was increased to $1,300. A three-point engine suspension was featured in 1904. For 1905 Stevens-Duryea introduced a new five-passenger touring car, powered by a new four-cylinder T-head engine rated at 20 hp ALAM, priced at $2,500. In 1906 Stevens-Duryea introduced a massive, more powerful touring, built on a 138-in. wheelbase chassis, powered by six-cylinder T-head engine rated at 50 hp. The chassis suspension was by four fully elliptical springs, the drive was by bevel rear axle gears rotated by an exposed propeller shaft with a universal joint at each end of the propeller shaft. The six-cylinder-engine-powered Stevens-Duryea was available in a seven-passenger touring car priced at $5,000.

In 1906 the Stevens-Duryea was organized as an independent entity, with capitalization at $300,000. J. Frank Duryea was the vice president of engineering.

Since Stevens-Duryea did not aspire to set production records, they built some of the finest cars available on the market at that time. Touring car models were the mainstay, and a few roadsters powered by four- or six-

1906 Duryea Power Company exhibit. (Source: M.J. Duryea)

cylinder engines were in the line. Stevens-Duryea production continued the same model line-up for the next four years, averaging about 100 cars per year.

Typical of J. Frank Duryea's safety in engineering, the 1911 Stevens-Duryea innovated dual ignition, accomplished by a Bosch magneto, and an additional battery, coil, and distributor ignition system, with a separate set of spark plugs for each system. Another innovation was the tire-inflating pump driven by a spur gear from the forward end of the camshaft. The spur driving gear was held out of mesh when the pump was not required. The fuel gauge dial was placed directly beneath the front seats, and an auxiliary fuel tank was provided for emergency use.

	1911 Model X	1911 Model AA	1911 Model Y
Wheelbase (in.)	128	131	142
Price	$2,850-$4,000	$3,750-$5,000	$4,000-$5,150
No. of Cylinders / Engine	T-4	T-6	6
Bore x Stroke (in.)		4.25 × 4.75	
Horsepower	36.1 NACC	43.8 NACC	54.1 NACC
Body Styles	seven-passenger touring/limousine	seven-passenger touring/limousine/Berline/Landaulet, torpedo touring, runabout, roadster	seven-passenger touring/limousine
Other Features	dual ignition, tire-inflating pump	dual ignition, tire-inflating pump	dual ignition, tire-inflating pump

1910 Stevens-Duryea exhibit.

The 1911 closed car bodies were built by custom coach builders, among which were: Seaman, Brunn, Brewster, Cunningham, Merrimac and many others. The 1911 custom body types available included limousines, open-vestibule limousines, Berlines, and Landaulets. The town cars and limousines had adjustable rear seats, up and down and fore and aft.

While the Stevens-Duryea total production remained fairly constant, the prices gradually increased. With the well-known reputation of Stevens-Duryea quality, it was placed into the marketing niche of Locomobile and Rolls-Royce. In spite of the business downturn of 1910, Stevens-Duryea did not suffer a great sales loss, building and selling 116 cars in 1911, and 89 cars in 1912. A second manufacturing plant for Stevens-Duryea in East Springfield, Massachusetts, was completed in 1912. Business in 1913 and 1914 remained about the same, averaging 100 cars per year.

However, the outbreak of World War I in Europe affected all financial institutions in Europe as well as in the United States by tightening credit and making less money available for automobile purchases. Stevens-Duryea Automobile Company was in a stable financial position, with two plants free from any incumbrances, and had on hand approximately $1,500,000 value of service replacement parts available to 14,000 Stevens-Duryea owners of cars in use.

On January 11, 1915, the Stevens-Duryea Automobile Company notified its selling representatives that, owing to the financial depression and uncertainty of business, further production of Stevens-Duryea cars for 1915 would be discontinued, and the introduction of new models would be postponed indefinitely. By June 1915, the Stevens-Duryea Automobile plant and the J. Stevens Arms and Tool Company of Chicopee Falls,

1912 Stevens-Duryea roadster.

Massachusetts, were sold to the New England Westinghouse Company for the production of war materiel. Frank retired from the company but continued to manufacture and distribute service parts from the Springfield Stevens-Duryea plant, to maintain and service the 14,000 Stevens-Duryea cars in operation, through 1918.

1914 Stevens-Duryea roadster.

1914 Stevens-Duryea open-vestibule limousine.

1914 Stevens-Duryea Demi-Berline.

1914 Stevens-Duryea touring.

Stevens-Duryea Incorporated

After World War I, in July 1919, Ray S. Deering, his associates, and a few former Stevens-Duryea employees formed a consortium and purchased the Stevens-Duryea name, good will, patent rights, and the former Stevens-Duryea plant in Chicopee Falls, Massachusetts, from the New England Westinghouse Company, and reorganized it as Stevens-Duryea Incorporated. The product to be manufactured was the refined Stevens-Duryea Model E powered by an updated six-cylinder T-head engine, its output increased to 80 hp. Production began in late 1919, but by this time the nation as a whole and all businesses were caught up in the grip of the severe 1920 post-WWI depression with its out-of-control inflation. To cope with the inflation, the price of the Stevens-Duryea was increased, starting at $9,500 up to $10,175.

Deering, thinking only of plant growth, increased his burden by buying Baker, Rauch and Lang of Cleveland, Ohio, in 1920. He had a new plant built next to the Stevens-Duryea plant in Chicopee Falls, intending to build the Rauch and Lang taxicab there; but, as the economic depression worsened, a bank failure and lack of financial control during 1921 forced Stevens-Duryea Incorporated into receivership, which lasted almost a year and a half. On May 9, 1923, Harry G. Fisk of Chicopee Falls and Frank H. Shaw of Chicago were appointed receivers by the Superior Court in Springfield, Massachusetts. During this period Stevens-Duryea Incorporated sold 116 new cars and 92 reconditioned models.

The indebtedness of Stevens-Duryea Incorporated was reported to be less than $1,100,000 while the assets were approximately $4,000,000, of which $350,000 were in notes receivable, $2,000,000 in real estate, plants, and equipment, and the remainder in the form of unused material and finished cars. Deering stated that his company relied on New York and Chicago banks for financial support, but that the closing of one of the banks in January 1923 had cut off one of its principal avenues of bank credit. On September 25, 1923, the Superior Court granted permission to sell the Stevens-Duryea plant in Chicopee Falls, Massachusetts, to a consortium headed by Raymond M. Owen for the sum of $450,000. Owen had been formerly associated with the Reo Motor Car Company and the Owen-Magnetic car.

Under Owen's leadership, the Stevens-Duryea Model G was in fact an updated Model E, powered by a six-cylinder T-head engine developing 80 hp. It was available in seven-passenger touring, four-passenger sport touring, two-passenger roadster, four-passenger coupe, seven-passenger sedan, seven-passenger closed-vestibule limousine, seven-passenger open-vestibule limousine, seven-passenger town-brougham, and cabriolet, priced at $7,500 for the touring and $10,175 for the cabriolet. During 1921, 155 cars were built and

1921 Stevens-Duryea Model E touring.

sold, 100 in 1922, 65 in 1923, 87 in 1924, until sales dwindled to the point that, after 1925, Stevens-Duryea cars were only on order. By 1927 Stevens-Duryea cars would no longer be built.

What Became of Frank and Charles?

After 1918 Frank went into full retirement in Springfield, Massachusetts, and enjoyed life to the fullest. J. Frank Duryea died in his Springfield, Massachusetts, home on February 15, 1967, in his 98th year, having fulfilled his destiny.

After the failure of Charles' ventures, for many years he conducted the Questions and Answers Department of the *Automotive Trade Journal*, and had a consulting-engineering and patent-expert business. One of his avocations was "simplified" spelling, and he issued a considerable number of pamphlets. These were of handy pocketbook size and he always carried a supply with him, which he distributed liberally. Most of these were written to bolster his claim to the title of "Father of the American Automobile." He was also interested in currency reform, and advocated the use of a dollar of fixed purchasing power instead of a dollar whose purchasing power fluctuated with the value of gold. He was also a champion of the League of Nations and of the Prohibition Movement. His letters appeared in Philadelphia newspapers at frequent intervals.

It has been a puzzle to many that Charles E. Duryea, who made such a brilliant start in the automobile business, would have been so unsuccessful as a manufacturer in his later years. Throughout the period he lived in Philadelphia he was in somewhat straitened circumstances and he once confided to an *Automotive Industries* writer that he often did not know "where tomorrow's meals were to come from." However, he never bewailed his fate, nor blamed others for his reverses. On the contrary, he was usually of quite buoyant disposition. He took great satisfaction in whatever public recognition was accorded him for his pioneer work in automobile development.

It seems that by nature he was a type of pioneer and crusader, but that he lacked the elements essential to succeed in business. The Duryea car that won the *Times-Herald* contest was a business-like appearing vehicle, and the fact that it won indicated that it was mechanically dependable. Later vehicles built by Charles sometimes seemed to be regressions rather than advances. Of course in his early endeavors, it was Frank who did all of the physical work. Charles did not get that kind of performance from his later associates. Charles Duryea quietly passed away at his home in Philadelphia, on September 28, 1939, at age 76.

In retrospect, one can only ponder what the history of Duryea Power Company would have been had Charles E. Duryea been less of a visionary and possessed more of J. Frank's constructive mechanical ability. A poet once said, "The saddest words of lips or pen are these, 'It might have been.'"

(l to r) Henry, Clement, Jacob, Peter, and John M. Studebaker.
(Source: Studebaker Corp.)

Studebaker

Why, after 115 years, did Studebaker cease building fine vehicles? In a quotation from the Passing of King Arthur, *"The old order changeth making place for new." With aging plants and an older management with old-fashioned philosophies on business, the demise of Studebaker was inevitable. Studebaker could only stand on its reputation of building fine transportation vehicles longer than any company in the world.*

* * *

The Studebaker Brothers and their father before them were blacksmiths, woodworkers, coach builders, and wagon makers by trade. Over a period of two generations they developed and expanded the horse-drawn vehicle business until it reached worldwide proportions. During the 19th century Studebaker carriages were used by royalty and military leaders of foreign countries, as well as by presidents of the United States. At the turn of the century and for two decades afterward, farmers owning a Studebaker wagon or carriage were expressing the ultimate of success.

How it Started

John Studebaker was born in 1799 and lived in Adams County, Pennsylvania, where he pursued his trade as blacksmith and wagon maker. He earned the reputation of a skilled and conscientious workman. He built his own blacksmith shop and a brick home. In 1820 he married Rebecca Mohler, and they had six children in Pennsylvania and four more after they moved to Ashland, Ohio. He built his first Conestoga-type wagon in 1830, and built a new larger one in 1835, which they used in their travel to Ashland.

The five Studebaker brothers born to John and Rebecca were: Henry (born in 1826), Clement (1831), John Mohler (1833), Peter Everst (1836), and Jacob Franklin (1844). The boys were put to work in the fields at

Rebecca Mohler Studebaker (top left); John Studebaker (top right); the covered wagon built by John in 1830. (Source: Studebaker Corp.)

an early age, and in the blacksmith shop their father had established in Ashland. Their father urged upon them the importance of getting an independent start in life.

Studebaker home and blacksmith shop in Ashland, Ohio, 1835. (Source: Studebaker Corp.)

About 1850, Henry and Clement Studebaker departed from home and traveled by wagon to South Bend, Indiana. Clem taught school for a short period of time and also worked at blacksmithing for 50 cents per day. John and Jacob followed them to South Bend, while Peter became a merchant in Goshen, Indiana. Ultimately the brothers brought their five sisters and parents to South Bend. Their father lived until 1877, and their mother until 1889.

H. and C. Studebaker

On February 15, 1852, Henry and Clement Studebaker, possessing a total capital of $68 and two forges, established the firm of H. and C. Studebaker, blacksmiths and wagon builders. John joined them with the sincere intention of pursuing the blacksmith trade; however, his adventurous nature lured him to California. He and his brothers built a sturdy Conestoga covered wagon, which John traded to the wagon master of a wagon train for his fare and board. This wagon was one of the few of the wagon train that arrived at old Hangtown (now Placerville), California, in good condition in August 1853. After five months of untold hazards and hardship with only 50 cents in his pocket, John obtained employment by a friendly blacksmith named H.L. Hines.

While John went to California to dig gold, a friendly prospector convinced him to work at his trade of building wagons and wheelbarrows. He earned $10 per week working many nights until dawn, repairing stagecoaches and wagons and making wheelbarrows and miner's picks. He continued working for five years and became a partner of Hines, saving reportedly over $4,000.

Studebaker plant, 1858 (top); Studebaker buggy built 1857 (bottom). (Source: Studebaker Corp.)

Meanwhile, in South Bend, his brothers had built their first carriage in 1856, but they were in need of additional capital. The brothers appealed to John to return to South Bend and join them in their business. Upon John's arrival in South Bend in 1858, he wisely invested his money in the H. and C. Studebaker Firm by buying out the interest of his brother Henry, who decided to retire from business and engage in farming. The additional money rehabilitated the firm and subsequent success was assured.

John M. Studebaker was put in charge of the manufacturing department, a position he would hold for 45 years. During this time he personally supervised the manufacturing of nearly all the vehicles, and built into these products his sturdy and honest character. This established a worldwide reputation of confidence in Studebaker vehicles. In fact, owning a Studebaker wagon or carriage was a symbol of character and good taste. In later years, H.L. Hines was superintendent and stockholder. The combination of Clement and John was unbeatable, as they proceeded on their path to success. They went after the business of the migrating pioneers and wagon train operators who knew of John's record of performance and quality products. They built up a large business, so big, in fact, that they induced their brother Peter (a merchant in Goshen, Indiana) to sell out and join them. After joining his brothers, Peter was placed in charge of sales, and his first responsibility was to establish a distributing center for Studebaker wagons in St. Joseph, Missouri. St. Joseph was an organization point for wagon trains, and as business developed and expanded, Peter established Studebaker branches in Salt Lake City, San Francisco, and Portland.

John Mohler Studebaker. (Source: Studebaker Corp.)

In 1857 H. and C. Studebaker received their first order for war materiel in the form of an order from the United States Government for 100 military supply wagons to be delivered in three months. Studebaker built and shipped the complete order of wagons within 3 weeks. When war was declared between the industrial north and the agricultural south, ammunition, clothing, food, horses, saddles, harnesses, and wagons were immediately demanded by the United States Government. Studebaker received orders for great numbers of troop carriers, supply wagons, and ambulances, essential for the movement of troops. According to the Bureau of Census Report of 1870, H. and C. Studebaker was the most important source of supply. The demand was so great that new buildings had to be erected and new equipment provided.

Studebaker Manufacturing Company

When the Civil War ended in 1865, the reputation of Studebaker had widened to cover the eastern half as well as the western half of the United States, and the reliability of Studebaker products became world renowned. The business had grown to such an extent that on January 1, 1868, they decided to form a corporation. The Studebaker Manufacturing Company was incorporated under the statutes of the state of Indiana on March 26, 1868, with a capital of $75,000. Clement Studebaker was elected president, John treasurer, and Peter secretary. The funds needed for factory expansion and increased working capital were provided by the use of accumulated profits and bank loans.

On October 10, 1879, John Mohler Studebaker II was born. He was destined to lead Studebaker during their most glorious years.

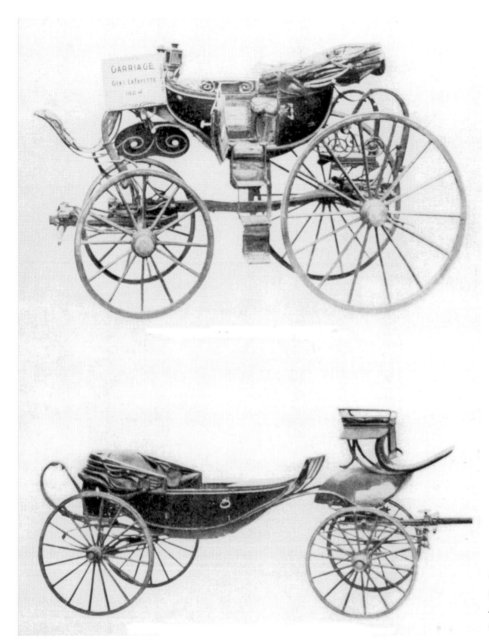

General LaFayette's carriage (top); President Abraham Lincoln's carriage (bottom). (Source: Studebaker Corp.)

The Studebakers increased business by building all types of horse-drawn vehicles, and Studebaker branch offices and dealerships popped up in practically every city and town where vehicles were sold. The sales organization developed on a large scale and prospered under the effective management of Peter Studebaker.

From Wagons to Cars

With the advent of the automobile in the late 1890s, it was inevitable that the Studebaker Manufacturing Company would become involved with the automobile business. In the spring of 1897, Studebaker began

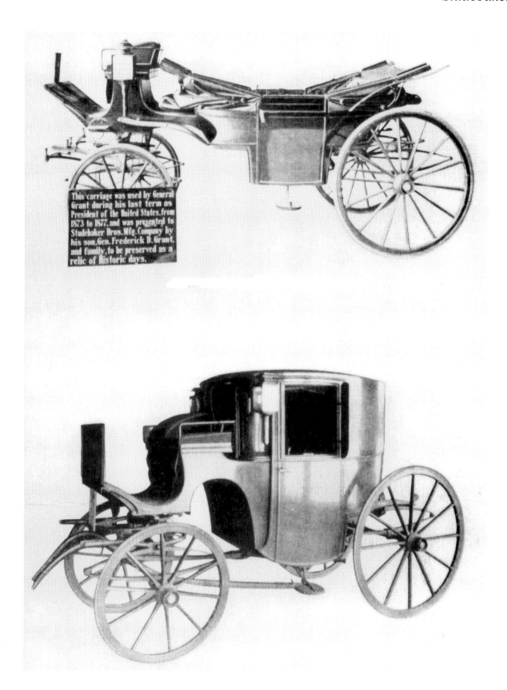

President Grant's carriage (top); President Harrison's carriage (bottom). (Source: Studebaker Corp.)

experimenting with horseless vehicles. In 1899 they started building bodies for electric-powered runabouts made by another company. Sadly, on November 26, 1901, Clement Studebaker died.

By 1902 Studebaker was building complete electric-powered runabouts and delivery vehicles, of which 20 were built that year and a total of 1,841 built and sold through 1912. In 1904 Studebaker began building gasoline-engine-powered vehicles (the chassis and engine supplied by Garford Motor Company), while the bodies were built and the cars were assembled in Studebaker factories in South Bend, Indiana. Under this arrangement, by 1911 a total of 2,481 gasoline-engine-powered cars and trucks were built and sold by Studebaker.

First Studebaker electric runabout, 1902. (Source: Studebaker Corp.)

1902 Studebaker gasoline car. (Source: Studebaker Corp.)

*1904 Studebaker gasoline car.
(Source: Studebaker Corp.)*

*1908 Studebaker seven-
passenger touring car.
(Source: Studebaker Corp.)*

*1908 Studebaker runabout.
(Source: Studebaker Corp.)*

1908 Studebaker limousine (top), touring car (second), tourabout (third), speed car (bottom). (Source: Studebaker Corp.)

Being assured of the future of the automobile, Studebaker management now perceived the necessity of embarking on a large-scale automotive venture, sufficient to employ its extensive manufacturing facilities and its worldwide sales organization. The Studebaker sales outlets at that time numbered approximately 4,000, which could market a moderately priced automobile in great quantities. It was therefore expedient to effect an agreement with the Everitt-Metzger-Flanders Company of Detroit, which at that time was being organized to manufacture vehicles of that character (see the Flanders chapter in this volume). The exclusive rights for the sale of E-M-F-manufactured cars were negotiated by Studebaker. Subsequently, Studebaker acquired substantial holdings in the E-M-F Company and, in March 1910, through the J.P. Morgan and Company, purchased the remaining stock and became sole owner of the E-M-F Company. The E-M-F Company was quite successful, building and selling 23,431 cars from August 4, 1908, through December 31, 1910, at which time it merged with the Studebaker Manufacturing Company.

The Studebaker Corporation

The Studebaker Corporation was organized and incorporated on February 14, 1911, under the laws of the state of New Jersey to acquire the assets, tradename, goodwill, and patent rights of the Studebaker Manufacturing Company and the E-M-F Company as of January 1, 1911, with a capitalization of $45,000,000. During early 1911 the cars were merchandised as the Flanders 20 and the Studebaker-E-M-F 30, but the names were changed to Studebaker 20 and Studebaker 30 for the 1912 model year. Walter E. Flanders stayed on as vice president and general manager of the Studebaker Corporation with a three-year contract. However, when it became known to the Studebaker management that Flanders had organized the Flanders Manufacturing Company on January 1, 1911, they felt that this was a conflict of interest. In August 1911 the Studebaker Corporation released Flanders from his contract, and Flanders rejoined his former partners, Barney Everitt and William Metzger, in their new business venture.

1910 Studebaker-E-M-F 30.
(Source: Studebaker Corp.)

On October 1, 1911, Albert R. Erskine was engaged by the Studebaker Corporation as treasurer, and appointed a member of the executive committee. Erskine had been vice president and member of the board of directors of the Underwood Typewriter Company.

Frederick S. Fish was a practicing lawyer in New Jersey, admitted to the bar in 1876. He was later a state senator, becoming president of the New Jersey Senate in 1887. Fish was appointed a director of the board and general counsel to the Studebaker Brothers Manufacturing Company in 1891. He later became chairman of the executive committee. He was also a son-in-law of John M. Studebaker.

On December 26, 1911, John M. Studebaker Sr. relinquished the presidency to become chairman of the board of directors of the Studebaker Corporation. Frederick Fish was elected president, Clement Studebaker Jr. first vice president, and Albert Erskine treasurer.

Frederick S. Fish (Source: Studebaker Corp.)

With the departure of Flanders, several organizational changes were made in early 1912. J.N. Gunn was appointed general manager replacing Flanders, W.S. Petit was named advertising manager, Charles Gordon was appointed factory manager, and Max F. Wollering remained as production manager. During 1912 Studebaker facilities expanded to the extent that, by May, Studebaker Corporation had twelve manufacturing plants in Detroit alone. Sales increased during 1912 to about 25,000 cars.

In November 1912 three new Studebaker lines of 1913 cars were introduced: Models 25 and 35 with four-cylinder engines, and Model E-6 (Studebaker Six) powered by a new six-cylinder engine. A roadster and touring car were available in the Model 25, a touring car, coupe, and sedan in the Model 35, and the Studebaker Six offered a touring car and limousine. All engines featured cast-en-bloc cylinder block design. The Studebaker Six and the 35 featured a new electric starting, charging, and lighting system by Wagner Electric Company of St. Louis, Missouri. The Model 25 had an acetylene primer to aid in starting the

1913 Studebaker seven-passenger E-6. (Source: John A. Conde)

1914 Studebaker Model SC. (Source: John A. Conde)

engine, and the fuel tank was located under the cowl. It also had fully elliptical rear springs, while the Six and 35 had three-quarter elliptical rear springs. The price range of the 1913 Studebaker models was $875 to $2,500.

In December 1913 Clement Studebaker Jr. retired as first vice president and was succeeded by Albert Erskine. James G. Heaslet was appointed vice president of engineering and production.

The 1914 Studebaker models introduced in October 1913 had a number of design changes, including moving the steering wheel and mechanism to the left side, and the gearshift control and hand brake lever to the center of the front compartment floor boards. The model availability was reduced to two: the Studebaker Four Model SC and the Studebaker Six Model EB. A new landau roadster was added to each model line. The new Studebaker Four replaced the Models 25 and 35, and was priced right between the former 25 and 35 price ranges. During 1914 the board of directors approved a profit-sharing plan for management personnel. Sales during 1913 and 1914 improved, with 32,504 cars sold in 1913 and 36,430 in 1914.

Albert R. Erskine. (Source: Studebaker Corp.)

On July 8, 1915, at a board of directors meeting, John M. Studebaker Sr. announced his retirement as board chairman. Frederick S. Fish was elected chairman of the board, and Albert R. Erskine was elected president of the Studebaker Corporation.

In 1915 closed cars were not offered, only the roadster and touring were available in both the Four and Six model lines. For 1916 the other body types were reinstated, including the coupe, sedan, and limousine. Sales for 1916 amounted to 65,885 automobiles, and the profits also were high. The net earnings rose from $2,534,042 in 1911 to $8,611,245 in 1916. Profits on war materiel amounted to only about 2%.

World War I

On February 6, 1917, Erskine sent President Woodrow Wilson the following telegram:

To the President of the United States, White House, Washington D.C. "Studebaker of course are at the disposal of the government. Any orders given will receive preference and clear right of way." A.R. Erskine President, The Studebaker Corporation.

President Woodrow Wilson's reply was:

February 6, 1917

My Dear Mr. Erskine:

Thank you for your generous assurances of your telegram today. I greatly appreciate your pledge of cooperation."

Sincerely yours,
Woodrow Wilson

The basic designs of the Studebaker cars continued through the end of 1917, consisting of four body types in the Studebaker Four, and seven body types in the Studebaker Six. A total of approximately 140,000 Studebaker cars were built and sold during the period of June 1915 through December 1917. Many Studebaker plants converted to war materiel production during the second half of 1917.

The design of the 1918 Studebaker models started in May 1917. The endurance testing began on September 15, 1917, and limited production was started in late December. However, introduction was delayed until February 1918, when Studebaker introduced three completely new lines of cars: the new Studebaker Light Four, Studebaker Light Six, and Studebaker Big Six. These cars were a complete departure from the carryover of the E-M-F and Flanders designs. The transaxle (combined transmission and rear axle) was discontinued and replaced by a new, selective, three-speed transmission and leather-faced aluminum cone clutch. These units were attached directly to the rear of the engine, and the drive was through an exposed propeller shaft and two universal joints to the new semi-floating rear axle.

1918 Studebaker Big Six.
(Source: Studebaker Corp.)

The engines of the 1918 models were newly designed with removable L-type cylinder heads. The bore and stroke of the Light Four and the Light Six were 3-1/2 in. by 5 in., and the bore of the Big Six was 3-7/8 in. and the stroke was 5 in. The bore and stroke dimensions of the six-cylinder engines would remain the same for many years. The Light Four was available in touring, roadster, and sedan body types. The Light Six offered a touring, four-passenger roadster, two-passenger roadster, coupe, and sedan. The Big Six was available in a seven-passenger touring model only.

World War I affected the production and sales of all cars in 1917-18. During this period Studebaker was producing ambulances, escort wagons, British tanks and track links, 4.7-in. cannons, gun carriages, shells and casings, and mine anchors. In 1918 Studebaker produced 18,270 automobiles and 58,830 horse-drawn vehicles.

In the spring of 1918 the British War Ministry requested that Studebaker design and build caterpillar-type tractors using the Light Six engine. Studebaker received an order for 5,000 of these tractors, and while a number were shipped before the Armistice, the order was subsequently cancelled in December 1918. By November 11, 1918, Studebaker was almost 100% into war materiel production.

On March 1, 1917, Studebaker had distributed wage dividends to hourly workers amounting to 5% of their pay for four years employed, 10% for the fifth year and beyond. On September 1, 1919, the wage dividends were increased and the workers were given vacations with pay. Stock purchase plans and pensions were offered, up to 25% of annual earnings.

Accounting records for 1919 showed that Studebaker received orders from the United States Government amounting to $30,979,416. Deliveries amounted to $16,909,820. Profit on war production was $979,699 or 5%. Car sales for the four previous years were: 1915 - 46,945 units, 1916 - 65,885 units, 1917 - 42,357 units, and 1918 - 23,951 units. Profits were as follows: 1915 - $9,000,000, 1916 - $8,600,000, 1917 - $3,500,000, and 1918 - $3,800,000.

World War I cannon and tow truck. (Source: Studebaker Corp.)

Post-War Prosperity

After the Armistice, Studebaker reconverted the plants and continued building the same car models and body types through 1919. All horse-drawn vehicles were discontinued, except farm trucks and wagons, which were then discontinued in 1920. For 1920 three new lines of cars were introduced and continued through 1921. The Studebaker Light Four was replaced by the new Studebaker Light Six. The former Light Six was replaced by the new Studebaker Special Six. The Studebaker Big Six was refined and a new four-passenger coupe and seven-passenger sedan were added in 1921. The 1922 and 1923 models remained basically the same as the 1921 models except for a few refinements. Sales amounted to 145,167 cars in 1923.

The 1924 Studebaker models had an improved appearance, and there was a marked difference in the radiator shell and hood shoulder lines between the Big Six, the Special Six, and the Light Six. The Big Six had a massive, nickel-plated radiator shell resembling the upper two-thirds of a hexagon. The Special Six had a nickel-plated radiator shell and hood shoulder line similar to that of Packard. The Light Six had a conventional, painted, rounded-corner radiator shell. Available in 14 body styles, the 1924 Studebakers were priced between $1,025 and $2,685.

1920 Studebaker Light Six Model EJ. (Source: John A. Conde)

1920 Studebaker Big Six. (Source: John A. Conde)

In September 1924 three new Studebaker lines of cars were introduced featuring balloon tires, engines with increased horsepower and full-pressure lubrication, automatic ignition timing advance, and the transmission mounted as a unit with the engine. Four-wheel mechanical brakes were offered as an option. The massive radiator hood, fender, and body were blended into smooth lines. The Studebaker Light Six was superseded by the Studebaker Standard Six. The open models, the Duplex Phaeton and Duplex Roadster, were fitted

1925 Studebaker Big Six brougham. (Source: John A. Conde)

1925 Studebaker Big Six coupe. (Source: John A. Conde)

1925 Studebaker Special Six roadster. (Source: John A. Conde)

1925 Studebaker Standard Six country club coupe. (Source: John A. Conde)

1925 Studebaker Standard Six duplex roadster. (Source: John A. Conde)

with a steel-reinforced permanent top. The side curtains were mounted on concealed rollers in the topside, which operated similar to a window shade roller, providing protection against weather.

Varnished, natural wood-spoke artillery wheels with demountable rims were standard. Disc wheels were used when four-wheel brakes were ordered. The wheels and balloon tires were of a generous size: 31 x 5.25-in. on the Standard Six, 32 x 6.20-in. on the Special Six, and 34 x 7.30-in. on the Big Six. Sales of the 1925 Studebakers amounted to a record 107,732 units domestically. During their 14 years (1911-25) of existence in the automobile industry, Studebaker earned a profit in excess of $150,000,000, of which 50% was retained in the business.

On April 7, 1925, Paul G. Hoffman, a successful Studebaker distributor for Southern California, joined the Studebaker Corporation as vice president of sales, succeeding H.A. Biggers. Hoffman was born in Chicago, Illinois, on April 26, 1891, and was educated at the University of Chicago, Rose Polytechnic Institute, and University of Southern California, earning his LLD and DBA degrees. He entered the automobile business at the Halladay Motor Company in Chicago in 1909. Hoffman became a Studebaker salesman in Los Angeles in 1911. He was made manager of retail sales in 1915, and by 1917 he was manager of the Southern California (Los Angeles) District. He served in the United States Army artillery division during 1917 and

1918. Upon release to inactive duty in 1919, Hoffman purchased the retail sales business in Los Angeles from Studebaker, becoming a Studebaker dealer on his own account. The Paul G. Hoffman Company was capitalized at $60,000 in 1919, and by 1925 the company assets exceeded $1,500,000.

In 1926 Studebaker refined the model lines and added two body types. The engineering and design groups were earnestly engaged in developing a new, lighter, and less-expensive companion car. Competition already had such companion cars. Max Wollering and Guy Henry were given the design and engineering responsibilities, while the European styling was done by Raymond Dietrich and his associates.

The new car, the 1927 Erskine, was announced at the Paris Automobile Salon on October 7, 1926. Two prototype models, a five-passenger sedan and a five-passenger touring car, were presented. The formal introduction of the 1927 Erskine was at the National Auto Show in New York in January 1927.

Paul S. Hoffman. (Source: Studebaker Corp.)

	1927 Erskine
Wheelbase (in.)	107
Price	$935-985
No. of Cylinders / Engine	L-6
Bore x Stroke (in.)	2.62 × 4.50
Horsepower	
Body Styles	five-passenger custom sedan, five-passenger touring, custom coupe with rumble seat, business coupe
Other Features	mechanical four-wheel brakes on wood-spoke artillery wheels

The Eight-Cylinder Engine

While it was hoped that the new Erskine with European styling would bolster Studebaker sales, the sales increased only 1,364 units, and Albert Erskine and Paul Hoffman noticed that the Big Six and Special Six were disproportionately outselling the lower-priced Studebaker models. They were also aware that the competition—Cadillac, Packard, Peerless, Stutz, Stearns, Jordan, Auburn, Diana, Chandler, and Hupmobile—were offering eight-cylinder engines. Thus, Erskine directed Max Wollering and his engineering staff to develop a new eight-cylinder engine. Wollering and Henry were pleased with the performance of the six-cylinder Studebaker engines, and vehemently objected to Erskine's order for a new eight-cylinder engine. Erskine easily resolved the problem by promptly discharging Max Wollering and Guy Henry on December 31, 1926. In January 1927 the model designations were changed from Standard Six to Dictator, from Special Six to Commander, and in September 1927 from Big Six to President.

On January 12, 1927, Harold S. Vance, vice president of Studebaker, announced the appointment of Delmar G. (Barney) Roos as vice president of engineering. Roos, who gained recognition through his photograph of the charging, horse-drawn, smoke-belching fire engine on the streets of New York, had been engaged in responsible engineering positions at Locomobile, General Electric, Pierce Arrow, and in 1925 and 1926 as

Delmar G. Roos. (Source: Studebaker Corp.)

chief engineer at Nordyke and Marmon. William S. James was named chief engineer of research, John A.C. Warner in charge of experimental and research laboratories, and Ralph DePalma (1915 Indianapolis 500 winner) as liaison engineer between research and experimental engineers.

Roos and his engineering staff "pulled out all the stops," and designed, engineered, built prototypes, and tested the eight-cylinder engine at full wide-open throttle for over 100 hours. The newly designed Studebaker in-line eight-cylinder engine came through with flying colors. The block and upper crankcase were cast integral, and the crankshaft was supported by five main bearings.

The plant rearrangement and tooling was accomplished in record time, so that the new eight-in-line engine and the newly styled Studebaker President Eight cars were able to make their debut at the National Auto Show in New York on January 7, 1928. Studebaker for 1928 had five series in their model line-up: the Erskine, Dictator, Commander, President Six, and President Eight.

1928 Studebaker President Eight

Wheelbase (in.)	131
Price	$1,985-2,250
No. of Cylinders / Engine	IL-8
Bore x Stroke (in.)	3.37 × 4.50
Horsepower	100 adv
Body Styles	4
Other Features	chromium plating on radiator shell, headlamps, and bumpers

1928 Studebaker President Eight. (Source: John A. Conde)

1929 Studebaker President Eight roadster, the 1929 Indianapolis 500 pace car. (Source: Indianapolis Motor Speedway Corp.)

Studebaker Acquires Pierce Arrow

In August 1928 the Studebaker Corporation bought controlling interest in the Pierce Arrow Motor Car Company (see the Pierce chapter in this volume). Pierce Arrow became a division of the Studebaker Corporation, with A.J. Chanter as vice president and general manager. Since Pierce Arrow was in the process of developing a V-12 engine, the Studebaker President Eight engine was used for the 1929 Pierce Arrow to supplement the six-cylinder engine during the interim period.

The new 1929 Studebaker models introduced on July 9, 1928, had a styling which was consistent with the trend of the times. On January 5, 1929, Studebaker introduced a new Commander model. In 1929 Studebaker enjoyed a record sales year, having built and sold 107,234 Studebaker and Erskine cars, and 8,386 Pierce Arrow cars.

	1929 Studebaker President Eight	1929 Studebaker Commander
Wheelbase (in.)	125/135	121
Price	$1,850-2,485	$1,685-1,850
No. of Cylinders / Engine	IL-8	IL-8
Bore x Stroke (in.)	3.37 × 4.87	3.18 × 4.25
Horsepower	109 adv	80 adv
Body Styles		
Other Features	higher single belt line, clamshell-type front fenders, bullet-shaped headlamps, cadet-type sun visor, wood-spoke wheels, and chromium-plated radiator, bumpers, and all bright metal parts	

Studebaker Faces Its First Losses

For 1930 most Studebaker models were refined and continued with basically the same technical specifications, except the Dictator model offered an optional, eight-in-line, L-head engine of 70 hp, at $200 extra. However, the new Dynamic Series 1930 Erskine model was introduced at the National Auto Show in New York on January 4, 1930. The new Erskine had a marked resemblance to the 1930 Studebaker Dictator Six,

1930 Studebaker Dictator Six landau sedan. (Source: John A. Conde)

Harold S. Vance. (Source: Studebaker Corp.)

which was introduced in June 1929. The radiator shell was tall, narrow, and peaked at the filler cap, and trilateral belt moldings in contrasting colors were used, giving an interesting effect. The Erskine models had dismal sales record though—24,893 in 1927, 22,275 in 1928, 25,565 in 1929, and 22,371 units in 1930—and were therefore discontinued in May 1930. Despite the many improvements and refinements, the 1930 Studebaker dropped sharply due to the deepening economic depression.

Sagging sales during 1931 became the chief concern of all automobile manufacturers, and many used various methods of sales promotion. For 1931 the Dictator Six became the Studebaker Six. The 1931 Studebaker, although greatly refined, could not increase the sales, thus, on March 24, 1931, Paul G. Hoffman, vice president, announced the appointment of Knute E. Rockne as manager of sales promotion. The appointment of Rockne did not interfere with his duties as athletic director for the University of Notre Dame. Sadly, seven days later on March 31, 1931, Rockne was killed in an airplane crash while flying to the west coast to keep a motion picture engagement.

In 1931 Studebaker Corporation organized the Rockne Motors Corporation (a subsidiary of Studebaker) with Harold S. Vance as president and George M. Graham as vice president. The Rockne car was introduced at the National Auto Show in New York in January 1932. The name Rockne was chosen in honor of Knute Rockne, and it replaced the Erskine in the model line-up. The Rockne was built in the Studebaker plant on Piquette Avenue in Detroit.

1932 Rockne

Wheelbase (in.)	114
Price	$685-840
No. of Cylinders / Engine	L-6
Bore x Stroke (in.)	3.25 × 4.12
Horsepower	72 adv
Body Styles	
Other Features	beautifully styled, handsome lines, agile, and lightweight

1932 Studebaker Commander Eight regal brougham. (Source: John A. Conde)

On December 23, 1931, Erskine and Hoffman announced the appointment of Roy H. Faulkner as vice president of the Studebaker Sales Corporation of America. Faulkner was previously the president of the Auburn Automobile Company. Faulkner became associated with the Studebaker Corporation at the time of reorganization, inaugurating a number of subsidiaries within the Studebaker Corporation.

On December 10, 1931, Studebaker introduced four new 1932 model lines: the Studebaker Six, Dictator Eight, Commander Eight, and President Eight. The wheelbases were 114, 117, 120, and 135 in., respectively. The 1932 Studebaker styling was known as the French-type air-curve coachcraft, which was aesthetically pleasing as well as functional. It featured rearward-sloping windshield pillars, inside sun visors, sweeping, one-piece, clamshell-type fenders, and integral unit-body construction. The instrument panel had round-dial (aircraft-type) instruments and a V-type radiator grille. Startix automatic starting, centrifugal and vacuum ignition spark advance, free-wheeling, synchronized transmission, and duo-servo mechanical four-wheel brakes were standard equipment on all models.

In 1932 performance was considered tantamount, as several Studebaker racing cars competed in the 1932 Indianapolis 500, with Cliff Bergere finishing in a respectable third position. Other Studebaker cars gave a good account of themselves, as they finished 6th, 13th, 15th, and 16th. Unfortunately, because of the deepening of the economic depression, the 1932 dollar sales volume dropped from $64,406,858 in 1931 to $46,233,830 in 1932. This resulted in a net loss $8,686,983, the first loss recorded in Studebaker Corporation history.

In an attempt to have more operating capital, the Studebaker Corporation and the White Motor Company of Cleveland made plans for a merger in September 1932, with Studebaker acquiring 95.11% of the White Motor Company stock for $26,853,822. There were several organizational changes; however, the identity of White Motor Company remained intact as Ashton G. Bean remained as president of White Motor Company. The actual merger of the two companies was never completed because it was blocked by a group of minority stockholders.

Cliff Bergere driving a Studebaker in the 1932 Indianapolis 500. (Source: Indianapolis Motor Speedway Corp.)

The Receivership

In the meantime the Studebaker Corporation went into a friendly receivership on March 18, 1933. The reasons given for going into receivership were overextension of capital, decline of sales with a resulting loss of $8,686,983 by Studebaker Corporation in 1932, a net operating loss of $3,618,000 reported by the White Motor Company, and a net loss of $3,032,430 suffered by Pierce Arrow Motor Car Company. These staggering losses were compounded by the National Bank Holiday declared by U.S. President Franklin D. Roosevelt on March 5, 1933, which dried up all sources of available capital. Studebaker halted production immediately.

When the Studebaker Corporation went into receivership, Harold S. Vance, Paul G. Hoffman, and Ashton G. Bean were named receivers by Judge Thomas W. Slick of the Federal Court. The appointment as receivers necessitated their resignation as directors of the Studebaker Corporation. John M. Studebaker Jr., Edward N. Hurley, and George H. Kelly were elected to succeed them on the board. The receivership did not affect the White Motor Company nor the Pierce Arrow Motor Car Company operations, but it did include the Rockne Motors Corporation.

The Studebaker plants reopened on March 21, 1933, following the shutdown necessitated by the Bank Holiday. On April 25, 1933, the board of directors, chaired by Frederick S. Fish, authorized an increase in common stock from 2,500,000 shares to 3,125,000 shares. This action provided the Studebaker Corporation with additional operating capital of approximately $1,750,000. The 1933 Studebaker car lines, as well as the Studebaker Pierce Arrow, White, and Indiana truck lines, enjoyed healthy increases in sales.

When the receivership took place, A.R. Erskine resigned from the board of directors, and was no longer active in management affairs. He despaired exceedingly, not only because of the gigantic loss sustained in 1932 (the only loss experienced by Studebaker up to that time), but also the thought of receivership. Tragically, on July 1, 1933, Erskine died by his own hand at his home in South Bend, Indiana. Ironically, the receivership was brief, lasting less than two years.

In September 1933 the Pierce Arrow Motor Car Company was re-established as an independent enterprise, unaffected by the Studebaker receivership. The Studebaker holdings of Pierce Arrow were reported to have been purchased by a group of Buffalo bankers, businessmen, and Arthur J. Chanter at a price of $1,000,000 and other considerations. Chanter was elected president and R.H. Faulkner vice president of sales of the new Pierce Arrow Motor Car Company.

Recovery and Expansion

During the receivership period, the Studebaker overhead and selling expenses were drastically reduced by approximately 50%. This allowed Studebaker to retool for the 1934 models, paying for the retooling out of the corporate cash on hand. By September 30, 1933, Studebaker had $3,000,000 on hand and up to $4,000,000 to $5,000,000 in quick assets.

For 1934 the Studebaker models were redesigned in a streamlined fashion, available in three car lines at reduced prices: the Studebaker Dictator Six (superseding the Rockne) at $645 to $695, the Commander Eight at $845 to $895, and the President Eight at $1,045 to $1,095. The 1934 models featured many innovations, among them the tumble-home transverse curvature of the body upper panels. The front fenders extended farther down in front, concealing the chassis components, and had a shallow trough on the inner edge to meet the hood side panel. The hood extended farther back to overlap the cowl up to the windshield pillar. The President Regal Sedan featured a built-in recessed rear trunk.

1934 Studebaker Dictator Six convertible roadster. (Source: John A. Conde)

1934 Studebaker Commander Eight convertible roadster. (Source: John A. Conde)

1934 Studebaker Commander Eight custom sedan. (Source: John A. Conde)

In January 1934 Studebaker introduced a streamlined model known as the Land Cruiser at the National Auto Show in New York. In February 1934 Studebaker added a new 3- to 4-ton truck model to their truck line, powered by a six-cylinder, L-head engine of 110 hp. In June 1934 Studebaker announced the first-series 1935 Dictator Six, and production began in July. The Land Cruiser model was added to the Dictator line. The 1935 President and Commander models were introduced in December 1934, featuring the Planar independent front suspension system, which was also added to the Dictator models. This type of suspension had a single transverse spring for support at each front wheel, allowing free movement of each front wheel. The President and Commander models had an 18-leaf spring, while the Dictator spring had 14 leaves. The 1935 models sported a new radiator shell and grille, and featured hydraulic four-wheel brakes and a transmission overdrive unit.

On November 17, 1934, with $5,500,000 in new money available to wipe out debts, a plan for reorganization was submitted to the United States Federal Court. Vance, Hoffman, and Bean, who had been receivers, were named trustees of the Studebaker Corporation in February 1935. After three days of hearings, the Federal Court gave approval of the Studebaker Corporation reorganization plan. Under this plan, Harold Vance became chairman of the board and Paul Hoffman was named president. On April 16, 1935, Ashton Bean became chairman of the board of White Motor Company. However, on July 20, 1935, Bean died of a heart attack at his Elyria, Ohio, home.

In 1935 Studebaker found itself with no less than a 38-model line-up. Therefore, for 1936 they offered only 12 models in two series: the Studebaker Dictator Six and the Studebaker President Eight, introduced on November 1, 1935; the Commander line was discontinued. The 1936 models physically had only minor changes from the 1935 models. The styling changes included a new, sloping, V-type, two-piece windshield, hinging the front doors at the windshield A-pillar, and horizontal hood louvres. The price range was $665 to $795 for the Dictator Six and $965 to $1,065 for the President Eight.

On August 15, 1936, Frederick Samuel Fish, who had remained a member of the board until his retirement in 1935, died at his home in South Bend, Indiana. In March 1937 Studebaker Corporation announced a profit of $2,187,783 for 1936, and a domestic sales total of 67,835 units, the greatest since 1929.

The 1937 Studebaker models were announced on September 1, 1936, with production starting September 4th and deliveries on September 15th. The model line-up was confined to two series: the Studebaker Dictator Six and the Studebaker President Eight.

	1937 Studebaker Dictator Six	1937 Studebaker President Eight
Wheelbase (in.)	116	125
Price	$665-795	$965-1,065
No. of Cylinders / Engine	L-6	IL-8
Bore x Stroke (in.)	3.25 × 4.37	3.06 × 4.25
Horsepower	90 adv	115 adv
Body Styles	7	6
Other Features	automatic hill-holder	aluminum-alloy cylinder head, automatic hill-holder

The styling appeal of the 1937 Studebaker was enhanced by the special treatment of the radiator grille and the new one-piece hood hinged at the rear. Horizontal grids on the radiator grille extended rearward, with four louvres along the upper portion of the hood side panels. The headlamps were mounted higher on the sides of the radiator housing (shell), and the one-piece front fenders were fared to reduce wind resistance. The frames were of box-section side rail and X-member construction. The front suspension was the Planar-type individual wheel suspension on the President Eight models and optional on the Dictator Six models. The rear axle was the Hypoid type allowing the body floor to be lowered 3-1/16 in. Automatic overdrive was standard on the President Eight and optional on the Dictator Six. All Studebaker engines were equipped with a Fram cartridge-type oil filter that required only seasonal oil changes, spring and fall. Delco-Remy electrical equipment with a shunt wound generator was used on all President Eight models, while Auto-Lite electrical equipment was used on the Dictator Six models. The 1937 Studebaker domestic sales amounted to 70,948 cars.

On May 24, 1937, work was resumed at full capacity after the United Automobile Workers got an exclusive contract with the Studebaker Corporation, claiming a 100% unionized shop. The temporary shutdown was caused by a walk-out on May 19th, when union workers refused to work with non-union workers or those who failed to pay their dues.

Studebaker Corporation announced their 1938 model lines on September 15, 1937.

	1938 Studebaker Six	1938 Commander Six	1938 President Eight
Wheelbase (in.)	116.5	116.5	122
Price	$875-1,365	$900-1,365	$1,120-1,555
No. of Cylinders / Engine	L-6	L-6	IL-8
Bore x Stroke (in.)	3.31 × 4.37	3.31 × 4.37	3.06 × 4.25
Horsepower	90 adv	90 adv	115 adv
Body Styles	4 plus business coupe	4	custom coupe, club sedan, cruising sedan, convertible sedan
Other Features			

Among the major changes in the 1938 Studebaker models was the 6-in. increase in body width, allowing the front seat to be 55-1/2 in. wide and the rear seat 47-1/4 in. The gear shifting was accomplished by means of a vacuum-powered 5-in. lever on the lower edge of the instrument panel. This arrangement not only cleared the front compartment floor of the obtrusive gearshift lever, but also eliminated the bulge in the floor by rotating the transmission case 90 degrees to lower the height. Overdrive was still optional equipment.

Style-wise, the car body contours were made smoother. The hood side panels were without louvres, and the fared headlamps were mounted on the inside of the front fender crown. The headlamp lens shape repeated

1938 Studebaker President club sedan. (Source: Studebaker Corp.)

the form of the radiator grille. Indeed the 1938 Studebakers were handsome cars. Unfortunately, the business recession of 1938 took its toll in loss of sales; domestic sales dropped to 41,504 units. On April 2, 1938, Harold E. Churchill was promoted to chief research engineer.

The 1939 Studebaker Commander Six Series 9A and the Studebaker President Series 5C were introduced on September 22, 1938.

	1939 Studebaker Commander Six Series 9A	1939 Studebaker President Series 5C
Wheelbase (in.)	116.5	122
Price	$875-965	$1,035-1,460
No. of Cylinders / Engine	L-6	IL-8
Bore x Stroke (in.)	3.31 × 4.37	3.06 × 4.25
Horsepower	90 adv	115 adv
Body Styles	4	4
Other Features		

While the engines and chassis remained basically the same as the 1937 models, the styling was changed considerably. The front end had two chromium-plated grilles with vertical louvres placed on the radiator splash apron, one on each side of the louvre-less radiator housing (shell). The hood side panels had two longitudinal moldings located just below the continuation of the belt side molding. The headlamps were fared and blended into the leading portion of the front fenders.

The Studebaker Champion

In early 1939 Delmar G. Roos resigned from Studebaker to join Joseph E. Fields, president of Willys-Overland Motors Company. Roos was succeeded by William S. James, former research engineer of Studebaker. Just before his departure, Roos designed the new engine, chassis, and body structure of the forthcoming Studebaker Champion. Raymond E. Loewy created the splendid body design and styling. Today's designers could have profited from Roos' design of an economical, efficient, roomy, and good-handling small car (now referred to as a compact).

Studebaker left no stone unturned in publicizing the announcement of the new Studebaker Champion. One phase of the publicity was the appearance of R.E. Cole, vice president of engineering, Raymond Loewy, designer, and W.S. James, chief engineer, on the CBS National Radio network feature broadcast of "Americans at Work." Lowell Thomas and Ted Husing were the co-moderators.

Apparently, the Champion's unique introduction, its splendid fuel economy, and the quality of Studebaker cars must have pleased the buying public, because sales of Studebaker cars in 1939 amounted to approximately 84,660 cars on a calendar year production of about 92,200 units. The sales performance placed Studebaker in 8th place in industry car sales, with the Champion accounting for 55% of its sales. Thus, Studebaker was able to show a profit of $1.31 per share of stock, and the working capital increased from $8,978,479 on December 31, 1938, to $12,952,934 on December 31, 1939.

Studebaker entered the 1940 model season confidently with three lines of cars, the Champion, Commander, and President models, and production began on August 25, 1939. While the styling of the front end remained unchanged, much had been done to improve the appearance. The Champion radiator grilles of chromium-plated die-cast zinc had narrower vertical blades, while the Commander and President had a small, square, grid pattern. The sealed beam headlamps were fared into the front fenders, and the alligator-type hood hinged at the rear had a release inside the driver's compartment. A single, horizontal, stainless-steel molding adorned the hood side panel.

The wheelbases remained unchanged, but the Commander and President rear tread was increased to 61 in. to allow wider rear seat cushions. The engines remained the same as those for the 1939 models. The 1940 Studebaker Champion, identified as Model 2-G, added coupe delivery and sedan delivery commercial models to the line-up, priced at $685 and $735, respectively.

By January 15, 1940, Studebaker had 3,118 dealer outlets in the United States. For the second year in a row, Studebaker won the Gilmore-Yosemite Fuel Economy Awards in all three classes.

The new Studebaker Champion Model G was first announced on March 11, 1939, via the CBS National Radio Network, and production started on April 1, 1939.

1939 Studebaker Champion Model G

Wheelbase (in.)	110.5
Price	$625-765
No. of Cylinders / Engine	L-6
Bore x Stroke (in.)	3.00 × 3.87
Horsepower	78 adv
Body Styles	three-passenger coupe, five-passenger club sedan, five-passenger four-door cruising sedan
Other Features	Planar independent front suspension system

The Champion featured full-size interior room with riding accommodations for six grown adults in the sedan models, and for three adult passengers in the coupe models. The bodies were of all-steel construction and insulated against noise and heat. The styling features included headlamps fared into the front fenders, concealed door hinges, door checks, and rotary door latches. Due to the low silhouette and ease of entry, running boards were omitted. The Custom models were upholstered in Bedford Cord and the Custom Deluxe models were upholstered in wool broadcloth.

With the extremely low weight of only 2,375 lb, an engine compression ration of 6.5 to 1, an ignition advance of both vacuum and centrifugal, the Champion models easily attained 30 mpg. During 1939 a Studebaker Champion established an AAA-sanctioned fuel economy record of 27.25 mpg average for the entire 6,144-mile distance from the San Francisco Golden Gate Exposition to the New York World's Fair in Long Island City, New York.

1940 Studebaker Champion coupe. (Source: John A. Conde)

With the Champion in the low-priced category, the Commander in the medium-priced group, and the President in a class by itself, Studebaker had a prosperous year in 1940, selling 102,281 cars domestically, on a calendar year production of about 117,091 units. In addition, truck sales amounted to 1,207 trucks.

In June 1940, William S. Knudsen, president of General Motors Corporation and on leave of absence to serve as director of the President's Advisory Commission, appointed J.D. Biggers, president of Libbey-Owens-Ford Glass Company, and Harold S. Vance, chairman of the board of Studebaker Corporation, as his aides.

On September 6, 1940, the new 1941 Studebaker models were announced: the Champion Model 3-G, the Commander Model 11-A, and the President Model 7-C. The wheelbase of the Champion remained at 110 in., while the Commander wheelbase was increased to 119 in. and the President wheelbase was increased to 124.5 in. The Champion engine stroke was increased to 4 in. and the resulting output was increased to 80 hp. The Commander horsepower was 94 and the President output was 117 hp.

Planar front suspension was continued on all models, with semi-elliptical rear spring suspension. Except for the lengthened wheelbases and the increase in tire cross-section dimension to 6.25 in. on the Commander and 7.00 in. on the President models, the rest of the specifications remained the same as those of 1940.

Among the many styling improvements were the omission of the radiator center grille, and the side grilles on the fender hood side extension (catwalk) were broader with numerous, narrow, vertical, chrome-plated bars. With the omission of the radiator center grille, the juncture of the hood side panels at the front was concealed and adorned by a vertical stainless-steel molding sporting the new Studebaker "S" insignia. Two body side moldings originated at the front center molding, continued to the rear, and converged at the rear quarter panel. The area between these two moldings was finished in a contrasting color tone.

The Land Cruiser sedan models introduced in 1934 were reinstated in the 1941 Studebaker Commander and President models. These sedans had no quarter windows, and the rear doors were hinged at the body center B-pillar. The Land Cruiser models had two-tone solid-color interiors, and offered two-tone exterior color finishes. Running boards were omitted on all models, and a stainless-steel molding covered the full length of the body sill.

In mid-1941 the Studebaker Corporation introduced the latest of Raymond Leowy's styling creations: a sedan coupe model in the Commander and President Skyway lines. This particular body style was devoid of all exterior, bright trim moldings, except for the chromium-plated parking lamps atop the front fenders, and stainless-steel window reveal moldings. The Skyway sedan coupe also featured a shatterproof, one-piece, curved windshield.

World War II

Studebaker Corporation prospered during 1941, building and selling 114,331 cars domestically, on a calendar production of 119,325 cars and 5,078 trucks. Sales dollar volume exceeded $100,000,000, allowing Studebaker to enjoy a sizeable net profit and be able to pay dividends regularly. During 1941 Studebaker Corporation also negotiated a number of contracts with the United States Government for the manufacture of aircraft engines, military cars, multiple-drive trucks, and amphibious vehicles.

The 1942 Studebaker models were announced on September 1, 1941, and consisted of the Champion, Commander, and President series. The 1942 models were a carryover of the 1941 models except for the changes in the chromium-plated, zinc, die-cast, front grille that extended across the entire front end, with parking lamps located at each end of the grille on the Champion models. Optional fog lamps were placed at the ends of the grille on the Commander and President models, with the parking lamps atop the front fenders. The grille consisted of narrow, horizontal bars, supported by less-conspicuous vertical bars, creating a matrix grid of small squares.

The Studebaker Weasel in World War II.
(Source: Studebaker Corp.)

Studebaker turbojet engine production. (Source: Studebaker Corp.)

A total of 26 body types were available, of which six Skyway Editions were offered: three on the Commander and three on the President chassis. The technical specifications were the same as those for the corresponding 1941 models; the price range was from $745 to $1,275. While production started in August 1941, it was halted on January 31, 1942, due to the conversion of all manufacturing facilities to production of military trucks, aircraft engines, and amphibious vehicles. Thus, the total cars built and sold in 1942 amounted to only 4,662 cars and 394 trucks. Production and sales of civilian cars and trucks did not begin again until late 1945.

During World War II, Studebaker built 197,678 multiple-drive, heavy-tonnage trucks, and 15,000 Weasels (tough amphibious vehicles), which were favorites of all World War II GIs because they could go anywhere. Prime Minister Winston Churchill commented, "You Americans can do anything." He had asked for a snow vehicle. Studebaker was given the assignment, and 180 days later the Weasel had been engineered, tested, and put into production. This versatile vehicle, carrying troops and ordnance, could propel itself and travel on water, rivers, swamps, mud, sand, ice, snow, grassy turf, hard roads, or bare rock, regardless of the incline angle. It rendered invaluable service to United States and Allied forces in all theatres of war.

Studebaker also mass-produced 63,789 Wright Cyclone radial aircraft engines for the United States Air Force's Flying Fortress B-17 Bombers.

Post-War Developments

During World War II, in order to make deliveries of war materiel to the military on a timely schedule, Studebaker agreed to terms in the labor contract (to avoid work stoppages) that would not allow the corporation to be profitable during peacetime operation. With outdated plants and equipment, and a pampered workforce, Studebaker could not achieve quality nor productivity.

After V-J Day in August 1945, Studebaker converted its plants to civilian production as quickly as possible. To expedite car production, the Studebaker management decided to offer only the 1946 Studebaker Champion Skyway models (a refined version of the 1942 models) as the only cars produced between December 1945 and March 1946. These cars, offered in four body types, priced at $1,005 to $1,095, represented a good value, and 19,775 were built and sold domestically during this period.

Studebaker

The 1947 Studebaker models, introduced in the spring of 1946, were truly the only new American cars with brand new engineering and styling. Hailed as Raymond Loewy's styling masterpiece, the body sides were flat with a single roll at the belt line. The rear fender contours were pressed into the quarter panels, a trend pioneered by Packard on the 1941 Clipper, which the industry would follow.

The roof and upper structure had a light and airy motif. The rear windows curved around the roof quarters. A model called the Starlight coupe was introduced and was, at times, jokingly called the "Glass Backward Studebaker." The 1947 Studebaker models were offered in three model lines.

	1947 Studebaker Champion	1947 Studebaker Commander	1947 Studebaker Land Cruiser
Wheelbase (in.)	110	119	123
Price		$1,445 - 2,235	
No. of Cylinders / Engine	L-6	L-6	L-6
Bore x Stroke (in.)	3.00 × 3.87	3.31 × 4.37	3.31 × 4.37
Horsepower	80 adv	94 adv	94 adv
Body Styles		10	
Other Features			

During the 1947 calendar year, 102,123 Studebaker cars were built and sold domestically, for a total of 140,399 cars for the 1947 model year (1946-1947).

The 1948 Studebaker introduced in the fall of 1947 consisted of the same three model lines. The Land Cruiser, Commander convertible and coupe, and Champion convertible and five-passenger coupe had a one-piece, curved windshield. All other models had a two-piece (V-shaped) flat windshield. A booming economy and a seller's market allowed Studebaker to build and sell 143,120 cars domestically, on a calendar year production of 186,516 units, resulting in a healthy profit in 1948.

The 1949 Studebaker continued virtually unchanged, except the Commander engine stroke was increased to 4-3/4 in. giving a piston displacement of 245.6 cu. in., developing 100 hp. The U.S. Office of Price Administration allowed Studebaker to increase their prices slightly, for a price range of $1,588 to $2,467. Sales were brisk, amounting to 199,460 cars domestically, resulting in a record profit in 1949.

The 1950 Studebaker was announced in mid-1949 and consisted of three model lines: the Champion, the Commander, and Land Cruiser sedan. The most significant change was the aircraft hood and front grille treatment, often referred to as the "bullet" nose. The rear quarters had a vertical tail and directional lamps. A new automatic transmission by the Borg-Warner Corporation was available as an extra-cost option. A total of 21 body styles was offered in a price range of $1,420 to $2,330. Sales were exceptionally good, selling 268,229 cars in 1950.

The 1951 Studebaker announced in mid-1950 was available in the same three lines. A new, 120-hp, valve-in-head, V-8 engine was introduced on the Commander and Land Cruiser. The engine of 232.6-cu.-in. displacement had a bore of 3-3/8 in. and a stroke of 3-1/4 in. The Champion L-head, in-line, six-cylinder engine developed 85 hp. The styling remained the same as for the 1950 models, except for the paint schemes, and the one-piece, curved windshield used on all models.

1950 Studebaker Champion custom sedan. (Source: Studebaker Corp.)

A Downward Turn

When the United States was engaged in the Korean War, Studebaker was called upon to build a great quantity of J-47-S25 jet aircraft engines for the giant bombers of the United States Air Force. The war also caused material shortages. Thus, the domestic sales dropped to 205,514 cars. Because of the antiquated manufacturing plants, Studebaker profits dropped.

The 1952 Studebaker model line-up remained the same, except the aircraft front end gave way to a more conventional, narrow, two-piece grille. Three manual transmissions were standard, with overdrive and automatic transmission as optional extra-cost items. The Korean War caused further restrictions on material availability. The chromium-plated trim was without the layer of copper between the nickel and chromium plating. Twenty body types were offered at prices from $1,735 to $2,550. Sales dipped further to 157,902 cars domestically, and profits dropped.

The 1953 Studebaker cars were announced in late 1952, having completely new styling by Raymond Loewy. All models were lower with a new hood and front-end lines featuring a gently sloping front grille area. Two horizontal wedge-shaped grilles graced the front end, providing a styling trend that would be copied 20 years later. The sporty Starliner hardtop model featured a one-piece wrap-around rear window.

The low silhouette made the Studebaker models look much longer. The styling improvement helped sales climb to 161,257 cars domestically. But the increased sales were not enough to offset Studebaker's rising operating costs, which caused Studebaker to sustain a loss in 1953.

The 1954 Studebaker, introduced in the fall of 1953, had substantially no changes from the 1953 models except for the vertical bars in the front grilles. The new Conestoga two-door station wagon was priced at $2,295 for the Champion and $2,555 for the Commander model.

Studebaker-Packard Corporation

On June 22, 1954, it was announced that a merger agreement between the Packard Motor Car Company and the Studebaker Corporation was signed by James J. Nance for Packard and Harold S. Vance for Studebaker. The merger created the Studebaker-Packard Corporation, incorporated in Michigan and Indiana on October 1, 1954. Harold Vance was elected chairman of the board and James Nance was named president and general manager. (See the Packard chapter in this volume.)

The merging of the two corporations progressed smoothly, with most officers retaining their positions within their respective divisions. Paul G. Hoffman retained the position of executive vice president of the Studebaker Division. Studebaker finished the 1954 model year with domestic sales of 95,914 units, sustaining an eight-digit operating loss.

The 1955 Studebaker consisted of three model lines: the Champion, Commander, and a reinstated President series. The new Studebaker President Speedster hardtop made its debut at the National Auto Show in New York in January 1955. The Studebaker President was powered by the larger, 259.2-cu.-in., V-8 engine of 185 hp. The styling remained the same except for a bolder, broader bumper and front grille outer housing treatment. A bright metal dihedral "V" adorned the grille inlet opening. The sedan models had wrap-around windshields. Available in 25 body styles, priced at $1,740 to $3,255, the Studebaker domestic sales hovered at 95,761 units, despite the combined effort of both Studebaker and Packard sales outlets. Needless to say, the Studebaker-Packard Corporation suffered a severe financial loss in 1955.

The 1956 Studebaker had a noticeable Packard influence, particularly in the Golden Hawk, which had a pleasing, vertical, gently sloping front grille, and was powered by a 352-cu.-in. displacement Packard V-8 engine of 275 hp. The extensive and costly promotional effort and concentrated advertising did not improve the sales, as they dipped to a dismal 76,402 units domestically, while the industry enjoyed a boom year in 1956.

1956 Studebaker Golden Hawk coupe. (Source: John A. Conde)

The Studebaker-Packard Corporation losses were so great, and the operating capital was so depleted in 1956, that it was only the financial help from the Curtiss-Wright Corporation that prevented the Studebaker-Packard Corporation assets from being sold at a receiver's sale. The Curtiss-Wright Corporation was able to do this as a tax write-off. The departure of James Nance on August 1, 1956, was coincident with the Curtiss-Wright financial deal. The Detroit Packard plant was closed, and Packard and Clipper components and materials were sold off as surplus. Studebaker and Packard operations were consolidated in South Bend, Indiana.

In 1958 Harold Churchill, vice president of engineering, was named president of the Studebaker-Packard Corporation. Churchill introduced the Studebaker Lark in late 1958. The Lark sold well, permitting Studebaker to sell a total of 133,382 cars domestically in 1959, earning a profit for the first time in over six years.

Churchill retired in 1961 and was succeeded by Sherwood Egbert. Egbert introduced the Studebaker Grand Tourismo hardtop and the Avanti in 1962. It was felt that the Packard name was hurting Studebaker sales, so it was dropped from the corporate name in 1962. Because of ailing health, Egbert resigned in 1963. The Studebaker production was limited to the Hamilton, Ontario, Canada plant with Gordon Grundy as general manager.

By 1966 Studebaker had diversified into other products, namely: STP, Paxton supercharger, Wagner Electric, and several others. Car production ceased in 1967, and the Hamilton, Ontario, plant was closed. Nathan Altman and Leo Newman, former Packard dealers, bought the Avanti interests in 1965 and organized the Avanti Motors Corporation. In 1982 the Avanti Motors Corporation was sold to the AMW Incorporated.

1929 Elcar. (Source: Indiana Historical)

The Pratt Family and the Elcar Motor Car Company

Fred Pratt and his sons, William and George, used sound business practices to build a good product along with honest merchandising. Fay Sears predicted in 1926, "that as long as a small manufacturer realized his limitations, he could co-exist among industrial giants." While the prediction could have been valid and very noble at the time, it could not hold true during the cruel Depression that followed soon afterward. Elcar became another industrial casualty, along with many other fine car builders.

* * *

Kicked Into a New Field

Frederick B. Pratt was born circa 1820. He owned a hardware store in Elkhart, Indiana. In 1873 he became very interested in the new horse-drawn buggies of the time. In order to examine the new buggies, he visited a nearby buggy dealer's display. He became quite engrossed and as he knelt to measure the buggy's dimensions, the annoyed dealer landed a swift boot to Pratt's posterior and ordered him off the property.

Fred Pratt became perturbed, but not angry. However, he was heard to mutter, "I'll show him, I will build a better buggy." Apparently Fred's bruised pride forced him to make good on his challenge. Shortly after the incident, he informed his family of his decision to build buggies, which meant giving up their well-established hardware business. While Fred's eldest son, William, questioned the wisdom of the decision, Fred's confidence and business sense prevailed, and it proved to be a wise decision "To build a better buggy and do it differently."

Elkhart Carriage and Harness Manufacturing Company

The firm was known as the Elkhart Buggy Company, with Fred and William Pratt as partners. Fred's younger son, George, joined the company in 1882. In 1888, when stock was sold in the company, it became known as the Elkhart Carriage and Harness Manufacturing Company, at which time Otis B. Thompson came aboard. It was about this time that the company went national in its merchandising, selling direct by catalog, following the example of Sears, Roebuck and Company and Montgomery Ward & Co. The Pratt's stressed the point of being pioneers in catalog selling.

It is quite possible that the company letterhead helped their business. In the section across the top were photographs of the two factories, one on Beardsley Avenue at Michigan Street and the other at Pratt and East Streets. Printed across the top was the statement: "As far as 5,000 miles away you can buy as safely as if you were here." On either side of the statement were imprinted the names of the officers: William B. Pratt, president, and George B. Pratt, vice president and treasurer.

During this time the Elkhart Carriage and Harness Company was very prosperous; not only did they employ 500 people, they built and sold about 10,000 buggies a year. William Pratt was a man with a human touch who could remember the name of each of the 500 employees in the plant. He was interested in them and their welfare, and for that reason he often gave them a raise when there was a new baby in the family because another mouth to feed would cost more money. His concern for his employees made him a very popular man.

The Elkhart Carriage and Harness Company built not just any kind of buggy, they offered Phaetons, stanhopes, surreys, and cabriolets, as well as spring wagons. Their illustrated catalogs were over 100 pages, and they shipped buggies to customers all over the United States. Customers could look over the Pratt buggy

Early Pratt buggy. (Source: Carriage Monthly)

carefully, and compare it to those displayed by local buggy dealers. The shipping charges were modest, and the buggies were remarkably low-priced.

Feeling that the business was successful and safely in the hands of his sons, Frederick retired in 1891. He remained active in business, civic, and fraternal organizations.

The Horseless Carriage

The Elkhart Carriage and Harness Company was indeed successful, but a new form of transportation was presenting a challenge: the "horseless carriage." The Pratt brothers experimented with a gasoline-engine-powered buggy starting in 1906. They were not satisfied with the prototypes, until 1909 when they developed and built a fine touring car, powered by a 30-hp, four-cylinder, gasoline-fueled engine. Being realists, William and George Pratt realized that the price of the car would not lend itself to selling by direct mail. Hence, a concentrated drive was made to establish dealers to sell their car. They did, however, continue to sell their carriages, buggies, wagons, and harnesses by mail.

In 1911 the Pratts dropped Elkhart from the name of the car because there was another car being built across town called the Crow-Elkhart. To avoid confusion, the name "Pratt" became the identity of the car from 1911 through 1915. The 1911 Pratt car was offered in touring, roadster, and limousine body types, priced at $1,800 to $2,000. In 1912 a demi-tonneau was offered in place of the limousine. From 1913 through 1915, Pratt concentrated on touring car and roadster models. However, the prices escalated to $1,950. During 1915 a six-cylinder-engine-powered car was added, priced at $2,150 and $2,250 for the roadster and seven-passenger touring, respectively.

The Elcar

In October 1915 the company was reorganized as the Elkhart Carriage and Motor Car Company, to manufacture a new, less-expensive car called "Elcar." The 1916 Elcar was introduced in late 1915.

	1916 Elcar
Wheelbase (in.)	114
Price	$795
No. of Cylinders / Engine	4
Bore x Stroke (in.)	3.50 × 5.00
Horsepower	34 adv
Body Styles	roadster, touring
Other Features	

Elkhart Motor Car Co. works. (Source: Indiana Historical)

For 1917 Elcar added a four-passenger "cloverleaf" roadster priced at $845 and a five-passenger sedan priced at $995.

In early 1917, after the declaration of war on Germany, Elkhart Carriage and Motor Car Company was requested by the United States Government to manufacture ambulance bodies. The Pratt brothers promptly discontinued the manufacture of horse-drawn vehicles, and concentrated on ambulance body production along with reduced automobile production. The Pratts kept car production limited to three body types in 1918.

	1918 Elcar D-4	1918 Elcar D-6
Wheelbase (in.)	116	116
Price	$1,095-1,625	$1,295-1,795
No. of Cylinders / Engine	4	L-6
Bore x Stroke (in.)	3.50 × 5.00	3.25 × 4.50
Horsepower	35 adv, 19.6 NACC	40 adv, 25.35 NACC
Body Styles	touring, roadster, sedan	touring, roadster, sedan
Other Features		

After the signing of the Armistice on November 11, 1918, Elcar went into full production of 1919 passenger cars; however, there was a slight increase in car prices.

	1919 Elcar H-4	1919 Elcar D-6
Wheelbase (in.)	116	116
Price	$1,175-1,625	$1,375-1,795
No. of Cylinders / Engine	L-4	L-6
Bore x Stroke (in.)	3.50 × 5.00	3.25 × 4.50
Horsepower	37.5 adv, 19.6 NACC	40 adv, 25.35 NACC
Body Styles	touring-roadster, sportster, sedan	touring-roadster, sportster, sedan
Other Features		

The sedan windshields were sloped and the body had staggered doors; the right door in the middle served both front and rear seats. The left side had a front and a rear door.

For 1920 Elcar continued with the same model line-up, with further increases in prices due to the inflationary period after World War I. The prices ranged from $1,395 to $1,595 for the four-cylinder models, and $1,595 to $2,195 for the six. During 1920 Elcar boasted of the Elcar-Lycoming engine in the four, and a new Continental Red Seal 7-R engine in the six. Delco electrical equipment, Stromberg carburetor, Salisbury rear axle, and Muncie transmission were listed as features, and a Boyce Moto-Meter was standard equipment. Considering the post-WWI depression, Elcar sales continued reasonably well.

The same model line-up continued for 1921, with a coupe being added to the Elcar four, priced at $2,095. The prices continued to climb, ranging from $1,495 for the four, and $1,795 to $2,395 for the six. The chassis wheelbase was increased to 117 in.

Most Attractive Line in Popular Priced Field

Never before have cars been built at Elcar prices with the style and finish —the all round quality and beauty—embodied in the new Elcar models.

Here is a car that absolutely **satisfies** the buyer with the most discriminating tastes. The Elcar is the very picture of grace, with deep, lustrous finish and luxury of appointments that you seldom find in cars under $2,000. The Elcar is even better than it **looks**.

Each and every detail bears the closest scrutiny. The motors are wonderful. Each part is built 150% strong. The spring suspension is wonderfully efficient. It's a good looking, classy performing, long lasting car of low upkeep. A combination that can't be beaten.

See the Elcar at the Chicago Show
The new models will be on display

GREATEST FAMILY CAR
In America at Its Price

Stylish, up-to-date design embodying the latest ideas in comforts and conveniences. Double cowl. Hand pads to match upholstery. Latest type one man top of 1919 design. Four cylinder model has powerful long stroke Elcar-Lycoming motor. Develops 37½ H.P. at 2,100 r.p.m. Six cylinder model has famous Red Seal Continental 3¼x4½ inch engine. Develops 40 H.P. at 2,100 r.p.m. Two unit electrical system. 116 inch wheel base. Complete in every detail.

4-cylinder model....$1175
6-cylinder model....$1375

The reproductions on these two pages suggest but do not do justice to the beauties of the Elcar. When you **see** the Elcar, you will marvel at the quality appearance that is reflected in every detail, from the way it is put together to its brilliant, high-grade finish.

A Real Opportunity, Dealers

The Elcar will be the VALUE car at the Chicago show. Elcar growth the past three years has been the sensation of the auto industry. Several attractive territories are without Elcar representatives, and offer real opportunities. See us at the Chicago show and let's talk it over. If you are not going to be at the show write or wire at once.

Elkhart Carriage & Motor Car Co.
781 Beardsley Ave.   **Elkhart, Ind.**

When Writing to Advertisers, Please Mention Motor Age

January 23, 1919 — MOTOR AGE

Stylish Touring Roadster

A very popular type of body, intermediate between the touring car and the roadster. Much favored by small families, and by young men who like a car of good capacity without the family car appearance. The car has four doors, forward compartment same as in touring car. Rear compartment uncommonly roomy for car of this type. In the rear a very commodious carrying space is provided. Made on both four and six cylinder chassis.

4 cylinder Touring
   Roadster..........$1175

6 cylinder Touring
   Roadster..........$1375

Smartest Sportster on Wheels

The Elcar Sportster, first introduced during the season of 1918, won immediate attention by its jaunty lines and original design. It is a car that "stands out," yet thoroughly practical because of Elcar construction and economy of service. The body is custom built, deeply upholstered, and richly finished in coach blue and olive green, with maroon, moleskin, beige and olive brown as optional colors. The beveled body-edges and square doors give an exceedingly smart appearance. Bevelled plate glass rear windows, extra long steering column, wire wheels, nickeled outside door handles, all add to the general effect. Price is same as for regular models.

4 cylinder..........$1175

6 cylinder..........$1375

A Luxurious Sedan To Fit Moderate Incomes

The Elcar Sedan provides plenty of room for five with all the comfort appointments you look for in a high grade car of this type. It is solid Sedan construction from the sills up —not a Sedan top on a touring body or imitation-Sedan in any sense. Built to withstand road stresses without developing squeaks or rattles. Large plate glass windows lower into the doors and sides of the car to admit "all the air there is" when the weather is fine. Doors are staggered, the right-hand door being in the center, while the left-hand door is placed forward for the convenience of the driver and shaped to conform to the slope of the windshield. The forward seats are of individual type with passageway between so that occupants may enter from either side of car.

4 cylinder......$1625

6 cylinder......$1795

When Writing to Advertisers, Please Mention Motor Age

Elcar Motor Car Company

On July 2, 1921, William and George Pratt retired and sold their interest in the Elkhart Carriage and Motor Car Company to a group of automotive executives formerly associated with the Auburn Automobile Company. Fay B. Sears was elected president and A. Michael Graffis was named vice president of engineering. G.W. Bundy and W.H. Denison also joined the company. Shortly afterward the name of the company was changed to Elcar Motor Car Company.

The 1922 Elcar prices remained high, with the four-cylinder models priced at $1,145 to $1,645, and the six-cylinder prices ranging from $1,595 to $2,495. During 1922 the management realized that the pricing was too high, so in 1923 the Elcar cars were priced realistically, $965 to $1,425 for the four and $1,395 to $1,995 for the six.

Late in 1922 Elcar entered the taxicab manufacturing field by obtaining an order for 1,000 cabs from the Diamond Cab Company of New York City. The five- to seven-passenger taxicab models were priced at $2,100 for the four-cylinder models and $2,450 for the six-cylinder models. Elcar continued building taxicabs for the remainder of the company's existence.

During 1923 Elcar production and sales were approximately 2,000 cars, and 1924 sales amounted to about 1,300 cars. A five-passenger sport sedan was added to the six-cylinder model line-up, priced at $2,195.

The Eight-Cylinder

In January 1925 at the National Automobile Show in New York, the new eight-cylinder-engine-powered Elcar was introduced.

1925 Elcar 8-80

Wheelbase (in.)	127
Price	$2,165-2,865
No. of Cylinders / Engine	IL-8
Bore x Stroke (in.)	3.12 × 4.25
Horsepower	65 adv, 31.25 SAE
Body Styles	7
Other Features	four-wheel hydraulic brakes, wood-spoke artillery wheels mounted with 30 x 6.00 balloon tires

The styling was attractive, in keeping with the trends of the time, yet distinctive. It had a massive nickel-plated radiator shell shaped in the form of an octagon with the bottom truncated. The vertical hood side louvres were arranged in two groups. The window line molding merged with the lower belt molding at the cowl. Wood-spoke artillery or steel disc wheels were available. Approximately 2,000 cars were built and sold in 1925. The four-cylinder and six-cylinder models also were continued, available in 13 body styles and priced at $995 to $2,195.

Some changes were made to the eight-cylinder model for 1926.

ELCAR
A WELL BUILT CAR

ELCAR Four Cylinder Touring

FOURS and SIXES
The Complete Profit Line for Dealers

DEALERS
Here is the price range that insures
PROFITS
—
FOURS
$965.00 to $1425.00
SIXES
$1395.00 to $1995.00
[f. o. b. factory]

Ask about
ELCAR TAXICABS
unusual proposition—quick turnover

New Elcar—new opportunity. Now is the time to "get set" on a real profit basis, selling the right car at the right price. Eliminate the annual question, "What car will I sell this year?" Alert dealers act—SIGN UP—while others hesitate.

The new Elcar line of Fours and Sixes has been developed along highest engineering standards. There is a market for Fours and Sixes which, though reasonably priced, are quality built by quality builders.

The new Elcar line gives the dealers' salesmen bigger leeway—wider field of prospects—more speed in closing sales.

ELCAR SIX is equipped with 8-R Continental Red Seal Motor and other units of equal worth.

ELCAR FOUR has improved Lycoming-Elcar motor. Elcar bodies are all Elcar-built insuring style and satisfaction.

Write TODAY for money-making facts.

ELCAR MOTOR COMPANY, Elkhart, Indiana
Builders of Fine Vehicles Since 1873

1926 Elcar 8-81

Wheelbase (in.)	127
Price	$2,095-2,765
No. of Cylinders / Engine	IL-8
Bore x Stroke (in.)	3.25 × 4.50
Horsepower	70 adv, 33.8 SAE
Body Styles	7
Other Features	

Three body types priced at $1,295 to $1,675 were offered in the Elcar 6-65 models, and three body styles were available in the Elcar 4-55 models priced at $1,095 to $1,395. The four-cylinder model was dropped after 1926. Sales increased somewhat during 1926 because of the financial boom, but not enough to affect the condition of the company.

In December 1926 Elcar entered the low-priced, eight-in-line field through the introduction of the new 1927 Elcar 8-82.

1927 Elcar 8-82

Wheelbase (in.)	123
Price	$1,595-1,870
No. of Cylinders / Engine	IL-8
Bore x Stroke (in.)	2.75 × 4.75
Horsepower	70 adv, 24.2 SAE
Body Styles	
Other Features	

The other two series of the 1927 Elcar line-up were the 6-70 and 8-90, built on 117- and 127-in. wheelbase chassis, respectively. The 6-70 had a 52-hp, six-cylinder Continental engine, while the 8-90 was powered by an 84-hp, eight-in-line Lycoming engine. Ten body styles were offered, priced at $1,295 to $2,065.

1927 Elcar. (Source: Indiana Historical)

During 1927 Elcar stressed quietness of operation (the shockless chassis) accomplished by the use of felt insulation and the incorporation of Bel-Flex rubber and fabric composite spring shackles. The 1927 body styling was basically the 1926 design with appearance improvements. Round, tubular, front and rear bumpers were offered. Sales continued to be modest, but the Elcar management believed that a small manufacturer could make the most of their opportunities as long as they knew their limitations.

On January 7, 1928, at the National Automobile Show in New York, Elcar introduced the new, lower-priced Elcar 8-78.

1928 Elcar 8-78

Wheelbase (in.)	123
Price	$1,395
No. of Cylinders / Engine	IL-8
Bore x Stroke (in.)	2.75 × 4.75
Horsepower	62 adv, 24.2 SAE
Body Styles	4
Other Features	

A total of 20 body styles were offered in Elcar's entire line-up on three chassis wheelbase lengths of 117, 123, and 132 in. Powered by a 52-hp, six-cylinder engine and three Lycoming eight-in-line engines of 62, 70, and 84 hp, the 1928 Elcar was priced at $1,295 to $2,565. The styling was distinctive, with the window belt molding and the lower belt molding joined by a curved vertical molding at the cowl. This vertical molding delineated the hood and cowl from the rest of the body panels. The vertical hood side louvres were grouped with five at the front and ten in the rear portion of the hood side panel. Despite the proliferation of body types, Elcar sales did not improve in 1928.

On November 3, 1928, Elcar made many bold changes in their 1929 model offerings, including a further reduction in the Model 75 price.

1929 Elcar 75

Wheelbase (in.)	117
Price	$995-1,195
No. of Cylinders / Engine	L-6
Bore x Stroke (in.)	2.87 × 4.75
Horsepower	60 adv, 19.8 SAE
Body Styles	7
Other Features	internal expanding Lockheed hydraulic brakes, shock absorbers, and bumpers

Other 1929 Elcar models included the 95, 96, and 120. Models 95 and 96 had a 123-in. wheelbase, powered by an 80-hp, Lycoming eight-in-line engine. Model 120 was on a 134-in. wheelbase powered by a 130-hp, Lycoming eight-in-line engine. The 1929 Elcar model lines offered 26 body types, priced at $995 to $2,645. The 1929 styling was more conventional in keeping with other competitive makes. It had long, sweeping, full crown front fenders, peaked hood center sections, and a chromium-plated radiator shell. The bullet-shaped headlamps were carried on the front fender tie bar. Similarly shaped cowl lamps were mounted on a circular molding at the cowl-hood juncture. The window belt and the lower belt joined into a single molding. Wood-spoke wheels mounted with balloon tires were standard.

An Abrupt End

Even with a proliferation of models available, "Black Thursday," October 24, 1929, brought an abrupt and steep decline in all automobile sales. On October 26, 1929, Elcar announced their new 1930 models, consisting of four lines: the Model 75 powered by a six-cylinder Continental engine, Models 95 and 96 powered by a 90-hp, Lycoming eight-in-line engine, and the Model 130 powered by a larger Continental eight-in-line engine of 3-3/8-in. bore and 4-1/2-in. stroke. They were some of the finest cars ever offered, but the timing couldn't have been worse.

The basic 1929 styling was carried over into 1930. In all there were 32 car models and types available in four Elcar lines ranging in price from $995 to $1,995. The Models 96 and 130 had a four-speed transmission as standard equipment, which was optional at extra cost on the Model 95. Even though the 1930 Elcar lines were the finest cars Elcar had ever built, and they represented a great motor car value, the Depression had taken its toll. Sales plummeted, and by April 1930, Elcar was in financial trouble.

A final and desperate attempt to save Elcar Motor Car Company was made in 1930 with the proposed merger of Elcar and the Lever Corporation, but it was to no avail. Elcar ceased car production in 1930. They continued for a short time afterward to build taxis to fill their orders.

Joseph Moon

Joseph Moon never sought nor intended to establish outstanding sales records. Instead he provided the American driving public with the exciting Moon cars and the exotic Diana. Like other small, independent car manufacturers, Moon became a victim of the Great Depression.

* * *

Joseph W. Moon was born in Brown County in southwest Ohio on March 29, 1850. After early experience in selling buggies in the midwest, Joseph and his brother, John C. Moon, moved to St. Louis, Missouri, in 1882 and organized the Moon Brothers Buggy Company. In 1892 the Moon Brothers Buggy Company was dissolved and John became the president of the Landis Machine Company. Joseph established the Joseph W. Moon Buggy Company. He was very successful in building and merchandising buggies and carriages. He also trained many pioneers in the buggy business who became prominent in their own right, among them Joseph J. Cole. Shortly after the turn of the century, while attending various carriage conventions, Moon became aware that the automobile would replace horse-drawn vehicles.

Moon Buggy Switches to Automobiles

On October 4, 1905, the following notice appeared in *Horseless Age*: "Joseph W. Moon Buggy Company of St. Louis, Missouri, are erecting a building of 140 ft. by 120 ft. in floor plan, three stories high

Louis P. Mooers.

(total 16,800 sq. ft.) adjoining their present plant, which will be used as an automobile factory." Louis P. Mooers had resigned his position as chief engineer at Peerless Motor Car Company on April 5, 1905, and joined the Joseph W. Moon Buggy Company in early May 1905 to design and engineer the Moon car, which went into production in early 1906.

1906 Moon

Wheelbase (in.)	106
Price	$3,000
No. of Cylinders / Engine	L-4
Bore x Stroke (in.)	4.50 × 5.00
Hp	30-35 adv, 32.4 ALAM
Body Styles	five-passenger touring
Other Features	

Ignition was by jump-spark from a storage battery to quadruple coils on the dashboard. It had individual cylinders bolted to the upper crankcase and was water-cooled by a Briscoe radiator. The engine power was transmitted through a multiple-disc clutch, three-speed sliding gear transmission, and an exposed propeller shaft to the rear axle. The suspension was by two parallel springs of semi-elliptical design and two parallel fully elliptical springs at the rear. Steering control was by steering wheel and steering gear mounted on the right side of the chassis. The side-entrance touring car body was made of wood, well-finished and upholstered. Approximately 45 cars were built and sold in 1906.

The 1907 Moon car had basically the same engine dimensions and chassis features, but was in fact an entirely new car offering many innovations.

1907 Moon

Wheelbase (in.)	110
Price	$3,000-3,800
No. of Cylinders / Engine	4
Bore x Stroke (in.)	4.50 × 5.00
Hp	30-35 adv, 32.4 ALAM
Body Styles	roadster, five- and seven-passenger touring
Other Features	aluminum bodies

It was powered by a new Rutenber, four-cylinder, overhead camshaft engine. The valve mechanism was actuated by a train of gears and shafts. The styling was improved, similar to that of most expensive cars built in America. Unfortunately, the price structure of Moon cars, and the panic of 1907, limited the potential buyer market.

Moon Motor Car Company

On November 10, 1907, the Moon Motor Car Company of St. Louis was incorporated under the laws of the state of Missouri, with $175,000 capital. It had purchased the automobile department of the Joseph W. Moon

Buggy Company and conducted it as a separate organization. The officers of the new company were: Joseph W. Moon president, Stewart MacDonald (Moon's son-in-law) vice president, Alfred Moberly treasurer, George W. Shelp secretary, E.J. Moon assistant secretary, and Louis P. Mooers general manager and chief engineer. At that time they were located in the Moon buggy plant, but they intended to move into a factory all their own.

When Moon introduced his first car he did not proceed through the usual stages of motorized carriage, nor the runabout stages of design, but rather went full-force into a real automobile in 1906. It featured a four-cylinder engine, multiple-disc clutch, sliding gear transmission, shaft drive, and a full-floating rear axle. Mooers did his homework well at Peerless before coming to Moon. His performance became a tough act to follow after he left Moon in late December 1908. The quality of Moon construction was always the best, due to the guidance of Carl Burst and the pride of workmanship of the mid-America workers.

Early Problems

The problem at Moon Motor Car Company was not one of design nor quality, but rather the merchandising philosophy and lack of dealer representation. While Moon had dealerships in the major cities, medium cities were represented by multiple-line dealers with divided interests, and the small towns had no Moon dealers at all. It seemed that Joseph Moon was infatuated with catering to the elite clientele in New York, Philadelphia, and New England. However, this market was already crowded by established manufacturers such as Cadillac, Packard, Peerless, Pierce-Arrow, Lozier, and Winton, not to mention the imported Mercedes, Renault, and Rolls-Royce. The market in this price class was definitely limited.

The 1908 Moon cars had the same features as the 1907 models including the overhead-camshaft engine, semi-elliptical springs in the front, and three-quarter elliptical at the rear. A seven-passenger touring model was added to the line-up that included a five-passenger touring and roadster model. The prices were $3,000 to $3,500.

1908 Moon. (Source: John A. Conde)

1908 Moon.

For 1909 Moon continued with the basic 1908 model construction, except the engine was changed to a T-head design and the cylinders cast in pairs. A new seven-passenger limousine model was added, priced at $3,800. The Moon sales were modest due to the price structure and the fierce competition in that market.

During 1910 a serious attempt was made to reduce the entry-level prices by introducing the new Moon Model 30.

	1910 Moon Model 30	1910 Moon Model 45
Wheelbase (in.)	114	121
Price	$1,500-2,750	$3,000-4,000
No. of Cylinders / Engine	T-4	T-4
Bore x Stroke (in.)	4.25 × 5.00	4.75 × 5.00
Hp	28.9 ALAM	36.1 ALAM
Body Styles		5
Other Features	pistons had four rings and connecting rods had four-bolt attachment	

The Moon car sales improved somewhat, but for the next two years Moon sales would average between 800 and 1,200 units annually.

The Moon Models 30 and 45 were carried over into 1911, with the 30 available in five body styles priced at $1,500 to $2,750. The 45 was also available in five body styles priced at $3,000 to $4,000. Sales during 1911 were modest. For 1912 the wheelbase of the Model 30 was increased to 116 in. and that of the 45 to 123 in. The 30 was priced at $1,600 to $1,650 and the 45 at $3,000 to $4,800. A new Model 40 was added to the line-up, built on a 120-in. wheelbase, available as a five- or seven-passenger limousine, priced at $3,000 and $3,200, respectively. Despite the expansion into numerous body styles, sales did not increase considerably.

1910 Moon.

1910 Moon. (Source: John A. Conde)

Sales Pick Up

On September 26, 1912, the Moon Motor Car Company announced its 1913 models, consisting of the 39, 48, and 65.

	1913 Moon Model 39	**1913 Moon Model 48**	**1913 Moon Model 65**
Wheelbase (in.)	116	121	132
Price	$1,650-3,000	$1,650-3,000	$2,500
No. of Cylinders / Engine	T-4	T-4	T-6
Bore x Stroke (in.)	4.00 × 5.75	4.50 × 5.00	4.00 × 5.75
Hp	39 adv, 25.6 NACC	48 adv, 32.6 NACC	65 adv, 38.4 NACC
Body Styles	11	11	five-passenger touring
Other Features		Wagner electric starting motors	

1912 Moon.

All 1913 Moon engines were of the T-head design, the cylinders cast in pairs, with the Model 39 and 65 cylinder blocks interchangeable. The steering gear and wheel were now on the left side, with the change gear and hand brake levers in the center of the floorboard. With the proliferation of body styles, Moon Motor Car Company enjoyed a sales total of over 1,500 cars in 1913.

For 1914 the Moon model availability was reduced to two series: the 4-42 and the 6-50.

	1914 Moon Model 4-42	**1914 Moon Model 6-50**
Wheelbase (in.)	118	129
Price	$1,750-2,250	$1,750-2,250
No. of Cylinders / Engine	T-4	L-6
Bore x Stroke (in.)	4.25 × 5.00	3.75 × 5.25
Hp	42 adv, 28.9 NACC	50 adv, 33.75 NACC
Body Styles	7	7
Other Features		

The limousine body style was discontinued, and the fuel tank was moved to the rear of the car, employing a pressure fuel system. Even though the lowest-priced model price escalated to $1,750, the favorable styling and mechanical improvements kept the sales up to a satisfactory level and generated a modest profit.

In August 1914 the new 1915 Moon cars were announced, consisting of Models 4-38, 6-40, and 6-50.

1914 Moon.

	1915 Moon Model 4-38	**1915 Moon Model 6-40**	**1915 Moon Model 6-50**
Wheelbase (in.)	122	122	130
Price	$1,350	$1,575-2,250	$2,150-2,950
No. of Cylinders / Engine	L-4	L-6	L-6
Bore x Stroke (in.)	3.75 × 5.00	3.50 × 5.00	3.75 × 5.25
Hp	38 adv, 22.5 NACC	40 adv, 29.4 NACC	50 adv, 33.75 NACC
Body Styles			
Other Features			four-speed transmission

In mid-1915 a new Moon Model 6-30 was introduced, available in touring and roadster body types, priced at $1,195. The model line-up gave Moon a total of 16 body types, priced at $1,195 to $2,150. The new features included a Stewart vacuum tank fuel feed. The bodies were exquisitely styled and sales were good due to the introduction of the 6-30.

The 1916 Moon models were a continuation of the 1915 Models 6-30 and 6-40; the 4-38 and 6-50 were discontinued.

	1916 Moon Model 6-30	**1916 Moon Model 6-40**
Wheelbase (in.)	118	124
Price	$1,195-1,475	$1,195-1,475
No. of Cylinders / Engine	L-6	L-6
Bore x Stroke (in.)	3.25 × 4.50	3.50 × 5.00
Hp	30 adv, 25.35 NACC	40 adv, 29.4 NACC
Body Styles	4	4
Other Features		

Starting was accomplished by a Delco combination starting-charging unit. The bodies incorporated the Moon "tumble home" design, and the seven-passenger touring had disappearing jump seats. Due to the pricing structure, sales in 1916 were a strong 5,351 cars.

The 1917 Moon models were introduced on August 15, 1916, consisting of the Moon Models 6-43 and 6-66. The new features embodied in the 1917 models included the Delauny-Bellville-type double-cowl touring car bodies. The charging and starting were accomplished by a Delco two-unit system, employing a Bendix drive for the starting motor.

	1917 Moon Model 6-43	**1917 Moon Model 6-66**
Wheelbase (in.)	118	125
Price	$1,295-1,950	$1,395-2,350
No. of Cylinders / Engine	L-6	L-6
Bore x Stroke (in.)	3.25 × 4.50	3.50 × 5.25
Hp	40 adv, 25.35 NACC	66 adv, 29.4 NACC
Body Styles	8	8
Other Features		

The bodies of the 1917 Moon cars were very stylish incorporating a streamlined design. A crescent-shaped Moon mascot adorned the nickel-plated radiator cap.

The United States entry into World War I in April 1917 affected the availability of materials, and thus the production and sales of 1917 cars. Moon Motor Car Company could build and sell only 1,049 cars in 1917.

On September 24, 1917, the Moon Motor Car Company took over part of the Joseph W. Moon Buggy Company plant to increase the car engine manufacturing facilities. The Moon Motor Car Company announced the sensational, new Moon 1918 Model 6-36, along with the Moon 6-45 and 6-66.

	1918 Moon Model 6-36	**1918 Moon Model 6-45**	**1918 Moon Model 6-66**
Wheelbase (in.)	114	125	125
Price	$1,095	$1,575-2,650	$1,575-2,650
No. of Cylinders / Engine	L-6	6	6
Bore x Stroke (in.)	2.87 × 5.50	3.25 × 4.50	3.50 × 5.25
Hp	36 adv, 19.76 NACC	45 adv, 25.35 NACC	66 adv, 29.4 NACC
Body Styles	five-passenger touring	7	7
Other Features			

The Moon 6-36 touring featured a one-piece, plate glass windshield, a five-passenger double-cowl body, and a genuine walnut instrument panel. It was an exceptionally good buy; however, the restrictions of World War I limited 1918 Moon car production and sales to 1,154 cars.

1918 Moon sedan. (Source: John A. Conde)

For 1919 the Moon Motor Car Company confined its model lines to two chassis: the 6-36 and the 6-66. The 6-36 continued as a touring only, while the 6-66 offered a club roadster, cabriolet, Victoria, and a five-passenger sedan in a price range of $2,250 to $2,950. The Model 6-45 was discontinued. While the 1919 chassis specifications were the same as the 1918 models, the styling was changed considerably. The new "German-Silver" radiator shell had a truncated hexagon shape (similar to the Rolls-Royce). The hood was reshaped to align with the radiator shell and the horizontal body lines. The open models had the top rear curtain wrap around the quarter corner and terminate at the top bow in a rearward slanting position. The 1919 production and sales amounted to 2,496 cars.

Joseph Moon Dies

On February 11, 1919, Joseph Moon passed away at his home in St. Louis, Missouri, following a month's confinement in bed due to illness. He was 69 years old. Vice president Stewart MacDonald succeeded Moon as president of Moon Motor Car Company, and the policies of the company remained the same.

On November 6, 1919, the new 1920 Moon models were announced, consisting of two lines: Moon 6-48 (called the Victory model) and the Moon 6-68.

	1920 Moon Model 6-48	**1920 Moon Model 6-68**
Wheelbase (in.)	122	125
Price	$1,885-2,750	$1,885-2,750
No. of Cylinders / Engine	L-6	L-6
Bore x Stroke (in.)	3.25 × 4.50	3.50 × 5.25
Hp	48 adv, 25.35 NACC	68 adv, 29.4 NACC
Body Styles	roadster, touring, coupe, sedan	roadster, touring, coupe, sedan
Other Features		

The styling remained the same as that of the 1919 models. The Moon 6-68 touring and sedan models had seven-passenger seating. The 1920 Moon car sales dropped to 1,501 units due to the depression that followed World War I.

In the fall of 1920, the 1921 Moon models were announced, which in fact were carryovers of the 6-48 and the 6-68. The specifications were the same as the 1920 models; however, due to the inflation that followed World War I, Moon had to increase prices accordingly. The 6-48, available in four body styles, was priced at $1,785 to $2,785, and the 6-68, offering three body types, was priced at $2,285 to $3,485. While production and sales dropped to 1,366 units in 1921, Moon was able to earn a profit of about $120,051.

With the announcement of the 1922 Moon models in the fall of 1921 came the debut of the new Moon 6-40.

	1922 Moon Model 6-40	**1922 Moon Model 6-48**	**1922 Moon Model 6-58**
Wheelbase (in.)	115	122	128
Price	$1,295	$1,785-2,785	$2,485
No. of Cylinders / Engine	L-6	L-6	L-6
Bore x Stroke (in.)	3.12 × 4.25	3.25 × 4.50	3.37 × 4.50
Hp	40 adv, 23.36 NACC	48 adv, 25.35 NACC	58 adv, 27.3 NACC
Body Styles	roadster and touring	5	seven-passenger touring and sedan
Other Features			

1922 Moon. (Source: George A. Moffitt)

The 6-48 was continued, but the new 6-58 superseded the 6-68. During 1922 Moon built and sold just under 10,000 cars and earned a profit of $913,883, due mainly to the introduction of the 6-40.

The 1923 Moon models announced in late 1922 consisted of two model lines: the 6-40 and the 6-58. The Moon 6-48 was discontinued. The styling and specifications remained the same as for 1922, except for minor improvements. Six body types priced at $1,295 to $1,885 were offered in the 6-40 line, and five body styles priced at $1,785 to $2,685 were originally offered in the Moon 6-58. On May 5, 1923, a new Sport Phaeton priced at $1,995 was added to the Moon 6-58. The new Sport Phaeton had full sport equipment, accommodating seven passengers. It was upholstered in Spanish leather, had a khaki fabric top, and a rear trunk. The car sported six Disteel wheels, including two spare wheels and tires carried up front between the front fenders and the front door openings.

On October 5, 1923, Moon Motor Car Company announced it was preparing to build and sell a new Moon Six, priced at just under $1,000. But the closest Moon came to that plan was a model priced at $1,295 for 1924 and $1,195 for 1925. The $1,000 car did arrive until the debut of the 1927 Moon 6-60 priced at $995.

The decision to limit the 1923 model lines to two series was a wise and profitable choice. Moon Motor Car Company built and sold 7,972 cars in 1923 and enjoyed earnings of about $932,107 for the first nine months of 1923.

The Line-up Expands

For 1924 Moon Motor Car Company expanded its model line-up to four series: the Series A, Newport, Metropolitan, and London Series. Despite the proliferation of models, Moon total production and sales increased to slightly more than 10,705 units in 1924, and the net income for the first nine months of 1924 amounted to $642,277.

1924 Moon.

	1924 Moon Series A	1924 Moon Newport	1924 Moon Metropolitan	1924 Moon London Series
Wheelbase (in.)	113	115	118	128
Price	$1,195-1,695	$1,495-1,915	$1,515-2,095	$1,985-2,540
No. of Cylinders / Engine	L-6	L-6	L-6	L-6
Bore x Stroke (in.)	3.12 × 4.25	3.12 × 4.25	3.25 × 4.50	3.25 × 4.50
Hp	50 adv, 23.36 NACC	50 adv, 23.36 NACC	58 adv, 25.35 NACC	58 adv, 25.35 NACC
Body Styles	4	4	3	Sport Touring and Petite Sedan
Other Features				

The 1925 Moon models were a carryover of the 1924 models but with the latest improvements, including balloon tires and hydraulic four-wheel brakes. The Series A was priced at $1,195, the Newport at $1,495 to $1,985, the Metropolitan at $1,515 to $2,540.

1925 Moon.

Howard (Dutch) Darrin. (Source: P.M.C. Co.)

On May 28, 1925, Stewart MacDonald, president of the Moon Motor Car Company, announced the formation of the Diana Motors Company to market the Diana Straight Eight built by Moon Motor Car Company. The officers of Diana Motors Company were: Stewart MacDonald president, Frederick Rengers vice president, Carl Durst vice president, and Stanley Moon (son of Joseph) secretary-treasurer. In Roman mythology, Diana was the goddess of hunting. She was said to have been the daughter of Jupiter and Latona, and the twin sister of Apollo. In art Diana is represented with bow and quiver of arrows on her shoulder, and a crescent moon on her head. Thus, the name Diana was chosen for the car.

The Diana car, when introduced on July 2, 1925, was indeed a dream car, expressing Howard (Dutch) Darrin's first success in America (with the opulence of the Belgian Minerva). The "Darrin" Packard convertibles of the late 30s and 40s are still one of the most sought-after classics, even today. The new Diana was built on a 125.5-in. wheelbase chassis, powered by a Red Seal Continental, L-head, straight-eight engine. It also featured a Warner three-speed transmission, Borg and Beck clutch, Hotchkiss drive with mechanics' universal joints, Lockheed hydraulic four-wheel brakes, and balloon tires. It was a well-built car and styled attractively, but, it is possible that America was not quite ready for the appearance of the Darrin-designed, convoluted hood radiator and shell. During 1925 production and sales of Moon and Diana cars amounted to 12,964 units, generating record profits.

New Merchandising Plan

Moon Motor Car Company increased its production capacity by 41% for 1926. However, the model line-up was limited to the Diana Eight, Moon Series A, and the Moon London models. On January 11, 1926, president MacDonald announced the inauguration of a new wholesale plan of merchandising. Under the new plan, cars would be sold to the Moon and Diana dealers directly by Moon Motor Car Company in New York,

MOON-DIANA
Announce
Startling New Prices

Because 1925 was the greatest year Moon Motors has ever had —*Because* Moon-Diana Sales showed more than a 36 per cent increase over the previous year—*Because* an increased demand calls for a 41 per cent increase in Moon-Diana production scheduled for 1926—These are the chief reasons for announcing the following new and startling prices effective January 12th, 1926.

New Low Prices For Moon Six		*New Low Prices For* Diana Eight	
De Luxe Coach	$1295	Roadster	$1695
Roadster	$1395	Phaeton	$1695
4-Door Standard Sedan	$1445	Cabriolet Roadster	$1995
Cabriolet Roadster	$1545	2-Door Brougham	$1795
De Luxe Sedan	$1595	4-Door Sedan De Luxe	$1995
F.O.B. Factories		F.O.B. Factories	

MOON MOTOR CAR COMPANY
DIANA MOTORS COMPANY
Stewart MacDonald, *President.*
St. Louis

It's a great thing to be connected with a successful company

1926 Moon Series A.
(Source: John A. Conde)

Chicago, and Philadelphia. Along with that announcement came the news that Moon Motor Car Company had purchased Quinlan Motors Company of Chicago and Milwaukee and that negotiations were underway for the acquisition of Moon Motor Car Company of New York. The intent of this method of wholesale merchandising was to obtain better dealer representation and increased sales.

Shortly after the new model announcement on January 12, 1926, Moon Motor Car Company announced startling price reductions, and the merchandising concentrated on the Moon Six (Series A) priced at $1,295 to

1926 Diana roadster.
(Source: Brown Brothers)

1926 Diana. (Source: John A. Conde)

$1,595 and the Diana Eight priced at $1,695 to $1,995. Still suffering from lack of dealer representation, Moon and Diana sales dwindled during 1926.

For 1927 the new Moon to be priced below $1,000 that was promised on October 5, 1923, finally arrived with the introduction of the new Moon 6-60 priced at $995 to $1,195.

1927 Moon Model 6-60

Wheelbase (in.)	110
Price	$995-1,195
No. of Cylinders / Engine	L-6
Bore x Stroke (in.)	2.87 × 4.75
Hp	47 adv, 19.76 SAE
Body Styles	7
Other Features	

The Moon Series A continued with the same specifications as the corresponding 1926 models priced at $1,195 to $1,595. The Diana Eight prices and specifications remained the same as the 1926 model with minor improvements; however, the 1927 Moon and Diana sales plunged further.

For 1928 the Moon "Aerotype" styling of the Model 6-72 was completely new with a more conventional hood and radiator shell of that period, peaked in the middle. The cowl sides followed the military-style European trend, with the parking lamps mounted on a bracket of the cowl surcingle molding. The Diana Eight continued into 1928, but later in the year its hood and radiator were restyled with Aerotype styling and it became the Moon Aerotype 8-80.

	1928 Moon Aerotype 6-72	1928 Moon Aerotype 8-80
Wheelbase (in.)	120	126
Price	$1,395-1,545	$2,195
No. of Cylinders / Engine	L-6	L-8
Bore x Stroke (in.)	3.37 × 4.00	3.00 × 4.75
Hp	72 adv, 27.3 SAE	80 adv, 28.8 SAE
Body Styles	6	sedan and convertible club coupe
Other Features		

Burst Takes Over

On September 4, 1928, Carl W. Burst, vice president and works manager of Moon Motor Car Company, was elected president, succeeding Stewart MacDonald who became chairman of the board, a newly created position. Helm Walker was appointed vice president in charge of sales on October 27, 1928.

On January 5, 1929, the Moon Motor Car Company introduced their stylish new line of 1929 cars. The name Windsor White Prince was given to the eight-cylinder-engine-powered models. (The name was chosen due to the popularity of the Prince of Wales, Duke of Windsor, and later King Edward VIII of England.) Engine power on the Windsor White Prince 8-82 and 8-92 was transmitted through a Warner "High-Flex" four-speed transmission.

	1929 Moon Windsor White Prince 8-82 and 8-92	1929 Moon Model 6-72
Wheelbase (in.)	125.5	120
Price	$1,745-2,345	$1,495-1,695
No. of Cylinders / Engine	L-8	L-6
Bore x Stroke (in.)	3.00 × 4.75	3.37 × 4.00
Hp	88 adv, 28.8 SAE	72 adv, 27.3 SAE
Body Styles	12	9 (Standard line or Royal line)
Other Features		

The Moon 6-72 was introduced at the same time as a companion model to the Windsor White Prince models. On April 1, 1929, the Moon 6-72 was superseded by the Windsor 6-72 and Windsor 6-77.

The Dark Clouds Appear

While the nation as a whole hailed 1929 as a boom year, and most other car manufacturers enjoyed record sales, Moon Motor Car Company was languishing in sagging sales and heavy company losses. Then, came "Black Thursday," October 24, 1929. While the former years at Moon Motor Car Company, with its ups and downs, were rather peaceful, that was not to be during 1930, when conditions became very turbulent.

The dark clouds of gloom and doom began to accumulate starting on April 25, 1929, when Archie M. Andrews announced the "Ruxton Front-Drive" car to be produced by New Era Motors Incorporated. In New York on May 22, 1929, New Era Motors Inc. announced its officers and directors: Archie M. Andrews president, W.H. Muller vice president and designer, and William V.C. Ruxton partner. A demonstration of the Ruxton car was given for newspaper and trade magazine representatives at the Columbia Yacht Club in New York City on May 28, 1929.

1926 Moon and Diana cars being loaded on barges in St. Louis.

In stressing the design characteristics and color effects, it was disclosed that they were credited to Joseph Urban, a nationally known interior and theatre set designer. At that time Andrews did not disclose who or what company would build the car (because he himself did not know). Temporarily, Andrews arranged for the building of the Ruxton car in the Kissel and Gardner plants on a contract basis. (Fortunately for Gardner, Russell E. Gardner was able to work out from under that contract with New Era. See his story in this volume.) Unfortunately, Moon Motor Car Company was not that lucky.

On November 25, 1929, Carl Burst announced the acquisition of The New Era Motors Inc., and the takeover of the manufacture of the Ruxton front-wheel-drive automobile from Gardner Motor Company, which had been building it on a contractual basis. The deal was effected through an exchange of stock and additional financing through the issuance of new stock, said to have been already underwritten.

To give the impending maelstrom the initial spin, the New Era Motors' interests in the Moon Motor Car Company held a special directors' meeting on April 12, 1930, at which they ousted the former officers and selected their successors to take office immediately. The new officers chosen were: W.J. Muller president,

Helm Walker vice president, J.E. Roberts second vice president, R.P. Kolwitz secretary, and F.E. Welch treasurer. The ousted officers were Carl W. Burst, W.D. Semenway, H.W. Klemme, and Stanley Moon. The New Era Motors group boasted that it had obtained 239,400 of the 350,000 shares of the company.

On August 23, 1930, F.W. Ayres was appointed vice president and general manager of Moon Motor Car Company, succeeding J.E. Roberts who had resigned. On November 15, 1930, the Moon Motor Car Company, makers of the Windsor and Ruxton automobiles, was placed in temporary receivership. Attorney Seneca G. Taylor was appointed temporary receiver by Judge Hogan. The petition declared that the company was solvent with total assets of $1,123,000 against general liabilities and accounts payable of no more than $697,000. The Moon Motor Car Company was sold to New Era Motors Inc. in April 1930. Shares of the company rose from $3.75 per share to $16.50 per share in the spring of 1930, but were selling at $1 per share in November. On December 31, 1930, the Moon Motor Car Company was in the process of liquidation under the direction of Seneca G. Taylor, federal court receiver.

Russell Gardner

Even in terminating their father's business, Russell Jr. and Fred Gardner could hold their heads high. Not only did the Gardners give Chevrolet and General Motors a boost in the manufacture and sales of Chevrolet cars, but they pioneered the decentralization of automobile manufacturing methods. They built and sold over 100,000 fine cars bearing the Gardner badge, upholding the excellent tradition of Russell Gardner Sr.

* * *

Russell E. Gardner was born in Gardner, Tennessee, on March 16, 1860, the son of W.H. Gardner, owner of the largest sawmill in the town founded by Gardner's ancestors, with a population of about 1,000. The sawmill worked mainly in hickory timbers from which spokes and wagon wheels were made. Russell's formal education was cut short when he was employed full time at his father's mill at the age of 10. It was only natural that he learned as he went along about wagon-making.

After nine years of employment in his father's mill, Russell decided to explore other fields of occupation. He heard of a spoke and wheel company in Galion, Ohio, and one day in the fall of 1879 he presented himself at the wheel company office and said he was ready to go to work. He was given a job as a clerk, and his earnest attitude and hard work earned him a promotion. Before the end of two years he was secretary of the company with all his savings invested in the company stock.

The wheel company was fairly prosperous, but eventually Gardner became convinced that the company was not operating under the strictest business ethics. He decided to leave the company, but he could not find a purchaser for his shares of company stock. One day in Galion, he met a fellow who owned a carload of wagon wheel hubs for which he had no use. Gardner offered to trade his stock certificates for the hubs even up. They shook hands on the deal and thus, Gardner became the sole owner of wagon wheel hubs, for which there was little demand in Galion, Ohio.

Because there was little hope of business success in Galion, Gardner boarded the train for Columbus, Ohio. After arriving, while sitting in the hotel lobby, he noticed an advertising calendar of Wabash Railroad which proclaimed to be the "Banner" line of the country. The name "Banner" struck his fancy and gave him an inspiration. At the same time he glanced out the window and saw a parade of horse-drawn vehicles. Putting his thoughts together he decided to build buggies and to call them Banner. However, he had no capital because all his savings were tied up in several thousand wheel hubs, so he searched for someone who would swap with him.

Banner Buggy Company

Gardner was able to find a man who had some wagon wheels that he would be willing to trade for the wheel hubs. The swap was made and Gardner had the components to start the buggy business. He went to the bank and borrowed some money, using the wheels as collateral. Thus, the Banner Buggy Company started in Columbus, Ohio, with Russell Gardner as the sole owner. He started in a small way, with a one-story factory in which he made a few buggies every week. He kept the books, and handled the orders and the correspondence. About the only thing he didn't do was to actually build the buggies. He prospered greatly, and 1890 marked the upward swing toward business success and wealth.

But those were weary years for the young man from Tennessee, who found himself at the head of an enterprise of astounding magnitude. By 1894 he was 34 years old and had accumulated a fortune of about $150,000. He decided to sell his factory and return to his native Gardner, Tennessee, but a few weeks after

1890 Banner Buggy.
(Source: Carriage Monthly)

his return to his home town he discovered that his temperament did not take kindly to loafing. With $150,000 at his disposal he became the president of the only bank in Gardner, Tennessee.

But one bank was not enough for him, so he purchased the controlling interest in the bank at Dresden, Tennessee. Within two years he owned and was president of two more banks in Martin and Greenville, Tennessee. However, after two years of banking, Gardner was convinced that banking was not his cup of tea. So, in 1896, he went to St. Louis, Missouri, and within a week was a buggy manufacturer again.

Gardner Carriage Company

Gardner started in a small plant on Third and Chouteau Streets. His first year output was 8,000 buggies, but by 1910 they were building 100,000 buggies annually. The Gardner Carriage Company was very successful, to the extent that they gave the Imperial Wheel Works of Flint, Michigan, an unprecedented order for 1,000,000 buggy and carriage wheels. In spite of the apparent success of the buggy and carriage business, and the bad experience he had with an attempt to build a horseless carriage in 1904, Gardner was convinced by 1915 that the automobile was replacing horse-drawn vehicles.

Chevrolet Motor Company of St. Louis

Knowing William C. Durant since the days of Durant-Dort Carriage Company, Gardner offered his plant facilities in 1915 to build the Chevrolet car. A mutual agreement was reached and Gardner was granted a license to build the Chevrolet car for all the territories encompassed by the Mississippi River basin. This was a profitable arrangement for Gardner, and he wisely discontinued production of horse-drawn vehicles. This arrangement was also favorable to Durant, as it provided manufacturing facilities for the ever-expanding Chevrolet car sales. Gardner, along with many former associates of Durant, willingly helped him in the growth and expansion of Chevrolet. The Chevrolet Motor Company of St. Louis was established to manufacture the Chevrolet car. Capitalized at $1,000,000, it was financed entirely by Russell Gardner's own capital and St. Louis investors.

During the years 1915-18, the Gardner-operated Chevrolet plant was remodeled and updated to have an efficient production capacity of 40,000 cars per year. On June 8, 1918, the Chevrolet plants operated by Gardner were sold to the General Motors Company, which by this time had absorbed the Chevrolet Motor Company.

The success of Chevrolet car sales, outgrowing the plant capacity, prompted General Motors Company to build a new larger plant in the St. Louis area. With the completion of the new General Motors Chevrolet plant in 1919, General Motors offered to sell the original plant back to Gardner. The terms were satisfactory and Russell Gardner repurchased the original Gardner plant on November 20, 1919.

Gardner Motor Company

The Gardner Motor Company was incorporated in St. Louis, Missouri, in 1919, with Russell E. Gardner, Sr., chairman of the board, and his sons Russell E. Gardner, Jr., president, and Frederick Gardner vice president. On October 23, 1919, the new Gardner car was announced. In spite of the 1920 depression following World War I, Gardner built and sold almost 3,000 cars in 1920.

Russell Gardner, Sr., and Russell Gardner, Jr. (Source: John R. Gardner)

	1920 Gardner
Wheelbase (in.)	112
Price	$1,125
No. of Cylinders / Engine	L-4
Bore x Stroke (in.)	3.50 × 5.00
Horsepower	35 adv, 19.6 NACC
Body Styles	touring, roadster
Other Features	

During 1921-24 Gardner continued to build the same car, adding a sedan, and changing the price from year to year: 1921 - $895-1,595; 1922 - $1,095-1,795; 1923 - $965-1,395; 1924 - $995-1,445.

In 1925 Gardner offered three car lines: a four-cylinder, a six-cylinder, and, introduced on December 17, 1924, a new eight-cylinder model. Gardner built and sold 5,034 cars in 1925.

	1925 Gardner	**1925 Gardner**	**1925 Gardner**
Wheelbase (in.)	112	125	125
Price	$995-1,595	$1,395	$1,995
No. of Cylinders / Engine	L-4	L-6	IL-8
Bore x Stroke (in.)	3.68 × 5.00	3.25 × 4.50	3.12 × 4.50
Horsepower	43 adv, 21.7 SAE	57 adv, 25.35 SAE	65 adv, 31.3 SAE
Body Styles	7	touring	touring, brougham
Other Features			

*1925 Gardner eight-cylinder.
(Source: NAHC)*

The four-cylinder was dropped, and only sixes and eights were offered in 1926 and 1927. In 1926 Gardner built and sold 6,253 six-cylinder models and 7,239 eights. Priced at $1,395-2,495 in 1926, and $1,395-2,295 in 1927, they were available in twelve body types, featured four-wheel mechanical brakes and wood-spoke wheels with balloon tires. In 1926 Gardner had 1,100 dealer sales and service stations.

For 1928 Gardner concentrated on eight-cylinder models, consisting of the 75, 80, 85, and 90 series. On September 8, 1928, Gardner introduced the new 1929 models, consisting of three lines: 120, 125 and 130 series. Gardner was able to advertise the lowest-priced eight.

	1928 Gardner 75	**1928 Gardner 85**	**1928 Gardner 95**
Wheelbase (in.)	122	125	130
Price	$1,195-1,695	$1,695-2,695	$2,095-2,495
No. of Cylinders / Engine	IL-8	IL-8	IL-8
Bore x Stroke (in.)	2.75 × 4.75	2.87 × 4.75	3.25 × 4.50
Horsepower	65 adv, 24.2 SAE	74 adv, 26.45 SAE	115 adv, 33.8 SAE
Body Styles	6	4	4
Other Features			

	1929 Gardner 120	**1929 Gardner 125**	**1929 Gardner 130**
Wheelbase (in.)	122	125	130
Price	$1,395-1,595	$1,695-1,855	$2,195-2,395
No. of Cylinders / Engine	IL-8	IL-8	IL-8
Bore x Stroke (in.)	2.75 × 4.75	2.87 × 4.75	3.25 × 4.50
Horsepower	65 adv, 24.2 SAE	85 adv, 26.45 SAE	115 adv, 33.8 SAE
Body Styles	4	6	5
Other Features			

*1929 Gardner in-line eight.
(Source: NAHC)*

The 1929 Gardner models were pleasingly styled, with a new peaked hood having two rows of louvres in the side panels, a new chromium-plated radiator shell with vertical radiator shutters. The bullet-shaped headlamps were chromium-plated, as well as the front fender tie bar and the headlamp supports. The cowl parking lamps were also chromium-plated, mounted on a chromium-plated band across the cowl.

The front fenders were long, sweeping, clamshell-styled, and the bumpers were the two-bar-type chromium-plated. The body styling was contemporary, having a cadet-type visor on the closed models. Wood-spoke wheels with balloon tires were standard, and wire-spoke wheels with the spares mounted in the front fender wells were optional at extra cost. The 120 and 125 series were equipped with Watson Stabilizers and a central chassis lubrication system.

In May 1929 there were negotiations for the possible merchandising of a light car built by Gardner, through the Sears, Roebuck and Company mail order and chain stores. On May 22, 1929, Russell Jr. and Fred Gardner announced the detailed plans for the Annual Gardner Trophy Race to be held in St. Louis on May 28 and May 30, 1929. In addition to the Gardner Trophy Silver Cup, the Gardners were donating $10,000 in prize money. During 1929 Gardner sales and profits were favorable, until "Black Thursday," October 24, 1929. This caused all plans with Sears merchandising to be dropped. Also, Gardner had arrangements to manufacture the Ruxton front-drive car for Archie Andrews' New Era Motors. However, due to the collapse of New Era Motors, this plan also fell through.

For 1930 Gardner carried over the 1929 models whose styling and features were well in advance of their time. During 1930, industry car sales plummeted to two-thirds of what they were in 1929, and operating loss mounted. On April 2, 1930, Gardner Motor Car Company consolidated its sales and engineering staffs with Moon Motor Car Company. F.H. Rengers, former Gardner sales manager, was named sales manager for the consolidated sales organization. However, Russell Jr. and Fred Gardner had the intelligence not to have their company "sucked up" by Archie Andrews' giant financial "vacuum cleaner," the New Era Motors.

On January 4, 1930, the Gardner "Front-Drive" car was announced and shown at the National Automobile Show in New York. It was built on a 133-in. wheelbase chassis, powered by a six-cylinder, 80-hp, L-head Lycoming engine. The Front-Drive Gardner car had a low Baker-Raulang sedan body. The intent was to merchandise it in the $2,000 price range; however, it never got beyond the prototype show-car stage.

On March 24, 1931, Russell Jr. and Fred Gardner offered four proposals at the stockholders meeting for the disposal of the Gardner manufacturing interests. At that time the Gardner stock on the New York Stock

1930 Gardner "Front-Drive." (Source: NAHC)

Exchange was quoted at $1.75, compared to $30 per share in 1920. On November 25, 1931, a dividend of 70 cents per share of Gardner stock was authorized by Circuit Court Judge Rosskopf, at the request of Russell Jr. The assets were reported at $592,213 against comparatively small liabilities. The distribution of dividends amounted to approximately $210,000. In liquidating their business, the Gardner stockholder actually got more money, since the Gardner stock by that time was listed by NYSE at 50 cents. The Gardner stock was taken off the NYSE on December 9, 1931.

1931 Gardners. (Source: NAHC)

Louis Clarke

Louis Clarke and his brothers, John and James, grew the Autocar Company and established its world-renowned reputation in the trucking industry. When they sold their Autocar stock in 1928, the same meticulous care in the design and manufacture of Autocar trucks continued. Autocar set the standard, and the most complimentary thing the competitors could say about their own product was that "it was almost as good as an Autocar."

* * *

Louis S. Clarke, his brothers John S. and James K., and their father Charles J. Clarke, were prominent steel producers who operated iron foundries and steel mills in Pittsburgh, Pennsylvania, and the many surrounding communities. Their principal steel users were railroads, locomotive manufacturers, and steel railroad car builders.

After the tragic and disastrous flood in Johnstown, Pennsylvania, on May 30, 1889, the Clarkes set out to restore and rebuild the town. They not only rebuilt their foundries and steel mills, they led in the restoration of the entire community by providing funds, their own physical effort, and the workforce from their own plants.

From Pittsburgh Motor Vehicle Company to Autocar

During the 1890s, the Clarkes decided to branch out into automotive transportation. They established the Pittsburgh Motor Vehicle Company in 1897, and for three years they built and tested experimental motorized

1897 Pittsburgh Motor Vehicle motorized tricycle. (Source: NAHC)

1899 Pittsburgh Motor Vehicle motorized four-wheeler. (Source: NAHC)

tricycles and four-wheel models. On November 1, 1899, the Pittsburgh Motor Vehicle Company announced the purchase of their business by Autocar Company, which would continue to manufacture motor vehicles on a large scale after January 1, 1900. They listed their address as the Autocar Company, Third Avenue and Ferry Street, Pittsburgh, Pennsylvania.

The officers of the company were: Louis S. Clarke, president; John S. Clarke, treasurer; William Morgan, secretary; and James K. Clarke, factory manager. Their father, Charles J. Clarke, furnished financial support. Their first production models, a runabout priced at $825 and a phaeton priced at $900, made their debut at the National Auto show in New York on November 3-10, 1899.

In late 1899, the Autocar Company purchased a 3-1/2-acre tract of land at Swissvale, Pennsylvania, located near Homestead, and awarded a contract to the Lytle Brothers of Wilkinsburg, Pennsylvania, for the building of a plant. The plant was to be completed by March 1, 1900, and was expected to employ up to 300 men. However, with the sudden and untimely death of Charles Clarke in mid-January, the brothers decided to move their manufacture to Ardmore, Pennsylvania, west of Philadelphia, perhaps because of the terrain of the mountainous region around Pittsburgh. The new plant at Swissvale was built and sold on completion.

In 1901 the first Autocar vehicles were built in the new plant in Ardmore. It was a four-wheel vehicle with a 70-in. wheelbase chassis, supported by parallel fully elliptical springs in front and three-quarter elliptical springs at the rear. The frame and running gear were made of heavy-gauge steel tubing, with all joints copper-brazed. The compensating gear was encased in a dustproof housing, lubricated by oil. However, the drive sprocket, chain, and rear axle sprocket were openly exposed to road dirt.

The chassis was powered by a water-cooled, 5.5-hp opposed two-cylinder engine. (The Clarke brothers claimed only 5 hp, while automotive writers exaggerated to 6 hp.) The axles were supported by tangent-laced wire spoke wheels having ball bearings. The wheels were mounted by 2.5-in. cross-section single-tube tires. The runabout body was amply large enough to seat two adults, and was also furnished with a top and storm apron. The price of the standard runabout was $800, and the vehicle including a Goddard top, storm apron, and panel seat was $875.

1900 Autocar. (Source: NAHC)

1901 Autocar. (Source: NAHC)

The Autocar Company adopted the policy of selling their vehicles entirely on their merits, refusing to accept cash bonuses for earlier delivery. They gave a liberal guarantee with their machines, and employed a full-time traveling mechanic (now called service representative) to visit towns where Autocar vehicles were located to handle the repairs and give valuable technical advice.

This field service arrangement was perhaps the first in the automobile industry. Having the traveling mechanic enabled the Clarkes to promptly discover the shortcomings of the vehicle, namely, the excessive wear and breakage of the drive chain. This prompted Louis Clarke to quickly change the method of drive mechanism. About two dozen Autocar vehicles were built and sold in 1901. For 1902, Autocar had a dustproof exposed shaft drive to the bevel gear rear axle, the first in the automotive industry.

	1901	1902	1903/4
Wheelbase (in.)	70	70/76	70/76
Price	$800-875	$1,100	$900-1,400
No. of Cylinders / Engine	2	2	2
Bore x Stroke (in.)	4.00 × 4.00	4.00 × 4.00	4.00 × 4.00
Horsepower	5.5 adv, 12.8 ALAM	8.5 adv, 12.8 ALAM	10 adv, 12.8 ALAM
Body Styles		runabout/tonneau	runabout/tonneau
Other Features		dustproof exposed shaft drive to bevel gear rear axle	

The 1902 Autocar models, introduced in June 1901, also included such innovations as the engine located longitudinally, under a sloping hood up front, with the crankshaft centerline in line with the sliding gear transmission, propeller shaft, and bevel drive pinion in the rear axle. Water-cooling of the engine was accomplished by a finned copper coiled-tubular radiator up front. The suspension was by four parallel three-quarter elliptical springs. The axles were supported by artillery-type wood-spoke wheels of 30-in. tire diameter. Another innovation was the left side steering, either by a lever or steering wheel.

1902 Autocar. (Source: National Museum of American History, Smithsonian Institution)

1902 Autocar. (Source: Autocar Company)

The 1903 and 1904 Autocar vehicles remained unchanged, except the rear entrance tonneau model replaced the dos-a-dos model of 1902. The engine output was increased to 10 hp through mechanical refinements. The runabout Model X priced at $900 and the tonneau Model VIII priced at $1,400 continued through 1905.

The long-awaited four-cylinder-engine-powered 1905 Autocar was introduced in the fall of 1904. The engine was located up front under a hood, and the power was transmitted through a dry-disc clutch, three-speed sliding gear transmission, and exposed propeller shaft to the bevel-gear rear axle. The chassis frame was of the armored-wood type, with the U channel steel section lined with hickory wood beams. The suspension was by four parallel semi-elliptical springs and Midgely steel-spoke wheels mounted with Fisk heavy-duty tires. The service brakes were of the external contracting type on drums of the rear wheels, and the emergency brake was an external contracting type on a drum at the rear of the transmission. Wheel steering was by a pinion and bevel gear sector unit mounted on the right frame side member.

Pioneers of the U.S. Automobile Industry, Volume 2: The Small Independents

1904 Autocar. (Source: NAHC)

1905 Autocar. (Source: NAHC)

	1905	1906 X	1906 XII
Wheelbase (in.)	96	76	100
Price	$2,000	$1,000	$2,400
No. of Cylinders / Engine	4	2	F-4
Bore x Stroke (in.)	3.50 × 4.00	4.00 × 4.00	3.50 × 4.00
Horsepower	16-20 adv, 19.6 ALAM	12 adv, 12.8 ALAM	24 adv, 19.6 ALAM
Body Styles	side-entrance tonneau	runabout	side-entrance tonneau
Other Features			

1906 Autocar. (Source: NAHC)

1910 Autocar. (Source: NAHC)

The 1906 Autocar models were introduced in late December 1905. They consisted of two model lines: the type X runabout powered by a two-cylinder engine and the type XII powered by a new larger four-cylinder F-head engine. In the new F-head engine design, the exhaust valves were located in pockets adjacent to the cylinder, while the inlet valves were arranged centrally above the pistons in the cylinder head. The exhaust valves were actuated directly by the camshaft lobe through a tappet, while the inlet valves were actuated by rocker levers and pushrod from the tappet raised by the camshaft lobe. This type of valve mechanism provided a 4-hp greater output with the same cylinder dimensions.

The 1907 Autocar models, the type XV runabout and the type XIV touring, were introduced in January 1907. The styling was similar to the 1906 models, but the wheelbases were longer.

1910 Autocars. (Source: Autocar Company)

	1907 XV	**1907 XIV**
Wheelbase (in.)	81	112
Price	$1,200	$3,000
No. of Cylinders / Engine	2	F-4
Bore x Stroke (in.)	4.00 × 4.00	4.00 × 4.50
Horsepower	14 adv, 12.8 ALAM	30 adv, 25.6 ALAM
Body Styles	runabout	touring
Other Features		

For 1908, the Autocar models were refined 1907 types. The price of the XV runabout remained at $1,200, but the price of the XIV touring was reduced to $2,750. A new limousine model was added to the type XIV model line, priced at $3,500. The basic 1908 model designs were continued through 1909 and 1910; however, the prices fluctuated.

The Move to Commercial

During 1908, the Clarkes were approached by many merchants, coal dealers, and oil refiners who asked them to build commercial vehicles with the same durability as the Autocar passenger vehicles. The Clarke brothers graciously accommodated them and, in fact, by 1910 found the commercial vehicle business more lucrative than passenger cars. Thus, Autocar discontinued passenger car production and concentrated on building commercial vehicles. Autocar used their "engine under the seat" design at first on their truck models to give a 3.5-ton load capacity on a 114-in. wheelbase chassis platform. This particular truck model became very popular with retail coal dealers and petroleum refining companies.

The Autocar trucks sold so well on their merits that the sales exceeded the production capacity. The Clarkes were very careful not to overbuild their facilities. Competitive truck manufacturers were unable to attract

1910 Autocar trucks.
(Source: Autocar Company)

loyal Autocar truck buyers. The service, reliability, and endurance of Autocar trucks available in many sizes and capacities, powered by four-cylinder and six-cylinder engines, became world famous.

By 1927, Autocar offered trucks of the conventional design, with the engine up front, as well as their successful short-wheelbase models with the engine under the seat. This was the forerunner of the COE (cab over engine) designs of competitive truck manufacturers.

In 1928, the Clarke brothers sold their stock in the Autocar Company to Prince and Whitely, New York bankers. In early 1929, Prince and Whitely split the stock three-for-one making 200,000 shares, of which Prince and Whitely retained 80,000 shares.

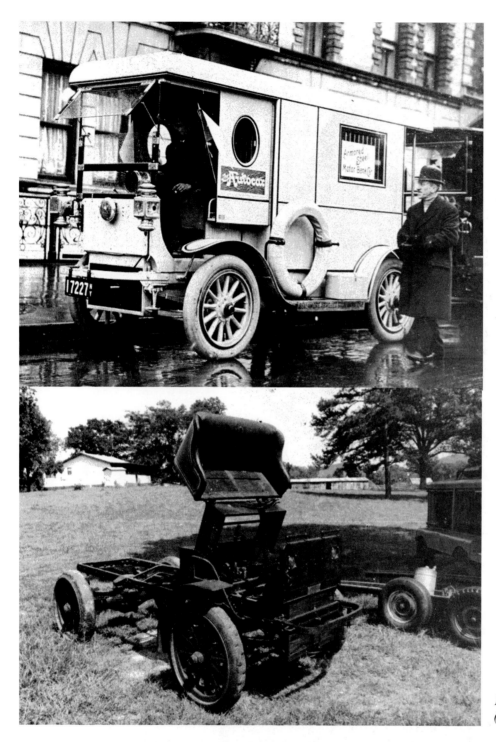

*1910 Autocar trucks.
(Source: Autocar Company)*

Then came "Black Thursday" on October 24, 1929. Prince and Whitely sold their Autocar and other holdings to the Prince and Whitely Investment Trust. In the fall of 1930, Prince and Whitely went into bankruptcy, which did not affect the Investment Trust. Shortly thereafter, the trust was reorganized and renamed the Phoenix Investment Trust.

In 1933, Wallace Groves, a New York investment banker, obtained control of Phoenix Investment Trust, and along with it the 80,000 shares of Autocar Company stock. While this did not give him control of Autocar, it

1911 Autocar truck chassis. (Source: Autocar Company)

was the largest block of stock owned by any Autocar Company stockholder. Early in 1934, Groves purchased 40,000 shares of Autocar stock from Walter Janney, a Philadelphia investment banker. Janney had a financial interest in the Autocar Company for many years and was a member of its board of directors. This gave Groves 120,000 shares of the 200,000 shares outstanding, and control of the Autocar Company.

At the annual stockholders meeting on March 9, 1934, Groves spoke optimistically of Autocar's future in the truck industry, and endorsed the management and policies of President Robert P. Page. Groves was elected chairman of the board of directors and Janney was re-elected to the board.

1929 Autocar dump truck. (Source: Autocar Company)

1932 Autocar tanker. (Source: Autocar Company)

In September 1935, the Autocar Company entered the four-wheel-drive truck field. They used Timken-Wisconsin front-drive axles and constant-velocity universal joints at the wheels. Transfer cases and auxiliary transmissions were also made by Timken-Wisconsin and were located amidships in rubber mountings. All three models were powered by larger Autocar Blue Streak six-cylinder engines and had four-speed transmissions and hydraulic brakes.

	1935 Truck Model 4DF	1935 Truck Model 4N	1935 Truck Model 4S
Wheelbase (in.)	159	165	165
Price	$5,000	$6,000	$7,000
No. of Cylinders / Engine	6	6	6
Bore x Stroke (in.)	4.00 × 4.75	4.50 × 4.75	4.50 × 5.25
Horsepower	94 adv, 38.4 SAE	101 adv, 48.6 SAE	124 adv, 48.6 SAE
Body Styles			
Other Features	3-ton rating	4-ton rating	6-ton rating

During World War II, the Autocar Company put all of its personnel, engineering, and manufacturing facilities at the disposal of the United States government and consistently met all production schedules on time. Autocar Company built many trucks of all designs for the United States armed forces, including all-wheel-drive and rear-wheel-drive models. Autocar trucks became the G.I.'s favorite when heavy hauling was needed.

1949 Autocar truck chassis. (Source: Autocar Company)

1959 Autocar tractor-trailer. (Source: Autocar Company)

In 1953, the White Motor Company purchased the assets, name, and good will of the Autocar Company, and continued to build Autocar trucks. Through mergers and acquisitions, the White Motor Truck Corporation became the Volvo-White Motor Truck Corporation. Autocar trucks are currently being purchased on special order.

Louis S. Clarke passed away in Palm Beach, Florida, on January 6, 1957.

Charles Clifton. (Source: Smithsonian Institution)

George Pierce and Charles Clifton

From bird cages to bicycles and motorettes, to manufacturing one of the finest cars built in America—that was the destiny of George Pierce and Colonel Charles Clifton. In retrospect it seems that Pierce Arrow Motor Car Company could have survived a few years longer had they not hastily separated themselves from the Studebaker Corporation in 1933. Moreover, it might have changed the history of the automobile industry during the hasty mergers of the independents in the 1950s. But, that was not to be.

* * *

George N. Pierce was born in Friendsville near Waverly, New York, in 1846. After high school he entered the Bryant and Stratton Business College. He began his business career in 1872 when he became a partner in the firm of Heintz, Pierce, and Munschauer Company, manufacturers of bird cages and refrigeration (ice) boxes. In 1881 Pierce married Louise H. Day; they eventually had two sons and four daughters.

In the late 1880s Pierce severed his partnership with Heintz and Munschauer, and in 1890 he organized a new enterprise under the firm name of George N. Pierce Company, for the purpose of manufacturing bicycles and tricycles. Before the turn of the century he was one of the leading American bicycle manufacturers.

Charles Clifton was born in Buffalo, New York, on September 29, 1853. He obtained his military title through twelve years of service (1881-1893) in the New York National Guard.

At age 18 Clifton began his active career with the Sidney Shepard and Company (a hardware firm) in Buffalo. After 1876 he was associated with the Erie Railroad, Buffalo Grape Sugar Company, Colgate

Gilbert Company as general manager, and with the Bell, Lewis and Yates Coal Mining Company as treasurer. Colonel Clifton married Grace Gorham on January 22, 1891. They had a son, Gorham, and a daughter, Alice.

The Start of a Lasting Relationship

In 1897 Pierce was fortunate to engage Col. Clifton as secretary and treasurer of the George N. Pierce Company. Clifton sensed the impending collapse of the bicycle industry and prevailed upon Pierce to experiment with automobile manufacture. They engaged Charles L. Sheppy as experimental engineer in 1898. Their first experiments showed the weight disadvantage and other shortcomings of American-built gasoline engines available at that time, so Clifton and Pierce agreed on the use of a DeDion Bouton engine (of French design) to power their motorette, benefiting from their experience in bicycle manufacturing. Their first motorette was completed in November 1900, and the second in May 1901.

Pierce Motorette

In 1901 Pierce and Clifton were fortunate enough to lure automotive engineer David Fergusson away from the E.C. Stearns Company of Syracuse, New York, and Toronto, Ontario, Canada, to become chief engineer for the George N. Pierce Company. Fergusson's experience in bicycle manufacturing certainly complemented that of Pierce and Clifton. The major part of 1901 was spent testing the two prototypes. Satisfied with their performance, the George N. Pierce Company was able to start production of the Pierce motorette in late 1901. The first motorette was identified as the Series D, powered by an air-cooled gasoline engine of 2.75 hp.

The Pierce motorette was constructed as a simple, light, and strong vehicle of American design, making use of the French experience, and Pierce's own experience in bicycle manufacturing. The DeDion Bouton single-cylinder gasoline engine had a bore of 2.913 in. and a stroke of 2.992 in. The engine and gear-change unit were carried at the rear of the vehicle on a brazed, tubular, steel underframe. The body was protected from road vibration by its unique hinged mounting and fully elliptical springs, front and rear. Steering was by a tiller lever on a steering post in front of and between the two passengers. The throttle, spark, and speed change levers were also located on the steering post.

Pierce sales headquarters were set up in Buffalo, New York City, and Boston for the New England states. Banker Brothers of Philadelphia was the general agency for the state of Pennsylvania. It is interesting to note that John N. Willys, who would become an automotive giant in his own right, in 1898 bought the Elmira Arms Company, a sporting goods store in Elmira, New York. They handled firearms and ammunition, but Willys added other merchandise, and specialized in selling Pierce bicycles profitably. As soon as Pierce began building the motorette, Willys was established as a Pierce automobile dealer, selling two motorettes during his first year. This relationship lasted many years (see the Willys chapter in Volume 3).

Expansion and Development

During 1902 the engine size was increased to 3.147-in. bore and 3.156-in. stroke, and the output was increased to 3.5 hp. These engines were water-cooled, and the models were identified as Series E and F.

Pierce Great Arrow touring car. (Source: Automotive Industries)

Approximately 187 motorettes were built during the balance of 1901 and 1902, 24 of which were the Series D.

The Series D was continued for 1903, and a new runabout Series G, stanhopes Series H, K, and L, and a new tonneau Series J were added to the model line-up. The runabout shared the 3.5-hp engine with the motorette, while the stanhopes were available with the larger 6-hp and 8-hp engines. The new tonneau sported a two-cylinder DeDion engine of 15 hp, mounted vertically up front under a hinged, sloping hood. The cooling radiator was carried up front, ahead of the front axle.

The Series J frame was made of seamless steel tubing, thoroughly trussed and reinforced by four cross stays brazed to the frame side tubes. Suspension was by four semi-elliptical leaf springs. The rear axle was of the Panhard type, with power transmitted to the bevel gear drive by a shaft having two universal joints. Wheel steering was provided and all three forward speeds and reverse were selected by a single lever on the steering column. About 50 tonneau models were built in 1903, and the Pierce capitalization was increased from $280,000 to $315,000.

For 1904 the motorette and the runabout were discontinued; however, the stanhope Series M with the 8-hp engine and the tonneau Series J with the two-cylinder were continued. That year they also introduced the Pierce Great Arrow touring Series N, powered by a new, four-cylinder, T-head type engine having a bore of 3.937 in. and a 4.75-in. stroke, developing 24-28 hp. The engine was located up front under a hood, and the drive to the rear wheels was by an enclosed shaft and bevel gears. About 50 Pierce Great Arrow tourings were built and sold in 1904.

In 1905 the single-cylinder, 8-hp stanhope at $1,200 continued to be a popular vehicle. The four-cylinder engine powered three Pierce Great Arrow models.

	1905 Pierce Great Arrow Series N	1905 Pierce Great Arrow Series NN	1905 Pierce Great Arrow Series P
Wheelbase (in.)	100	104	109
Price	$3,500	4,000	$5,000
No. of Cylinders / Engine	T-4	T-4	T-4
Bore x Stroke (in.)	3.94 × 4.75	4.25 × 4.75	4.87 × 5.00
Horsepower	24-28 adv, 24.7 ALAM	28-32 adv, 28.9 ALAM	40 adv, 37.9 ALAM
Body Styles	King of Belgium tonneau	3 King of Belgium Tonneau (touring) models and five-passenger Victoria	seven-passenger Landaulet, seven-passenger Suburban, eight-passenger Opera Coach
Other Features			

The same fine engineering, best of materials, and precise skilled workmanship were found in all Pierce Great Arrow models. Approximately 450 Series N and NN, and 25 Series P cars were built and sold in 1905.

The 8-hp, single-cylinder stanhope on a 70-in. wheelbase, at $1,200, again was the price leader for Pierce in 1906. The Pierce Great Arrow consisted of two model lines; and about 700 of them were built and sold in 1906.

	1906 Pierce Great Arrow Series NN	1906 Pierce Great Arrow Series PP
Wheelbase (in.)	107	109
Price	$4,000-5,250	$5,000-6,250
No. of Cylinders / Engine	T-4	T-4
Bore x Stroke (in.)	4.25 × 4.75	5.00 × 5.50
Horsepower	28-32 adv, 28.9 ALAM	60 ALAM
Body Styles	5	4
Other Features		

Focusing on Automobiles

So rapidly did the automobile department of the Pierce organization grow, that in 1906 it was necessary to divide the firm into two organizations: the automobile branch remained the George N. Pierce Company, while the cycle branch became the Pierce Cycle Company.

Pierce Great Arrow Series NN and PP were continued for 1907; however, their wheelbase lengths were increased to 112 and 124 in., respectively. Even though Pierce had had their six-cylinder T-head engine under development for a long time, they did not introduce it until mid-1907 on the new Series Q.

	1907 Pierce Great Arrow Series Q
Wheelbase (in.)	135
Price	$6,500
No. of Cylinders / Engine	T-6
Bore x Stroke (in.)	5.00 × 5.50
Horsepower	60 ALAM
Body Styles	seven-passenger touring
Other Features	

1908 Pierce Arrow.

Approximately 100 Series Q touring cars were built and sold, and about 900 Series NN and Series PP were sold in 1907, despite the Wall Street panic on October 22, 1907.

The George N. Pierce Company manufactured four model lines during 1908: the Series NN, PP, and Q were continued, and a new Series S was introduced.

1908 Pierce Great Arrow Series S

Wheelbase (in.)	130
Price	$5,500 (touring)
No. of Cylinders / Engine	T-4
Bore x Stroke (in.)	4.25 × 4.75
Horsepower	40 adv, 28.9 ALAM
Body Styles	roadster, suburban, touring
Other Features	

While the Great Arrow designation was discontinued in 1909, the George N. Pierce Company proliferated with five model lines. The Series Q was superseded by Series UU, and the Series PP was continued on a 124-in. wheelbase chassis. Three new series were added to the model line-up.

	1909 Pierce Series SS	1909 Pierce Series T	1909 Pierce Series QQ
Wheelbase (in.)	119	111.5	130
Price	$3,700-4,650	$3,100-4,050	$4,800-6,100
No. of Cylinders / Engine	6	4	6
Bore x Stroke (in.)	3.94 × 4.75	3.94 × 4.75	4.50 × 4.75
Horsepower	37 ALAM	24.7 ALAM	48.6 ALAM
Body Styles			
Other Features			

1908-09 Pierce. (Source: Smithsonian Institution)

Pierce Arrow Motor Car Company

On January 20, 1909, at the board of directors meeting, George Pierce asked to be relieved of active management, and that Charles Clifton be elected president and chief operating officer. However, Pierce remained as chairman of the board of directors of the Pierce Cycle Company. The name of the George N. Pierce Company was changed to Pierce Arrow Motor Car Company. The cars and trucks built thereafter were called the Pierce Arrow.

One of Clifton's first moves as president and chief operating officer of Pierce Arrow Motor Car Company was to reduce the number of models available and discontinue the use of the four-cylinder engine. Thus, the 1910 model line-up included the Series SS with a six-cylinder engine of larger bore (4 in.) and the chassis wheelbase increased to 125 in. The Series QQ was continued with the same 48-hp engine, but the wheelbase was increased to 134.5 in. The Series UU engine bore was increased to 5.25 in., for a piston displacement of 714.36 cu. in., the largest of any production passenger car in America. The engine was rated at 66 hp, and the chassis wheelbase was 140 in. While this Pierce Arrow was the most powerful car available, it also had many pleasing design features. Two new body types were added: a new "protected" touring car, and a new close-coupled body with accommodation for four passengers, introduced as the Miniature Tonneau.

The protected touring car, while following the popular torpedo styling, did not carry the radical lines of that type, but instead was embodied in a rationalized form of high body sides and front doors. Designed for luxurious touring, it provided unusually comprehensive appointments not found in other cars. The luggage rack and tool box were removed from the running boards and gracefully placed at the rear end of the vehicle, supported by the gusset plates that reinforce the frame at that point. The protected touring, the miniature tonneau, landaulet, brougham, and runabout completed the body style line-up available on three chassis of 125-in., 134.5-in., and 140-in. wheelbase lengths. The runabout wheelbases were 6.5 in. shorter. This, in effect, offered the prospective buyer 15 cars to choose from.

A noteworthy change was made in the selective gear set, employing a telescoping form of direct speed clutch instead of the "craw-crab" clutch previously used. A foot accelerator pedal was introduced to supplement the hand throttle lever atop the steering wheel. A radical and commendable feature was the adoption of a special priming pump on the dash panel to inject fuel into the intake manifold, to facilitate cold-weather starting.

Identification of the 1911 Pierce Arrow model lines was by numbers corresponding to the engine horsepower ratings. The Models SS, QQ, and UU were carried over into 1911, renamed the Model 36, Model 48, and Model 66, respectively. The close-coupled, five-passenger sedan and protected touring were continued in the 48 and 66 model lines, priced at $5,000 and $6,000, respectively. The Model 66 featured a new stylish cowl, concealed door hinges, electric lighting, and an air starting system.

Pierce Arrow Trucks

But the big news in January 1911 was the debut of the Pierce Arrow 5-ton truck at the National Auto Show in New York. The experience of successfully building fine, dependable, passenger cars for ten years, combined with four years of design, development, testing, and experimentation with motor trucks, truly qualified Pierce Arrow to offer one of the finest motor trucks ever built.

The Pierce Arrow motor truck was built on a 156-in. wheelbase chassis, driven by a worm-drive rear axle, and was the first American-built 5-ton motor truck to possess such a feature. The final drive ratio of 8:1 was obtained by a single worm and worm wheel reduction, without the use of an internal gear and pinion at each rear wheel. The power was transmitted to the full-floating rear axle by means of a nickel-steel propeller shaft fitted with two universal joints. The chassis had an overall length of 20 ft. and a width of 7 ft. The platform in back of the driver's seat was 12 ft. 8 in. long, and 7 ft. wide. The truck was geared to operate at 13 mph with full load. The price of the chassis was $4,500 f.o.b. Buffalo, New York.

Pierce Dies

By 1911 Pierce Arrow had established itself as one of the finest and most desired cars in America. Approximately 2,200 cars and trucks were built and sold profitably in 1911. However, everyone in the Pierce Arrow organization was saddened by the sudden and untimely death of George N. Pierce from heart failure on the evening of March 23, 1911. At just 64 years of age, he was popularly known as one of the automobile industry's "grand old men."

Continued Growth

In September 1911, Pierce Arrow introduced the 1912 models. The wheelbase length of the Model 36 was increased to 127 in., the Model 48 remained at 134.5 in., and the Model 66 stayed at 140 in. All three chassis models had three-quarter elliptical spring rear suspension. The Model 66 engine was changed to 5-in. bore and 7-in. stroke.

The main features of the 1912 models were: a wider flush-sided body; the gear change and brake levers were placed within the body, inside the door; the concealed door hinges allowed the doors to open wide; the dash

1912 Pierce Arrow runabout. (Source: Borg-Warner)

was formed in the shape of the cowl, and inside were fitted two closets to facilitate storage of personal items; the spark coil for supplementary battery ignition during starting was placed between two closets; a spark switch was placed alongside the coil, permitting the driver to ignite the gas headlamps without getting out of the car; and a new type of windshield with a flexible composition flap allowed ventilation of the front passenger compartment.

While the changes in the 1913 Pierce Arrow were primarily mechanical, a letter had been added to the series identification. The Series 38 became the 38C on a 132-in. wheelbase chassis, the Series 48B was available on a 134.5-in. wheelbase chassis, the 48D on a 142-in. wheelbase chassis, and the Series 66A on a 147.5-in. wheelbase chassis. The addition of the Series 48D brought the total number of body types available to 35. The price range was $4,300 to $7,300.

The 1913 models featured a Disco self-starter and the Westinghouse electric lighting and charging systems. Full-pressure lubrication to all connecting rod bearings and the seven crankshaft main bearings was provided, thereby eliminating the gravity oil tank. A new carburetor of the double-jet type was used. Three

1912 Pierce 66 vestibule suburban. (Source: Smithsonian Institution)

sizes of six-cylinder engines were used: 38 hp with 339-cu.-in. piston displacement, 48 hp with 525-cu.-in. displacement, and 66 hp with 825-cu.-in. displacement. All engines had T-type combustion chambers, and the cylinders were cast in pairs, bolted to an aluminum crankcase.

A new 2-ton Pierce Arrow truck chassis was introduced in late September 1913, on a 150-in. wheelbase, powered by a conventional Pierce Arrow four-cylinder, T-type engine of 4-in. bore and 5.5-in. stroke, having a rating of 25.6 hp. The engine was pressure lubricated, and the drive was by the now-famous Pierce Arrow worm-drive rear axle. The Pierce Arrow 2-ton truck was in fact a smaller edition of their 5-ton truck and was developed at the request of the United States and French governments.

Hallmark Headlamps

The Pierce Arrow cars introduced in January 1914 included the debut of the most famous Pierce Arrow hallmark, the trumpet-shaped, front-fender-mounted headlamps. The designer of this outstanding headlamp styling was Herbert M. Dawley, who had the foresight to patent this design. This unique feature (although somewhat modified later) remained the Pierce Arrow identity for the life of the company. Although stanchion-mounted, conventional headlamps were offered as an option, it is doubtful that many, if any, Pierce Arrow cars were sold with this option.

The 1914 Pierce Arrow cars appeared in three chassis forms, each in two wheelbase lengths (the runabouts in the 38C-2 and the 48B-2 had the shorter wheelbase chassis).

	1914 Pierce Arrow Series 38C-2	1914 Pierce Arrow Series 48B-2	1914 Pierce Arrow Series 66A-2
Wheelbase (in.)	127/132	134.5/142	140/147.5
Price	$4,300-5,400	$4,850-6,300	$5,850-7,000
No. of Cylinders / Engine	6	6	6
Bore x Stroke (in.)	4.00 × 5.50	4.50 × 5.50	5.00 × 7.00
Horsepower	38.4 NACC	48.6 NACC	70 NACC
Body Styles	7	6	11
Other Features	Westinghouse starting, lighting, and charging system		

In October 1914, Pierce Arrow was awarded a contract for 300 2-ton trucks by the French government.

World War I

Except for lowering the entire car height by 3 in., the 1914 Pierce Arrow car models were carried into 1915. In fact, all models remained basically the same through 1918, although the appearance improved each year by smoother body lines, one-man top on the touring models, and a better finish on all cars. During that time Pierce Arrow concentrated on trucks because of the military vehicle orders. In 1917, 2,532 cars and 5,171 trucks were built, and in 1918, 1,168 cars and 7,467 trucks were built and sold, even though Pierce Arrow proliferated the car body types to 56, priced at $4,800 to $8,000.

The energies of the Pierce Arrow Motor Car Company during 1917 and 1918 were consumed mainly by the growing military demands of the United States War Industries Board, and the building of commercial trucks.

1914 Pierce Arrow 48B touring. (Source: Smithsonian Institution)

Car production was greatly curtailed during the latter part of 1918, and it became impossible to obtain material for either passenger cars or civilian commercial trucks. The military trucks built by Pierce Arrow gave a good account of themselves by their performance in France during World War I.

In August 1918, the U.S. War Production Board entered into a contract with the Pierce Arrow Motor Car Company to manufacture a large number of Hispano-Suiza aircraft engines, with deliveries to begin in 1918. However, with the signing of the Armistice on November 11, 1918, the entire program of production of military vehicles was terminated, and the preparation for the production of aircraft engines came to an abrupt halt. The net profits for the Pierce Arrow Motor Car Company were $4,791,274 in 1917 and $4,273,171 in 1918. The profits from the sales of military vehicles were $1,161,803 in 1917 and about $1,200,000 in 1918, since the war profits were being regulated by the United States government. Following the Armistice, factory night work and overtime were eliminated, and they went to a 48-hour work week.

In September 1918 Pierce Arrow introduced two new series of 1919 cars. The appearance of the cars was greatly enhanced by the smoother lines in the cowl area. The body type availability was reduced to 24 models. The Series 31 on a 134-in. wheelbase chassis was powered by a 38-hp, six-cylinder engine. The Series 51 on a 142-in. wheelbase chassis was powered by the 48-hp, six-cylinder engine. The main feature of the 1919 Pierce Arrow cars was the adoption of the dual-valve system (two intake and two exhaust valves per cylinder) of T-head design. The cylinders were cast in pairs with detachable heads and bolted to an aluminum crankcase. The dual-valve arrangement increased the engine torque and acceleration ability.

Management Overhaul

In July 1919, as the result of a decision reached by the Pierce Arrow board of directors, G.W. Goethals and Company was placed in control of Pierce Arrow Motor Car Company operations. Col. Charles Clifton was elected chairman of the board, but the Goethals and Company installed Goethals partner John C. Jay as president and George W. Mixter as chief operating officer. Myron E. Forbes was elected vice president and treasurer, to succeed Walter C. Wrye, who had resigned. During the reorganization, David Fergusson resigned as chief engineer, and was succeeded by Charles Sheppy. Henry May, vice president of manufacturing, also resigned at this time, with Mixter taking over May's responsibilities.

The Pierce Arrow Series 31 and 51 were carried over into 1920, but the price escalated to a range of $7,500 to $9,700. Many automobile manufacturers struggled that year because of the business depression and inherent inflation. Pierce Arrow sales dropped only slightly thanks to buyer loyalty, and were fair considering the economic conditions, with earnings of $1,769,914. However, the management became uneasy and engaged in a drastic cost-reduction procedure. Thus, the 1921 models consisted of only one line: the Series 32 on a 138-in. wheelbase chassis, powered by a 38-hp, six-cylinder engine.

The 1921 Pierce Arrow Series 32 continued the dual-valve engine, but the six cylinders were cast en-bloc. Ten body types comprised the Pierce Arrow passenger car line, with the prices ranging from $7,500 to $9,000. A new three-speed transmission and a multiple dry-disc clutch replaced the former four-speed transmission and cone clutch.

The body lines of the 1921 cars were entirely different from the former styles, yet the stately dignity of the Pierce Arrow was retained and easily recognizable. The open models were long, low, and graceful, with the general outline consisting of a straight line from the radiator shell, back across the hood and cowl, with a gentle drop to the top of the doors, and carried on around the rear of the car. Two new distinctive body styles were added: a two-passenger runabout or roadster, and a new sporty six-passenger touring.

Forbes and Clifton in Charge

On May 14, 1921, the G.W. Goethals and Company retired from the Pierce Arrow management, and George Mixter was elected president. Mixter stayed on as president for only a short time, resigning on December 29, 1921. Myron Forbes, who had been vice president and treasurer since 1919, succeeded Mixter as president of Pierce Arrow Motor Car Company. At this time Pierce Arrow had about 3,200 employees.

Austerity prevailed at Pierce Arrow for 1922. The Series 32 designs became Series 33 for 1922, 1923, and 1924, with only very minor changes, with 19 body styles available for 1924. The domestic sales of 1,669 Pierce Arrow cars in 1923, and 2,078 cars in 1924, could not be considered a resounding success, with Peerless delivering about 9,000 cars during that same period, and Packard selling approximately 28,000 cars.

Going for the Middle

Forbes and Clifton were convinced that a single, expensive car line ($6,250 to $8,000) could not sustain the company. Thus, on July 31, 1924, the Pierce Arrow Motor Car Company announced a new, moderately priced car line, the Pierce Arrow Series 80. The chassis was powered by a new seven-main-bearing engine of L-head design.

1924 Pierce Arrow Series 80

Wheelbase (in.)	130
Price	$2,895-4,045
No. of Cylinders / Engine	6
Bore x Stroke (in.)	4.00 × 5.50
Horsepower	72 adv, 38.4 NACC
Body Styles	7
Other Features	mechanical four-wheel brakes, balloon tires, and torque arm on rear axle to prevent loading the rear springs on heavy acceleration and braking

1926 Pierce 80 roadster. (Source: Smithsonian Institution)

On June 24, 1925, Pierce Arrow added a five-passenger, two-door, closed model known as the Series 80 Custom Built Coach, priced at $3,150. The weight of the Series 80 phaeton was only 3,640 lb compared to 5,090 lb for the Series 33. The more expensive Series 33 was continued through 1925 without change.

Apparently Forbes' and Clifton's decision had merit, because Pierce Arrow sales for 1925 climbed to 5,231 units. On July 28, 1925, Pierce Arrow Motor Car Company had completed arrangements to retire all of its prior preference stock, and paid off the balance of its outstanding bank loans. They were again in a sound financial position.

Pierce Arrow successfully carried over the Series 33 and Series 80 into 1926, with 18 body types available in the Series 33, and 9 body styles in the Series 80. The price structure was $2,895 to $4,045 for the Series 80, and $5,250 to $8,000 for the Series 33. The economy of the United States was good in 1926, and Pierce Arrow accounted for the sales of 5,682 units.

In October 1926, Pierce Arrow announced the introduction of the new 1927 Pierce Arrow Series 36 Dual Valve Six, which superseded the Series 33. The Series 36 Pierce Arrow was lower, had more graceful body lines, and more luxurious body appointments. It featured the "Isotta Fraschini" mechanical four-wheel brakes, and pioneered the use of the Bragg-Kliesrath vacuum power brake unit, balloon tires, roller-sector steering gear, and lacquer exterior finish. The chassis wheelbase remained at 138 in., but the rear wheel tread was increased to 58 in. The characteristic Pierce Arrow trumpet-style headlamp housings built into the fenders were continued; however, drum-type headlamps mounted on brackets were available at extra cost (the author does not recall ever seeing a Series 36 Pierce Arrow with drum headlamps). The price range was $5,875 to $8,000. The Series 80 line was continued without change or price revision. The 1927 Pierce Arrow apparently appealed to the buyers, because despite the business downturn in 1927, whereby the automotive industry total sales dropped by 19%, Pierce Arrow sales increased by 3% to 5,836 units.

Sadly, on May 2, 1927, chief engineer Charles Sheppy died suddenly in Summerville, South Carolina. He was succeeded by John C. Talcott, who had joined Pierce Arrow in 1909.

In November 1927, Pierce Arrow introduced the new 1928 Series 81, superseding the Series 80. New body designs, improved performance, and the increased use of aluminum in the engine were featured. In addition to the aluminum crankcase, transmission case, and clutch housing were the forged aluminum connecting rods, aluminum pistons, and aluminum intake manifold, allowing the carburetor to be equidistant to all cylinders. The first known use of aluminum for the cylinder head allowed increased compression ratio without detonation, thereby providing better performance. The price structure was $2,900 to $3,550 for the Series 81. The Series 36 was continued without change or price revision. Pierce Arrow sold 5,736 cars domestically in 1928.

The Merger with Studebaker

In early 1928 Forbes and Clifton were aware that the small number of Pierce Arrow dealers, the price structure, and the now-ancient six-cylinder engines limited Pierce Arrow sales potential. Therefore, on June 11 and 12, 1928, Forbes and Albert R. Erskine, chairman of the Studebaker Corporation, met in New York City. While there they discussed proposals for the merging of the Pierce Arrow Motor Car Company and the Studebaker Corporation.

The directors of Pierce Arrow Motor Car Company met on Wednesday, June 13, 1928, to review the preliminary terms of the proposed merger. It should be remembered that the Studebaker Corporation earned $11,938,000 in 1927, and had a net income of $3,980,000 for the first three months of 1928. Pierce Arrow showed a deficit of $783,000 in 1927, and the arrears on preferred dividends totaled $4,400,000 as of January 1928.

Just a week later the entire automotive industry was saddened by the death of Colonel Charles Clifton on June 21, 1928, after an illness of only one week. He was 79 years old.

At a special meeting held on July 25, 1928, the stockholders voted for the reorganization of Pierce Arrow Motor Car Company, and consolidation with a new company to be formed in the Studebaker Corporation. The Studebaker Corporation invested $2,000,000 and received 230,125 shares of Pierce Arrow class B stock. Studebaker was able to throw the weight of its giant selling organization behind the new company. Albert Erskine was named chairman of the board, and Myron Forbes became president of the Pierce Arrow Motor Car Company division. A.J. Chanter, then manager of Studebaker branches, was assigned to work with the Pierce Arrow selling organization.

On August 7, 1928, the stockholders approved the reorganization plans, and Studebaker Corporation acquired control of Pierce Arrow.

While certain economies were put into effect immediately, Pierce Arrow designers and engineers were given the go-ahead for new eight-cylinder engines and new body styling. The six-cylinder engines were discontinued. In a miraculous fashion the Pierce Arrow Motor Car Company had two new lines of cars, with modern body styling, ready for the International Auto Show in New York in January 1929.

	1929 Pierce Arrow Series 133	**1929 Pierce Arrow Series 143**
Wheelbase (in.)	133	143
Price	$2,875-3,350	$3,750-5,750
No. of Cylinders / Engine	IL-8	IL-8
Bore x Stroke (in.)	3.50 × 4.75	3.50 × 4.75
Horsepower	125 adv, 39.2 SAE	125 adv, 39.2 SAE
Body Styles	8	5
Other Features	shatterproof glass all around, adjustable driver's seat, side and top cowl ventilators, Marshall seat cushion springs, and complete deluxe equipment	

The cylinders and upper crankcase were cast en-bloc, with the crankshaft supported on nine main bearings, having a Lanchester vibration damper at the front end. Thermostatically operated vertical radiator shutters, located inside the chromium radiator shell, regulated the engine coolant temperature. The engine shared certain design characteristics with the Studebaker President eight-cylinder engine. The engine power was transmitted through a Long Manufacturing Company torsional-vibration-dampening two-disc clutch, a three-speed transmission, propeller shaft with two Spicer universal joints to a semi-floating hypoid rear axle.

The 1929 Pierce Arrow body lines were new and decidedly more modern than previous cars, having a single belt line, omitting the external sun visor, and using vertical ventilating doors in the hood sides in place of louvres. During 1929 Pierce Arrow enjoyed sales of 8,386 units and a profit of $2,566,112 as compared to losses of $1,293,026 sustained in the preceding year. Then came "Black Thursday," October 24, 1929.

The Depression

While the Wall Street stock market crash did not immediately affect Pierce Arrow sales, it put the company on stricter cost control. On December 11, 1929, Erskine assured his distributors that Pierce Arrow trucks would be continued. The 1930 Pierce Arrow cars were announced in January 1930, designated as Series C, B, and A. The cars were offered on four wheelbase lengths, with the net effect of widening the price range in a downward direction.

	1930 Pierce Arrow Series C	1930 Pierce Arrow Series B	1930 Pierce Arrow Series A
Wheelbase (in.)	132	134/139	144
Price	$2,595-2,750	$2,975-3,750	$3,975-6,250
No. of Cylinders / Engine	IL-8	IL-8	IL-8
Bore x Stroke (in.)	3.37 × 4.75	3.50 × 4.75	3.50 × 5.00
Horsepower	115 adv, 36.45 SAE	125 adv, 39.2 SAE	132 adv, 39.2 SAE
Body Styles	3	9	5
Other Features			

On December 30, 1929, Myron Forbes resigned as president of Pierce Arrow Motor Car Company. His duties were assigned to Arthur J. Chanter, who was named first vice president and general manager at the

1930 Pierce Arrow. (Source: Studebaker Museum)

1931 Pierce Arrow. (Source: Studebaker Museum)

board of directors meeting in New York on January 7, 1930. The Pierce Arrow sales during 1930 dropped to 6,795 units, which should not be considered bad during the unfavorable business climate. The Series C, B, and A were carried over from September 2 to December 31, 1930, and were considered early production 1931 models.

In January 1931 Pierce Arrow came to the National Auto Show with three lines of cars powered by eight-cylinder engines, on four wheelbase lengths. The former Series C models were dropped from the line-up.

	1931 Pierce Arrow Salon Series 41	1931 Pierce Arrow Series 42	1931 Pierce Arrow Series 43
Wheelbase (in.)	147	142	134/137
Price	$6,250 (town car)		$2,685 (coupe)
No. of Cylinders / Engine	8	8	8
Bore x Stroke (in.)	3.50 × 5.00	3.50 × 5.00	3.50 × 4.75
Horsepower	132 adv, 39.2 SAE	132 adv, 39.2 SAE	125 adv, 39.2 SAE
Body Styles	5	9	3
Other Features			free-wheeling on the rear of the transmission

The bodies were almost completely new in appearance, construction, and luxury appointments. Body side panels, doors, and cowl sides were extended down below the frame side rail, leaving only a narrow splasher between the body sill and the running board. The clamshell front fenders were new with deeper crowns, and the fender inner shield had a strut effect in front concealing the shock absorbers and front suspension components. The cowl, dash, windshield header, and hinge pillar (A-pillar) were stamped in one piece. The Salon models had a provision of a screw knob-type adjustment for the rear seats as well as the front. The radiator shell was more massive and chromium-plated, as were the curved horns up front. The bumpers were a heavier, broader, chromium-plated single bar. The sales amounted to 4,522 units in 1931.

New Trucks

In January 1931 Pierce Arrow announced the introduction of a completely new line of medium-duty and heavy-duty trucks in five models. A special six-wheeler, for extreme heavy-duty service, was also announced and shown in the New York Pierce Arrow showroom during the Auto Show week. These trucks were offered in three wheelbase lengths each: 160-, 180-, 200-in. for the 2-ton and 5-ton capacity models; and 150-, 170-, and 190-in. for the 3-ton model. The 8-ton six-wheeler was offered in wheelbase lengths of 168 and 204 in.

The truck engines were of the L-head type of Pierce Arrow design, having piston displacements of 298-, 361-, 479-, and 611-cu.-in. capacity. The hp ratings were 70, 77, 103, and 130, respectively. The new lines of Pierce Arrow trucks carried all the features of the previous Pierce Arrow trucks introduced in September 1913, including the famous Pierce Arrow worm-drive rear axle. The engine lubrication was by force-feed, and the engines were protected against dirt and moisture by oil, fuel, and air filters. The radiators were of the built-up tubular core type with cast tanks supported by vertical steel members, and mounted on flexible cushions on the frame side rails. The power was transmitted through a twin-disc clutch and four-speed transmission to the worm-drive rear axle, by means of an exposed tubular propeller shaft and two enclosed universal joints. The merchandising of the Pierce Arrow trucks was aggressively handled through the S.P.A.R. Sales Company (Studebaker, Pierce Arrow, Rockne) formed in late November 1931, with Paul Hoffman as president. Hoffman was also vice president of the Studebaker Corporation.

Multi-Cylinders

With the "multi-cylinder" race that began in 1930, Pierce Arrow was not to be denied the leadership in the many fine cars that were designed and produced in 1932. In mid-November 1931, Pierce Arrow announced their new twelve-cylinder-engine-powered cars, Series 53, 52, and 51, and a new straight-eight-cylinder line of cars, Series 54, for 1932.

	1932 Pierce Arrow Series 53	**1932 Pierce Arrow Series 52**	**1932 Pierce Arrow Series 51**	**1932 Pierce Arrow Series 54**
Wheelbase (in.)	137	142	147	
Price	$3,785-4,250	$4,295-4,800	$4,295-4,800	$2,385-3,050
No. of Cylinders / Engine	12	12	12	IL-8
Bore x Stroke (in.)	3.62 × 4.00	3.62 × 4.00	3.62 × 4.00	3.50 × 4.75
Horsepower	150 adv, 50.7 SAE	150 adv, 50.7 SAE	150 adv, 50.7 SAE	125 adv, 39.2 SAE
Body Styles	12	3	seven-passenger sedan and enclosed drive limousine plus custom bodies by Brunn and LeBaron	12
Other Features		Bendix Startix automatic starting device		

Everything about the new 1932 Pierce Arrow lines of cars bespoke elaboration in riding comfort. The wheel treads were 58 in. front, and 61.5 in. in the rear. The frame side rails were of the double-drop box-section type through the mid-section of the car's length. The bodies were carried on lowered outrigger brackets, reducing the thickness of the body sills, and in effect lowering the entire car body. The styling was pleasing, with the broader belt line molding extending rearward from the chromium-plated, sloping, V-type radiator shell, through the front and rear doors, and gently sloping downward in the rear quarter panel. The mascot that adorned the radiator filler cap was a statuette of a crouched archer with bow and arrow, finished in chromium plating. The trumpet-shaped headlamp housings blended gently into the clamshell front fenders, which had fender well provisions for the spare tire and wheel assemblies.

The mechanical features of the new 1932 Pierce Arrow were outstanding: synchro-mesh transmission with helical gear teeth for silent second gear operation, and four double-acting shock absorbers that had adjustable relief valves controlled by a linkage and control lever on the instrument panel, for full ride control at the will of the driver. The eight-cylinder engine of 366-cu.-in. displacement had a nine main bearing crankshaft. The twelve-cylinder engine of the Series 53 had a displacement of 398 cu. in., and those of the Series 52 and 51 had 429-cu.-in. displacement. All Pierce Arrow twelve-cylinder engines carried their crankshaft on seven main bearings.

In a depressed industrial market where the total automobile industry car sales tumbled from 3,880,247 units in 1929 to 1,096,399 units in 1932, Pierce Arrow sales also dropped to 2,692 cars in 1932. Business conditions did not improve during 1932, but instead, the cruel Depression deepened.

Sales Continue to Fall

It was obvious why Pierce Arrow sales dropped in 1932: lack of dealer representation. In 1931 Pierce Arrow had 449 dealers, but by the end of 1932 the dealer count had dropped to 385. What was more alarming was that of that number, 344 of the dealers were handling multiple makes of cars, and only 41 were exclusively Pierce Arrow. By population groups, Pierce Arrow had only 46 dealers in cities of populations of 50,000 to 100,000.

In October 1932 Chanter announced the appointment of Roy H. Faulkner as president of the Pierce Arrow Sales Corporation, and vice president of the Pierce Arrow Motor Car Company. Faulkner was president of the Auburn Automobile Company before he became associated with the Studebaker interests in November 1931, when he was made vice president of the Studebaker Sales Corporation.

Pierce Silver Arrow

The production of the 1933 Pierce Arrow models started in November 1932, but the formal introduction came in January 1933 at the National Auto Show in New York. In the Salon section of the show, the star was the ultra-streamlined Pierce Silver Arrow. It had long, sloping rear quarters to overcome wind drag, an extremely sloping V-type radiator shell with shutters, and a two-piece, V-shaped windshield. The styling by Philip Wright and interior appointments reached a new pinnacle of luxury, comfort, and beauty. Five Silver Arrows were built and it stole the show wherever it was exhibited.

1933 Pierce Silver Arrow. (Source: Smithsonian Institution)

1933 Pierce Silver Arrow

Wheelbase (in.)	139
Price	$10,000
No. of Cylinders / Engine	V-12
Bore x Stroke (in.)	3.62 × 4.00
Horsepower	175 adv, 50.7 SAE
Body Styles	1
Other Features	capable of speeds up to 115 mph

While the car was not intended for large production, it reflected the capabilities of Pierce Arrow, and was the forerunner of future cars to be built by the automobile industry. The Pierce Silver Arrow pioneered the use of a pressed metal roof. It had shrouds to cover the rear wheels, and the spare tire and wheel were carried in a shrouded compartment in the front fenders, just behind the front wheels.

The other 1933 Pierce Arrow models consisted of two entirely new production lines, two salon models, and two custom model lines.

	1933 Pierce Arrow Series 836	1933 Pierce Arrow Salon Series 1236	1933 Pierce Arrow Custom Series 1242 and 1247
Wheelbase (in.)	136	136	142/147
Price	$2,385-2,975	$2,785-3,375	$3,650-7,200
No. of Cylinders / Engine	IL-8	V-12	V-12
Bore x Stroke (in.)	3.50 × 5.00	3.62 × 4.00	3.62 × 4.00
Horsepower	135 adv, 39.2 SAE	160 adv, 50.7 SAE	175 adv, 50.7 SAE
Body Styles			
Other Features			

The design and styling of the 1933 Pierce Arrow models were the epitome of functional design and good taste. The long, sweeping front fenders had the crown and skirt follow the outline of the front tire, while the nacelle-type headlamp housings blended gracefully into the top of the front fenders. The extremely sloping V-type radiator shell and shutters were pleasing to the eye and remained distinctive of the maker. The body rear lower panel was curved gracefully rearward giving a beaver tail effect. A handsome, functional, folding trunk rack adorned the rear end of the car. The chromium-plated tail, stop, backup lamps, license bracket, and fuel filler neck were assembled into a neat unit at the left rear.

Because of the splendid design, quality manufacture, and great value, sales of 2,152 units could be considered fair in a depressed national economy. Taking everything into account, Pierce Arrow management showed a net profit of $4,770 for the June quarter of 1933, compared to a net loss of $878,800 for the June quarter of 1932.

More Red Flags and Reorganization

Unfortunately, the friendly receivership of the Studebaker Corporation on March 18, 1933, and the untimely death of Albert R. Erskine, board chairman, on July 1, 1933, created much apprehension within the Pierce Arrow organization, even though the receivership did not affect Pierce.

Arthur Chanter was elected president of Pierce Arrow Motor Car Company in April 1933, and a plan for reorganization was worked out by Chanter and a group of Buffalo businessmen and bankers, many of whom were affiliated with Pierce Arrow in one way or another. The plan conceived was the purchase of Pierce Arrow Motor Car Company from the receivers as authorized by Federal Judge Thomas W. Slick. The purchase was made in September 1933 for $1,000,000 and other considerations. The funding was provided by Chanter and the group of Buffalo bankers and businessmen, and a board of directors was set up.

Chanter continued as president and had associated with him Roy Faulkner as vice president of sales, B.H. Warner vice president in charge of operations, K.M. Wise director of engineering, and M.C. Ewald secretary-treasurer. Faulkner was elected to the Pierce Arrow board of directors on November 3, 1933.

Publicity Stunts

A Pierce Arrow twelve roadster was stripped of the fenders, top, and windshield, and the engine was tuned to 207 hp, in preparation for a record run that the company hoped would boost car sales by emphasizing performance. On August 7, 1933, Ab Jenkins drove this roadster on a circular course laid out on the Salt Flats in Utah, and broke all records for 500 miles to 2,000 miles and 12-hour and 24-hour records by at least 5 mph. The runs were supervised by the AAA Contest Board, as Jenkins also established a new world speed record of 118 mph for 3,000 miles, and a one-lap record of 128.1 mph. Despite all these efforts, Pierce Arrow sales for 1933 amounted to 2,152 units, but the sales of all luxury cars also tumbled in 1933.

In January 1934, Pierce Arrow introduced three distinct lines of cars.

	1934 Pierce Arrow Series 1240	1934 Pierce Arrow Series 1248A	1934 Pierce Arrow Series 840
Wheelbase (in.)	139/144	147	139/144
Price	$2,795-4,495	$4,295-7,000	$2,795-4,495
No. of Cylinders / Engine	12	12	IL-8
Bore x Stroke (in.)	3.62 × 4.00	3.62 × 4.00	3.50 × 5.00
Horsepower	175 adv, 50.7 SAE	175 adv, 50.7 SAE	140 adv, 39.2 SAE
Body Styles		5 standard and 3 custom	
Other Features		Free-wheeling and Stewart-Warner power brakes	

The 1934 styling had a more pronounced streamlined effect, the hood sides had door-type louvres, the bodies were roomier, and they featured draftless window ventilation, and adjustable rear as well as front seats. The frame side rails were of full-length box-channel girder type, and the chassis featured hypoid rear axle drive gears. The Series 1248A carried the most formal and conservative body types, many of which were built by custom coach builders.

The outstanding body type of the 1934 models was the Silver Arrow, a five-passenger, four-door sedan. This particular sedan had all the features of the conventional sedan, plus many of the styling features of the Silver Arrow Show car, such as the downward-sloping rear end panels and the sloping roof quarter panels. The Silver Arrow model was priced at $3,495 for the eight and $3,895 for the twelve, both on a 144-in. wheelbase chassis.

In March 1934 a new lower-priced Series 836A was introduced.

1934 Pierce Arrow Series 836A

Wheelbase (in.)	136
Price	$2,195-2,295
No. of Cylinders / Engine	IL-8
Bore x Stroke (in.)	3.50 × 5.00
Horsepower	135 adv, 39.2 SAE
Body Styles	two-door brougham and club sedan
Other Features	

On August 22, 1934, Ab Jenkins drove a Pierce Arrow special-bodied car for a new world's 24-hour record. He drove 3,053 miles on the Bonneville Salt Flats in Utah at an average speed of 127.208 mph. Despite the phenomenal speed records and promotional effort, there was a constant shrinkage of the demand for cars in the fine-car field. Only 1,740 Pierce Arrow cars were sold in 1934. This caused the board of directors of Pierce Arrow Motor Car Company to apply to the U.S. Federal Court for permission to reorganize under the National Bankruptcy Law. A hearing was set for September 17, 1934.

The End Had Come

The 1935 Pierce Arrow models introduced in January 1935 were in fact a carryover of the 1934 models, except for more modernization in style and a conspicuous refinement of interior and exterior. The identification became Series 845 for the eight and Series 1245 and 1255 for the twelve-cylinder-engine-powered cars. The price structure was $2,795 to $3,495 for the eights, and $3,195 to $7,000 for the twelves. By the end of 1935 most Pierce Arrow branches were closed and the distributors in most cities became multiple-car-line dealers. Pierce Arrow sales slipped further to 875 units.

Again for 1936 the Pierce Arrow cars were a carryover of the 1935 models. The identification was changed to Series 1601 for the eight-cylinder models and Series 1602 and 1603 for the twelve-cylinder models. The main new styling feature was the horizontal ventilating doors in the hood sides, and the seven-passenger sedan had a built-in trunk, while the other body types had a folding trunk rack in the rear. The price schedule for the 1936 Pierce Arrow models was $3,195 to $5,295 for the eights and $3,695 to $5,795 for the twelves. An automatic overdrive was standard on all models, and an X-member had been incorporated in the frame

1936 Pierce Arrow brougham.

structure while the box section side members and tubular cross-members were retained. Despite the Pierce Arrow refinements, engineering innovations, and excellent workmanship, sales for 1936 were only 787 units.

In August 1936 a new division of Pierce Arrow was formed to handle the production and sale of tourist-type trailers (commonly known as house trailers). The features of the "Travelodge" (the tradename for the vehicle) included independent wheel suspension, Bendix hydraulic brakes, and Houdaille shock absorbers. The vehicle was built on an all-steel chassis and body frame covered by an outer shell of 18-gauge sheet aluminum.

The 1937 Pierce Arrow models announced in November 1936 were again a carryover from 1936. They were now identified as Series 1701, 1702, and 1703, respectively. Twenty-four body types were available, including the new Brunn Metropolitan Town Brougham, available with either the eight-cylinder or twelve-cylinder engine on a 144-in. wheelbase chassis. The price range was $3,195 to $5,795.

The sales of 1937 Pierce Arrow cars amounted to only 167 units; thus, in late December 1937, Arthur J. Chanter, on behalf of the Pierce Arrow Motor Car Company, filed a petition seeking reorganization under section 77B of the Bankruptcy Act, for the purpose of preserving the status quo. Federal Judge Knight set January 17, 1938, as the first hearing date.

At that hearing Pierce Arrow Motor Car Company was ruled insolvent by Judge Knight and he ordered the liquidation of the Buffalo automobile manufacturing firm. During the third week of February, the last chapter of the history of the once proud and great Pierce Arrow Motor Car Company was written by a five- and ten-cent sale in the quiet of the huge plant once alive with men at work. Pierce Arrow was sold piece-meal for the liquidation by the 1695 Elmwood Avenue Corporation. The bulk of the office furniture and supplies were purchased by former employees as mementos.

Finally, in March 1939, the application of A. Howard Aaron to be discharged as trustee of the now-defunct Pierce Arrow Motor Car Company was granted in Federal Court by Judge Knight. Thus was the conclusion of a great fine-car manufacturing firm.

Despite the demise of Pierce Arrow Motor Car Company in 1937, at no time did they lower their prestige, engineering excellence, or manufacturing quality. Even today, Pierce Arrow cars are sought as valuable possessions, thanks to the dedication and untiring effort of George N. Pierce and Colonel Charles Clifton.

1937 Pierce Arrow V-12.

Packard/ Joy/ Macauley and the Packard Motor Car Company

Joy in the 1899 Model A. (Source: P.M.C. Co.)

Macauley (left) and J. Packard. (Source: P.M.C. Co.)

Many might question, "How could a company that built cars of such magnificent beauty and enduring quality be condemned to oblivion?" In retrospect, it could be said that Packard's passing was unavoidable, and if the company had survived, it would have only prolonged the agony. If you were to "ask the man who owned one," he might reply that Packard accepted an honorable exit, in preference to an existence in a business environment in which success was marked by high profit, ostentation, gaudy exhibition in advertising, and makeshift substitution. James Packard, Henry Joy, and Alvan Macauley certainly would have had none of that.

* * *

James Packard

James Packard. (Source: P.M.C. Co.)

James Ward Packard was born on November 5, 1863, in Warren, Ohio. He entered Lehigh University in Bethlehem, Pennsylvania, in 1882, and graduated with honors in 1886 with a degree in mechanical and electrical engineering. His interest in design, experimentation, and invention asserted itself through the electrical and mechanical devices he designed and made while in college. After graduation, he was employed by the Sawyer-Mann Electric Company in New York City as a shop worker in the incandescent lamp department. He quickly rose to assistant superintendent of the factory. In 1893 he returned to Warren and, with his older brother William, founded the Packard Electric Company, engaged in the manufacture of electrical bulbs, transformers, cable, and wiring supplies.

The electrical business prospered greatly, and Packard, being interested in self-propelled vehicles, purchased a Winton car in August 1898. (See the Winton chapter in Volume 4.) Becoming exasperated with the mechanical difficulties he experienced with the car, he traveled to Cleveland, Ohio, to make a few suggestions to Alexander Winton. Winton, who had built his car in 1896 and made the first sale to a retail customer on March 21, 1898, was not about to take advice from an electrical manufacturer. With his patience wearing thin as the result of Packard's persistent suggestive remarks, Winton abruptly snapped, "If you are so smart Packard, why don't you make a car yourself?" Packard thought for a moment, and then replied, "I think I will."

Ohio Automobile Company

William Packard. (Source: P.M.C. Co.)

James and William, and two Cleveland craftsmen with horseless carriage experience, George L. Weiss and William A. Hatcher, formed a partnership of Packard and Weiss. They proceeded to build a horseless vehicle, completed on November 6, 1899, and tested it on the streets of Warren, Ohio. The Ohio Automobile Company was organized to build the car, and was incorporated under the statutes of the state of West Virginia on December 30, 1899, capitalized at $3,050,000.

The "Packard Carriage" was the center of interest at America's first automobile show in New York's old Madison Square Garden in January 1900. The car was a single-seat roadster with wire-spoke wheels. It was powered by a 9-hp, single-cylinder engine, and driven by a chain drive to the rear wheels. The steering was by a tiller; however, the engine had automatic spark advance as well as other features not found in other cars until years later, among which was an H-slot quadrant for the gearshift lever. Five Packard cars were built in 1899 to take care of the orders on hand.

1899 Model A. (Source: P.M.C. Co.)

1900 Model B. (Source: P.M.C. Co.)

The 1900 Packard basic model was continued with minor improvements, such as a foot-operated accelerator pedal and a dos-a-dos seat to make it a four-passenger vehicle. About 45 to 50 Packard cars were built and sold in 1900. The 1901 models featured steering wheel control, and a three-speed sliding-gear transmission with an H-slot gearshift lever quadrant. The engine was the same except the output was increased to 12 hp. Approximately 80 Packard cars were built and sold in 1901.

1901 Model B. (Source: P.M.C. Co.)

Sidney D. Waldon joined the Packard organization in 1902, filling a responsible sales position. The 1902 Packard Model F showed a marked improvement in design and styling, changing from the horseless carriage (or buggy) to that of a real automotive vehicle. Most new vehicles introduced by competitive manufacturers after 1902 resembled the 1902 Packard Model F. More attention was given to the comfort and functional aspects of the car. Two body types were offered in 1902: a two-passenger roadster and a five-passenger rear entrance. In 1902 Packard built about four experimental prototypes of the Packard Model G, featuring an opposed two-cylinder engine, but this model did not get into production. The Model F continued through 1903 without change, except for a slightly longer hood.

During the summer of 1903, starting on June 20th, Marcus Krarup and Thomas Fetch, driving a Packard Model F nicknamed the "Old Pacific," established a new transcontinental record run from San Francisco, California, to New York City, covering the distance in 53 days.

1902 Packard Model F. (Source: P.M.C. Co.)

Henry Joy

Henry Bourne Joy was born in Detroit, Michigan, on November 23, 1864. He graduated from the Phillips Academy in Andover, Massachusetts, in 1883, and the Yale Scientific School in 1884. His business career began as an office boy for the Peninsular Car Company, of which he later became assistant treasurer. From 1887 to 1889 Joy was actively engaged in mining operations in Utah. Subsequently, in 1890 he was involved with the development of transportation methods for the Detroit Street Railway System. He stayed there until 1898, when he was called to active duty by the United States Navy during the Spanish-American War. Joy was a member of the Michigan Naval Militia, and served as a Boatswain's Mate aboard the U.S.S. Yosemite.

Upon his release to inactive duty in 1900, Joy became an official of the Detroit Union Depot Company. He later served as a receiver of the Chicago and Grand Trunk Railway Company from 1900 to 1902. With the automobile industry growing in Detroit, it was only natural that Henry Joy became interested in the automobile business.

Henry Joy. (Source: P.M.C. Co.)

On May 1, 1902, Joy purchased a Model F Packard, with which he later made a record-breaking non-stop run between Boston and New York City. He was so impressed with the performance of the Packard car, and being of comfortable wealth, he knew he wanted to be part of the action at the Ohio Automobile Company. Joy went to Warren and convinced Packard of the need for additional facilities, and to accept reorganization with additional capital to be funded by prominent Detroit businessmen. In addition to Henry B. Joy, the Detroit men who provided the necessary funds were: Russell A. Alger Jr., Fred M. Alger, John T. Newberry, Truman Newberry, R.P. Joy, Charles M. DuCharme, D.M. Ferry Jr., Willard C. McMillan, Philip H. McMillan, and Joseph Boyer. James W. Packard, Henry B. Joy, Truman Newberry, George L. Weiss, Philip H. McMillan, Joseph Boyer, and Russell A. Alger were elected to the board of directors.

1903 Packard Model K.
(Source: P.M.C. Co.)

*1904 Packard Model L.
(Source: P.M.C. Co.)*

Packard Motor Car Company

The name of the Ohio Automobile Company was changed to Packard Motor Car Company in November 1902. James Packard was elected president, Henry Joy became general manager, and Sidney D. Waldon was appointed sales manager. William E. Metzger was appointed agent for Packard on April 1, 1903. Frederick W. Slack joined Packard in 1903 as a draftsman. Alvan T. Fuller became Packard's first distributor on December 30, 1903, in Boston, Massachusetts. By 1906 he established Boston's famous Motor Mart, having no less than 21 independent salesrooms. He merchandised eight makes of cars in addition to Packard.

In early 1902, Albert Kahn, then a young architect, was commissioned to design the world's first reinforced-steel concrete factory building at 1580 East Grand Boulevard in Detroit, the new Packard home. Upon completion of the plant in 1903, the Packard automotive operations and 247 employees were moved from Warren to Detroit. William D. Packard, George L. Weiss, and William A. Hatcher chose to remain in Warren with the Packard Electric Company plant. The Packard Electric business is still in existence today as a division of the General Motors Corporation.

The first Packard car to be built in the new Detroit plant was the Packard Model L, which definitely set the styling trend for the next few years. With the engine up front under the hood it no longer resembled the horseless carriage of a few years ago. This Packard model introduced the distinctive Packard creased shoulder line of the radiator shell and hood, which would remain the Packard identity for the next 44 years. The engine was a new, vertical, four-cylinder, L-head type, with the head cast integral with the cylinder block. Valves were accessible through the port plugs in the head portion of the cylinder block. The cylinder dimensions were 3-7/8-in. bore and 5-1/8-in. stroke, giving a piston displacement of 241.69 cu. in.

Gray Wolf

Prior to Packard's move to Detroit, a French-born engineer named Charles Schmidt joined the Ohio Automobile Company in April 1902. Schmidt had been employed by the Mors Works in Paris, France, migrating to

Charles Schmidt driving Packard Gray Wolf at Ormond-Daytona Beach, January 1904. (Source: P.M.C. Co.)

the United States in early 1902. Schmidt was responsible for the design of Packard's first four-cylinder-engine-powered passenger car, the massive and very expensive Model K, of which only a few more than two dozen were built. However, he developed and used the Packard Model K four-cylinder engine (4-in. bore, 5-in. stroke) and chassis for the design of the famous "Gray Wolf." By careful engineering and design, Schmidt was able to reduce the weight of the Gray Wolf to 1,310 lb.

The Gray Wolf was an exquisite, slender vehicle with a very narrow vertical snout without the flat vertical radiator, offering as little resistance to airflow as possible. Engine cooling was accomplished by longitudinally placed copper tubing following the body lines, just above the frame rail. Suspension was by semi-elliptical springs: one transverse spring at the front, and two placed longitudinally along the frame rails at the rear. The Gray Wolf was tested at the Warren, Ohio, Fair Grounds track. It was entered in the race at Warren on September 5, 1903, in which Schmidt was injured as the Gray Wolf crashed into the inner fence of the track.

On September 8, 1903, the Gray Wolf competed on the Grosse Pointe, Michigan, track, with Harry Cunningham driving (Schmidt's injuries would not permit him to drive). Cunningham easily defeated Barney Oldfield in Winton's Bullet as well as the other racing cars of the day. By January 1904 Schmidt's injuries were healed well enough for him to break all records at Ormond-Daytona Beach, Florida. Schmidt drove the measured mile in 46.6 seconds, and the 5-mile run in 4 minutes, 21.6 seconds. The records remained intact for many years. After the record runs, the Packard Gray Wolf was exhibited on a roped-off platform at the National Automobile Show in New York City.

Charles Schmidt returned to the Packard factory in Detroit to continue performing his responsibilities as chief engineer. In late 1904 the Gray Wolf was sold, and Packard did not compete again in racing until 1915. Charles Schmidt resigned from Packard in April 1905, moving to Cleveland, Ohio, after accepting the position of chief engineer at the Peerless Motor Car Company. Russell Huff, who joined Packard at Warren, Ohio, in 1902, succeeded Schmidt as chief engineer.

The 1905 Packard Model N continued with the same four-cylinder engine, but had many refinements and body innovations. A side entrance touring car with two side doors replaced the rear entrance tonneau. Two new closed models were introduced: the closed brougham and the limousine with open driver's compartment. These were the first known closed models built on a production basis. Approximately 400 Packard

1905 Packard truck chassis. (Source: P.M.C. Co.)

model Ns were built and sold in 1905. During 1905 Earle C. Anthony Incorporated became Packard distributors for the San Francisco and Los Angeles zones, covering the entire state of California.

Packard Trucks

The merchandising of passenger cars was definitely a seasonal spring and summer activity, causing high peaks and low depressions in production schedules and manpower requirements. Henry Joy saw the need to diversify the product line-up to include products that would sell in the winter months as well. Thus, during 1903 Packard engineers and designers came up with a number of commercial vehicle types. Strangely, even though Packard had already graduated to an in-line four-cylinder engine for passenger cars, the designers felt that since commercial vehicles would be operated at lower speeds, through greater numerical gear ratio only half the power of a limousine would be sufficient. Hence, the first production Packard trucks of 1.5-ton capacity emerged with a two-cylinder engine under the driver's seat. In 1908 Packard introduced a new 3-ton capacity truck with a vertical, four-cylinder L-head engine, with the cylinders cast en-bloc.

By early 1905 Joy and Huff felt confident enough in Packard trucks that they had the new trucks placed in the hands of users and potential buyers on a consignment basis. What a unique way of merchandising a new product! Needless to say none of the Packard truck users were willing to part with them once they appreciated the value of their service. They gladly paid $2,500 for each truck chassis. Packard continued to build trucks through 1922, when a decision was made to phase out truck production. A total of approximately 43,500 Packard trucks were built and sold during 1905-22.

New Passenger-Car Developments

For 1906 Packard introduced the Model S, which featured the new, larger, and more powerful T-head engine, with a bore of 4-1/2 in. and a stroke of 5-1/2 in., providing a displacement of 349.9 cu. in. Aside from the increased cylinder dimensions, the most important change from the 1905 Model N engine was that the inlet and exhaust valves were located in separate chambers on opposite sides of the engine. This allowed larger valves, improved breathing, and consequently more power from a given size of engine.

*1906 Packard Model S.
(Source: P.M.C. Co.)*

The 1906 Model S featured a high-tension magneto by Eisemann, which induced a low voltage increased by a coil located in a weatherproof box on the forward end of the running board. The high-voltage current was distributed in the correct sequence to the four cylinders by a timing distributor located on the magneto. In addition, auxiliary battery ignition was provided using a quadruple vibrator coil and a distributor of roller contact type. Two sets of spark plugs were used to accommodate the dual ignition.

The Packard Model S discarded the transverse front spring arrangement, which had for so long been a feature of Packard construction. The new front axle of tubular design arched downward in the center for engine clearance. The front and rear suspension was accomplished by parallel semi-elliptical springs. The Packard 1906 Model S was offered in five body styles: a touring, runabout, landau (replacing the brougham), limousine, and a Victoria model.

*1907 Packard Model 30.
(Source: P.M.C. Co.)*

During 1906 and subsequent years, the hexagon coined depression in the center of the hubcaps was painted black in production. When the car was reconditioned or the engine was rebuilt at the distributors' or the dealers' shops, the hexagon would be painted red indicating the services performed. This practice continued through 1912. The 1913 and subsequent models had the hexagon painted red in production. Production and sales amounted to approximately 725 units in 1906.

While the new Packard 30, introduced in August 1906, gave the first impression of being a refined 1906 Model S, it was, in fact, an entirely new model for 1907 that would create Packard history. The wheelbase was increased by 3 in. to accommodate the larger engine and longer hood. The rear suspension, while retaining the parallel semi-elliptical springs, eliminated the conventional rear spring frame horn, and provided an additional transverse semi-elliptical rear spring, giving the car a platform spring arrangement. This provided for a better ride, controlling roll and increasing handling stability.

1907 Packard Model 30

Wheelbase (in.)	108/122
Price	$4,200-5,600
No. of Cylinders / Engine	T-4
Bore x Stroke (in.)	5.00 × 5.50
Horsepower	30 adv, 40 ALAM
Body Styles	touring, limousine, landau, runabout (108 in.)
Other Features	folding top for touring and runabout
	acetylene headlamps with automatic lighters

Despite the price increase, and 1907 being considered a business panic year, the buying public apparently liked the Packard 30. Packard built and sold more than 1,125 cars in 1907, and the profits exceeded $1,000,000.

The 1908 Packard 30 models were a refinement of the 1907 models, with the wheelbase lengthened to 123-1/2 in., except the runabout wheelbase, which remained at 108 in. This model saw the introduction of Packard's famous bail-type radiator cap, allowing the cap to be tilted while adding water. A new close-coupled touring body was added and the Victoria model was discontinued. Even with a modest price increase Packard was able to build and sell over 1,300 cars during the 1908 model year.

For 1909 the Packard 30 models were further refined, with the major change being a new dry multiple-disc clutch, the addition of an inner wheel housing to the front fenders, and longitudinal running board splash

1908 Packard Model 30 touring car. (Source: P.M.C. Co.)

shields. The new 1909 Packard 18 was introduced on August 19, 1908, and featured the same style and quality of material and workmanship as that of the Packard 30. Approximately 1,500 Packard 30s and about 800 Packard 18s were built and sold in 1909.

1909 Packard Model 18

Wheelbase (in.)	102/112
Price	$4,300-4,400
No. of Cylinders / Engine	T-4
Bore x Stroke (in.)	4.06 × 5.12
Horsepower	26.3 ALAM
Body Styles	touring, limousine, landaulet, runabout (102 in.)
Other Features	

On September 9, 1909, Packard Motor Car Company was incorporated in the state of Michigan, capitalized at $10,000,000. During 1909 Ormond E. Hunt joined Packard Motor Car Company as engineering draftsman.

Alvan Macauley

For 1910 Packard 30 and 18 models were continued with minor changes. However, a major change in the Packard organization occurred when Sidney D. Waldon was elevated to vice president, and Alvan Macauley joined Packard Motor Car Company on April 11, 1910, as general manager.

Macauley started his business career as a practicing lawyer in the District of Columbia, and later in the state of Ohio. He joined the National Cash Register Company of Dayton, Ohio, as patent attorney in 1895, and later was elected to the board of directors. During his tenure at National Cash Register, Macauley figured out that some of the register mechanism could be applied to an adding machine. In 1902 he joined the American Arithometer Company at Main and Mullanphy Streets in St. Louis, Missouri. His progress was envied by Patterson, the president of American Arithometer Company, who in 1904 resigned and left the company, taking with him 52 key executives. This left Macauley in charge, since he was the only one who knew the adding machine business. During 1904, Macauley tried to obtain permission from the city of

1910 Packard Model 30 limousine. (Source: John A. Conde)

St. Louis to build a bridge across an alley to connect two buildings of the plant, but he was repeatedly turned down.

Macauley was able to get W.M. Burroughs and Joe Boyer of Detroit, who already had majority interests in the Chicago Pneumatic Tool Company and the Boyer Machine Shops, to invest in the adding machine business. After reorganization was completed, the name was changed to Burroughs Adding Machine Company, and in 1905 the operations and all personnel were moved into the newly erected plant in Detroit. Favorable business climate and labor market was given as the reason for the move. Alvan Macauley moved up in that organization steadily in the following five years to become general manager.

Macauley's executive ability was noticed by Henry Joy, and when Sidney Waldon was elected vice president, Joy convinced Macauley to Join the Packard Motor Car Company as general manager. Shortly afterward, Joy brought engineer Jesse G. Vincent to Packard, upon Macauley's recommendation.

Alvan Macauley. (Source: P.M.C. Co.)

Six-Cylinder Engines

After joining Packard in 1910, Jesse Vincent and his engineering staff worked feverishly to develop a satisfactory six-cylinder engine, because Packard 18 and 30 sales were declining. Pierce Arrow had had a powerful six-cylinder engine since 1907, and Peerless had their 50-hp, six-cylinder engine since 1908. Rear axle noises were prevalent in all cars at that time, hence, Vincent also directed some of the engineering efforts toward developing quieter rear axle gears.

By early 1911, the Packard engineering department had designed and developed a remarkable six-cylinder engine of T-head design with the cylinders cast in pairs of three, bolted to an aluminum crankcase. The

1911 Packard Model 18 runabout. (Source: P.M.C. Co.)

cylinder bore of 4-1/2 in. and a stroke of 5-1/2 in. made it possible for the engine to produce 48 hp. This engine featured full-pressure lubrication to 35 vital moving parts, and set the precedent for the automotive industry to follow.

To be sure the engine would be reliable, Joy, Huff, and E.F. Roberts, general superintendent, took a production Packard equipped with the new six-cylinder engine on a 4,000-mile endurance trip through the impassable Rocky Mountains and the plateaus of Wyoming. The test team left Detroit on May 15, 1911, and in their journey climbed mountains, forded streams, and passed through treacherous alkali sinks and marshes to the southern entrance of Yellowstone National Park, over terrain and in weather conditions never before encountered by a motor car. The test team returned to Detroit five weeks later, carrying with them their camping equipment, three adult passengers, fuel and oil, which brought the total weight of the vehicle to over 3 tons.

1912 Packard Model 30 touring. (Source: P.M.C. Co.)

The owner of a 1912 Packard Model 30 expresses himself. (Source: P.M.C. Co.)

1912 Packard Model 1-48 touring car. (Source: P.M.C. Co.)

The 1912 six-cylinder Packard Model 1-48 went into production in April 1911 and deliveries started in late May. The Packard Models 18 and 30 were continued along with the 1-48 through the 1912 model year. In August 1912 Packard introduced a smaller six-cylinder-engine-powered car, designated Model 1-38, having an L-head six-cylinder engine with a 4-in. bore and a 5-1/2-in. stroke, as a companion to the Packard 1-48. The Models 18 and 30 were discontinued after the 1912 model year. During 1912 Packard started using model series designation in place of yearly models. This practice continued through 1953. For 1913 the 1-48 was unchanged except the wheelbase was increased to 121-1/2 and 139 in., with the model designation changed to 2-48. The 1-38 continued through 1913 unchanged.

On March 8, 1913, Packard introduced the new Model 3-48 with left-hand drive and central gearshift and brake lever, and also a centralized control console over the steering column (the forerunner of the instrument panel). Packard also introduced electric charging, lighting, ignition, and starting, using the combination starter-generator manufactured by Delco. Spiral-toothed bevel driving pinion and ring gears were introduced in the rear axle. The spiral-toothed (or helical) bevel gears were, in effect, a combination of bevel gear and worm drive design principles. These new spiral-toothed bevel gear rear axles were the first in the industry, and practically eliminated all rear axle noise. The rest of the automotive industry followed Packard's design, some shortly afterward and some not until 15 years later.

The 1914 Model 2-38, introduced in September 1913, also changed to left-hand drive and central control levers and console. The engine exhaust manifold had to be moved to the right side, and the carburetor and intake manifold to the left side, to provide space for the steering gear. The year 1913 was profitable for

1913 Packard Model 2-48 touring. (Source: P.M.C. Co.)

Packard, building and selling more than 4,400 cars. On October 16, 1913, at the stockholders meeting, a dividend of 40% was declared. The board of directors also authorized increasing the capital from $10,000,000 to $16,000,000. The net income was $2,157,472 and the surplus was $1,198,784.

Lincoln Highway Association

On June 4, 1913, the Lincoln Highway Association was formed in Detroit, with the headquarters located in the Dime Bank Building, for the purpose of promoting a continuous, connecting, transcontinental highway reaching from the Atlantic Ocean to the Pacific Ocean. Henry B. Joy was elected president of the association. On June 14, 1913, Joy started on a fact-finding trip to the west coast, accompanied by Frank H. Trego, research engineer for Packard. The trip was made over the proposed route in a six-cylinder Packard car, with Joy driving and Trego making engineering notes as the trip progressed. The written notes included the various types of roads, their conditions, types of bridges and condition, and the improvements needed. After their completion of the trip, Joy and Trego submitted a report on their findings to the association, with recommendation for action and improvements. Roy D. Chapin was one of the prime movers of the Lincoln Highway project (see the Chapin, *et al.*, chapter in this volume).

Unfortunately very little action was taken by the Federal government to improve highway conditions, even though some states and counties made some attempt to eliminate the quagmires in their road systems within their boundaries. It wasn't until the 1920s that the need was noticed and acknowledged. On June 6, 1923, the Zero Milestone, a gift of the Robert E. Lee Highway Association (a volunteer organization) was dedicated in Washington D.C. with President Warren G. Harding making the principal address. The Zero Milestone is located in the Mall in Washington D.C., and marks the point from which the national system of highways would evolve. The President regretfully noted that as of that date, there was no single road from coast to coast that could be used unimpeded throughout the year. It was not until after World War II that billions of dollars would be spent in the United States to provide the finest network of super-highways in all directions. Contrary to the general belief that the fine highway system was the product of government funding, the funding of early highway construction was mostly by private capital.

Col. Jesse Vincent. (Source: P.M.C. Co.)

In 1913 Russell Huff resigned as chief engineer of Packard to join the Dodge Brothers, and was succeeded by Ormond E. Hunt in September 1913. The major changes for the 1914 Packard were: the increase in wheelbase length on the Model 2-38 to 140 in. for all 17 body types; the 2-38-cylinder blocks were now cast in a pair of three cylinders each instead of two cylinders per casting as before; the ignition system was changed to a Bosch Duplex system; and the starting, charging, and lighting systems were changed to Bijur manufacture, with the starting motor and generator being separate units. The Model 3-48 was continued through 1914, incorporating the same changes as the Model 2-38 and adding additional body types to fill out the model line-up.

In February 1915, Sidney Waldon, vice president of engineering, resigned from Packard to join the Cadillac Motor Car Company in the same capacity. Jesse Vincent succeeded him on February 17, 1915, and lost no time in furthering Packard's innovativeness.

On May 23, 1915, Packard startled the automotive industry with the announcement of the new Packard Twin Six. It was powered by a new twelve-cylinder engine, with two six-cylinder blocks set at a 60-degree angle on a single crankcase. A multiple dry-disc clutch was used, and the electrical system for starting, charging, and lighting was by a two-unit system manufactured by Bijur.

1915 Packard Twin Six Models 1-25 and 1-35

Wheelbase (in.)	125/135
Price	$2,600-4,600
No. of Cylinders / Engine	V-12
Bore x Stroke (in.)	3.00 × 5.00
Horsepower	88 adv, 43.2 ALAM
Body Styles	19
Other Features	transmission bolted directly to crankcase

A production of 7,500 cars was planned for the 1916 model year. The six-cylinder Models 38 and 48 were discontinued after the 1915 models.

On June 1, 1915, Alvan Macauley, vice president and general manager, was elected to the board of directors to succeed James Packard, who retired from the board. During the fiscal year ending August 31, 1915, Packard Motor Car Company enjoyed an increase in its surplus to $3,713,747 which represented an increase of 50% over the previous year. The total number of vehicles built and sold during this one-year period was 29,936, of which approximately 8,000 were trucks.

The first use of the new Chicago board speedway for testing manufacturers' vehicles took place on July 19, 1915, when Packard tested the new Twin Six for speed, economy, and endurance. The Twin Six touring, weighing 5,400 lb (with passengers), was run under AAA official supervision. With the top and windshield

1915 Packard Model 3-38 selected to pace the 1915 Indianapolis 500. (l to r) Joe Boyer Jr., Herbert Chase, James Corbett, E.H. Belden (in rear), Darwin Hatch, Frank Trego, Carl Fisher, Chester Ricker, W. R. McCulla (driver), and Sidney Waldon. (Source: P.M.C. Co.)

Packard/Joy/Macauley and the Packard Motor Car Company

up, and carrying five passengers, the Twin Six consumed 3 gallons and 13-1/2 ounces of gasoline for the 50-mile test run, averaging 13.3 mpg. In the speed tests, with Ralph DePalma driving, and another passenger, top down and windshield up, DePalma covered the 10-mile distance (five laps) in 8 minutes, 15 seconds, averaging 72.7 mph. In November 1915, Jesse Vincent drove a Twin Six chassis with a racing body on the famous Sheepshead Bay Speedway, covering the 2-mile lap in 1 minute, 10.52 seconds, at a speed of 102.25 mph. The chassis was standard stock, except for axle ratio and tune-up for maximum speed.

Aircraft Engines

World War I had been going on for over a year, and the value of aeroplanes to the armies had been amply demonstrated, so Joy gave orders for the design and development of light aircraft engines, knowing that the time would come when the United States Government would need a fleet of aeroplanes. Packard did this all on its own, not waiting to be asked to develop aircraft engines.

Experimental work had been going on for some time, and it was found by testing that the twin-six-type engine was best suited for aircraft power. Packard's first successful aircraft engine was the 299-cu.-in. V-12 with a bore of 2-21/32 in. and a stroke of 5 in. All the testing performed during 1914-15 was done in automotive vehicles at that time, since aircraft were not available. The 299 engine developed approximately 200 hp, and subsequently was succeeded by a larger V-12-cylinder aircraft engine of 905-cu.-in. displacement. The 299 and 905 were tested in racing cars, establishing many speed records. The 905 became the forerunner of the famous Liberty aircraft engine developed in 1917-18. The use and maintenance of the experimental garages at the Indianapolis, Chicago, and Sheepshead Bay Speedways led to the speculation that Packard had intended to compete in speedway contests. This was vehemently denied on March 24, 1916, by Vincent, emphasizing that the speedways were used only for testing experimental engines.

On May 12, 1916, Packard Motor Car Company acquired the Krit Motor Car Company properties that adjoined the Packard facilities, and built the necessary additional buildings for the manufacture of aircraft engines. During 1916 Joy arranged for the purchase of 600 acres of flat land in Macomb County, Michigan (near Mount Clemens), for the purpose of testing aircraft and engines. The hangars and support buildings were built, as well as the runways. The area was named Joy Aviation Field. In June 1916, Joy left Packard

(l to r) Liberty 1250, Packard 905, Packard 299 aircraft engines. (Source: P.M.C. Co.)

Motor Car Company to devote full time to the Lincoln Highway Project. Alvan Macauley was elected president of Packard Motor Car Company on June 15, 1916.

The second series Packard Twin Six models were introduced in August 1916. After successfully building and selling several thousand Twin Six models, Packard made new improvements in the second series to make the fine cars even finer. The engines, while basically the same as those of the first series, now had removable cylinder heads and a relocated engine coolant thermostat. The wheelbase of the Series 2-25 was increased to 126-1/2 in., while the 2-35 wheelbase remained at 135 in.

The overall height of the cars was reduced by changes in the frame and the use of 35 x 5-in. tires, bringing the running boards to within 16 in. of the ground. The rear fenders followed the curve of the rear tire, with the ends downward, smoothing out the entire line of the car. The second series saw the introduction of an entirely new car, a four-passenger runabout model with individual front seats that allowed a middle passageway for seating two passengers in the rear compartment. For 1917 Packard offered 21 body styles: 9 in Series 2-25 and 12 in Series 2-35. The price range was from $2,865 for the 2-25 touring to $4,915 for the 2-35 Imperial Limousine. Production and sales for 1917 were approximately 9,000 units.

On April 6, 1917, the United States Government declared war on the government of Germany and the Austro-Hungarian empire. While the facilities of Packard Motor Car Company were quickly occupied with aircraft engines and military trucks, the new designs of the Packard third series were already in place.

The new third series Packard Twin Six models were announced in August 1917, and the same basic design remained unchanged through 1919. The new Packard radiator and shell were narrower and higher, and the cowl was blended into the projection of the famous Packard shoulder line in a smooth continuation from the radiator shell, hood, cowl, body doors, and quarter panels around to the rear. The fenders were of the semi-crown type, made of a one-piece steel stamping. The elimination of the battery box from the running boards gave a clean effect of the running board side shields, portraying a lower, longer car.

The improved Packard Twin Six engine powered the 3-25 and 3-35 models of 128- and 135-in. wheelbase lengths. The Packard third series was offered in 21 body styles in 1917, 17 styles in 1918, and 15 in 1919. The price range was $3,265 to $4,915 in 1917, $3,450 to $5,600 in 1918, and $4,100 to $6,000 in 1919. Approximately 9,500 Twin Six models were sold during 1917-19. While sales were good and Packard could have sold all the cars they could build, government restrictions on material limited the output.

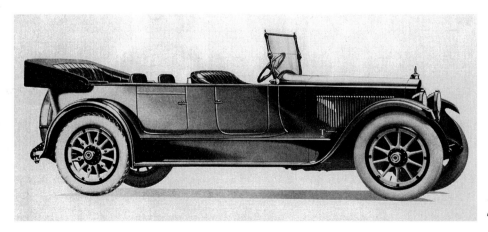

1917-20 third series Twin Six. (Source: P.M.C. Co.)

Liberty Aircraft Engine

In 1917 Joy Aviation Field was sold to the United States Government for a reported $1.00. The field was renamed Selfridge Field in honor of Lieutenant Thomas N. Selfridge, the first American military officer killed while flying. Because Packard Motor Car Company for over two years had been working on the development of an aviation engine in the 200-hp class, it represented the furthest advance in the United States toward the development of a satisfactory combat aircraft engine. Therefore, it was only natural that Jesse Vincent, Packard's vice president of engineering, and E.J. Hall, president of Hall-Scott Motor Company of Berkeley, California, were called by Colonel Edward A. Deeds to a conference in Washington D.C.

On May 29, 1917, Vincent and Hall were given the go-ahead to design and develop an American engine capable of 200 hp with eight cylinders, and 300 hp with twelve cylinders. Vincent and Hall organized a large force of skilled designers and draftsmen, with one team of designers in Washington and another team of detail draftsmen in Detroit. As soon as the design drawings were hurried from Washington to Detroit, and the detailing completed in Detroit, the drawings were given to the Packard Motor Car Company tool room where the corresponding parts were made.

The complete engineering organization had been staffed by engineers from the various automobile factories in Detroit, including Packard, Cadillac, Dodge Brothers, and even Pierce Arrow engineers from Buffalo, New York. So well did Vincent have the engineering organized, that on his return to Detroit on June 8, 1917, he found that Ormond Hunt (chief engineer of the carriage department) had already obtained the necessary steel billets from Cleveland, Ohio, for the making of the cylinders, and expedited all related work.

The work progressed so smoothly and rapidly that on July 3, 1917, the first sample eight-cylinder aircraft engine was delivered to the Bureau of Standards in Washington D.C. Additional samples of the eight-cylinder and samples of the twelve-cylinder aircraft engine followed quickly. The first engineering sample twelve-cylinder aircraft engine completed its first official 50-hour endurance run at 1:30 a.m. on August 25, 1917.

The first production twelve-cylinder aircraft engine was delivered to McCook Field, Dayton, Ohio, on Thanksgiving Day, November 29, 1917. As a matter of interest it may be noted that Vincent and Hall had originally designated this line of engines as the "U.S.A. Standardized" line, but a little later, Admiral Taylor of the U.S. Navy dubbed it the Liberty engine in one of the aircraft meetings. This name took so well that Vincent and Hall agreed to adopt the name change and changed all the titles on the drawings accordingly. The unusual design of the Liberty engine was that the cylinder banks were inclined to a 45-degree included angle between them. The valves actuated by overhead camshafts were inclined at a 15-degree angle. The cylinder bore was 5 in. and the piston stroke was 7 in., providing a displacement of 1,649.34 cu. in.

One important aspect of the Liberty engine program was that, while the money for the development work was appropriated, the usual bureaucratic procedures in Washington D.C. would have delayed the delivery of the engines for months. Thus, Colonel Deeds explained to Alvan Macauley that in order to make timely deliveries, Packard would have to purchase the needed materials through commercial channels. To this end he had requested Packard Motor Car Company to finance the job, so to speak. In other words, Packard was to buy and pay for the needed material, with the understanding that Packard would render an accounting and be reimbursed by the United States Government.

Macauley graciously agreed to the request, and for good measure gave the right-of-way not only to the services of the Packard Engineering department, but the entire factory facilities as well. Thus, without being asked, Packard curtailed automobile productions, and totally placed all its personnel and manufacturing facilities at the disposal of the United States Government for the building of Liberty aircraft engines.

The automotive industry as a whole was saddled with the Senate Military Affairs subcommittee, who was more anxious to find evidence of graft than to find a means of improving the aircraft production program. This environment, plus the appointment of many military officers to important aeronautical work, who were in no way familiar with their duties, added to the problems. (This was eventually brought to light by the Judge Hughes Investigation Report in late October 1918.) Despite the hampering and delays caused by bureaucratic blunders, the automobile industry (Packard and Lincoln, in particular) gave a good account of themselves in the quality manufacture and timely delivery of Liberty aircraft engines.

Of the approximately 22,000 Liberty engines on order by contract, Packard was to build 6,500, Lincoln 6,500, Ford 3,950, Marmon 1,000, and General Motors (Buick and Cadillac) 2,528. As of October 10, 1918, 10,151 Liberty aircraft engines were built: 3,965 by Packard, 2,824 by Lincoln, 2,010 by Ford, 1,144 by General Motors, and 208 by Nordyke and Marmon. During October 1918 the Liberty engine production was 5,603 engines, and would have been 10,000 per month by April 1919 if World War I had continued.

After the Armistice on November 11, 1918, the United States Government contracts were readjusted, which permitted the automobile companies to build engines to use up the material on hand.

Post-War Recovery

The year 1919 was spent realigning personnel within the Packard organization. Chief engineer O.E. Hunt resigned in November to accept the vice presidency of Hares Incorporated of Trenton, New Jersey. E.G. Gunn was promoted to chief engineer of the carriage department. George L. McClain moved up to chassis engineer of the carriage department. E.A. Moorehouse became chief engineer of the truck department. R.E. Chamberlain was promoted to assistant general sales manager of Packard Motor Car Co.

Despite the general optimism nationally that a business boom would follow the Armistice, Packard management wisely appraised the economic climate of intense inflation, rising costs, and unemployment that would follow the return of military personnel. The curtailment of truck production for the United States Government, and the increase in truck manufacturers, flooded the truck market. Packard management knew that the economic conditions would not sustain the present level of Packard Twin Six sales.

Thus, in early 1919, Packard Engineering and Design groups went full-steam ahead to design a new lower-priced Packard car. For the 1920 Packard Twin Six models, an electrically heated fumer or fuelizer was incorporated in the intake manifold to aid in fuel vaporization during cold-weather starting. While the 1920 sales of the Packard Twin Six models amounted to 5,193 units, the sales plunged to 1,310 cars in 1921, due to a full-scale economic depression nationwide.

Thus, in September 1920, Packard introduced a new six-cylinder model called the Single Six, a companion to the Twin Six.

1921 Packard Single Six

Wheelbase (in.)	116
Price	$3,640-4,940
No. of Cylinders / Engine	L-6
Bore x Stroke (in.)	3.37 × 4.50
Horsepower	52 adv, 27.3 NACC
Body Styles	five-passenger touring, two-passenger runabout, four-passenger coupe, and five-passenger sedan
Other Features	

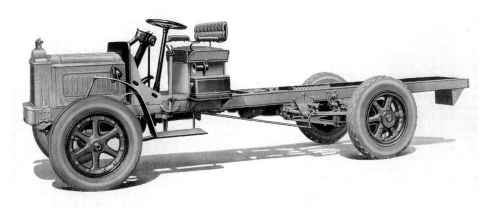

Packard was first to test pneumatic tires on 1921 trucks. (Source: P.M.C. Co.)

Price cuts took place during 1921, putting the Packard Single Six in the $2,350 to $3,350 price range. The introduction of the Single Six increased the Packard dealers in cities and towns that were formerly too small for Twin Six representation, and brought in new streams of car orders. Production and sales of the 1921 Packard Single Six Series 116 amounted to approximately 8,800 units; however, it was not enough to show a profit.

The catastrophic business recession of 1920-21 was being felt by the Packard Motor Car Company. On November 15, 1921, Macauley reported to the board of directors an operating loss of $987,366 for fiscal year 1921. This did not reflect any financial weakness of the company. In fact, after paying dividends on preferred stock and setting aside reserves for contingencies, the Packard Motor Car Company surplus amounted to $15,923,985, with $10,323,000 shown as cash on hand and readily marketable securities. The total assets of the Packard Motor Car Company at that time were $58,739,638 and liabilities were only $3,807,342. Stringent cost reductions in manufacturing expenses from wartime levels were put into effect in 1921.

Even though the national economy improved somewhat during 1922, and Packard built and sold 2,032 trucks, the decision was made to phase out truck manufacture. The 1922 Packard trucks were carried over into 1923.

On April 20, 1922, the new Packard Single Six Series 126-133 was announced. These new models were beautifully styled: long, sleek, and truly representing the kind of vehicle Packard buyers would want.

1922 Packard Single Six Series 126-133

Wheelbase (in.)	116/126/133
Price	$2,485-3,575
No. of Cylinders / Engine	L-6
Bore x Stroke (in.)	3.37 × 4.50
Horsepower	54 adv, 27.3 NACC
Body Styles	11
Other Features	

The Packard Single Six now apparently appealed to the buying public. While the Twin Six sales tumbled to 1,944 units in 1922, the Single Six took up the slack. A total of 13,382 cars were built and sold in 1923, but only 303 were Twin Six models. The first six months of the 1923 fiscal year showed a net profit of $4,435,559.

1922-23 Packard Single Six 126. (Source: P.M.C. Co.)

In January 1923 George M. Berry organized the Berry Motor Car Company of St. Louis, Missouri, to take over the wholesale end of the car distribution formerly handled by the Packard Motor Car Company of Missouri. Berry had been identified with the automobile business since 1902, and remained a Packard distributor until 1954 when he retired.

In May 1923 Packard entered three racing cars in the Indianapolis 500 Memorial Day Classic. The racing cars were designed by Jesse Vincent and built at the Packard plant. The drivers selected were Ralph DePalma, Dario Resta, and Joe Boyer. The cars were powered by a 122-cu.-in.-displacement, six-cylinder, overhead camshaft engine. The racing cars qualified for 3rd, 11th, and 13th starting positions by Resta, DePalma, and Boyer, respectively. Unfortunately, Boyer was forced out on lap 59 with a differential problem, DePalma's car suffered a cylinder head gasket failure on lap 69, and Resta also experienced a differential problem on lap 88. Thus, none of the Packard cars finished the race.

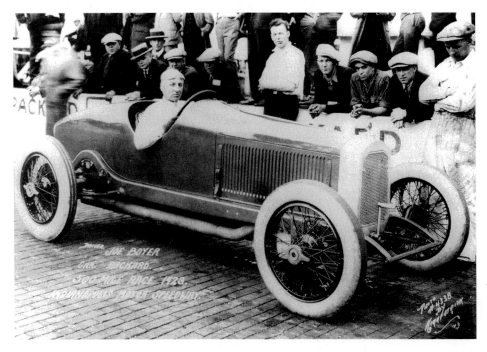

Joe Boyer in a Packard at 1923 Indianapolis 500. (Source: Indianapolis Motor Speedway Corp.)

Eight-in-Line Engine

While automotive publications carried the announcements two days earlier, on June 16, 1923, Packard Motor Car Company officially introduced the new Packard Single Eight, which replaced the Twin Six and was marketed along with the Packard Single Six.

1923 Packard Single Eight

Wheelbase (in.)	136/143
Price	$3,650-4,950
No. of Cylinders / Engine	IL-8
Bore x Stroke (in.)	3.37 × 5.00
Horsepower	85 adv, 36.4 NACC
Body Styles	9
Other Features	four-wheel mechanical brakes

The eight-in-line, L-head engine featured aluminum-alloy pistons with three rings, and had the crankshaft supported on nine main bearings in an aluminum crankcase. It had a Lanchester Vibration Damper on the forward end of the crankshaft, which helped to account for the Packard Single Eight's smoothness.

The styling was in good taste, similar to that of the Single Six, except the hood was about 10 in. longer to accommodate the bigger engine. The radiator was equipped with a Pines automatic Winterfront with horizontal louvres, and a Moto-Meter as standard equipment.

The Packard Single Eight was considered the pioneer of the eight-in-line engines. Even though Duesenberg introduced an eight-in-line-engine-powered car in November 1920, in the subsequent 16 years they built less than 600 passenger cars. Many competitive automobile manufacturers followed Packard's lead, introducing eight-in-line (straight-eight) engines in their car lines.

Equally important in the Packard Single Eight was the mechanical four-wheel brakes of the internal-expanding-shoe-type with ribbed (finned) brake drums. Four-wheel brakes were available on some European fine cars since 1920, and Duesenberg had hydraulic four-wheel brakes in 1921.

Packard domestic car sales amounted to 13,382 in 1923, and 14,220 in 1924. On top of this, Packard enjoyed a healthy export car market amounting to about 8,000 units annually, as many affluent Europeans preferred Packard to Europe's fine cars.

1923 Packard first series Single Eight. (Source: P.M.C. Co.)

On October 30, 1923, Alvan Macauley reported to the stockholders that as of the end of the fiscal year on August 31, 1923, Packard enjoyed a gross volume of business amounting to $55,670,464, resulting in a net profit of $7,881,878 after payment of stock and cash dividends in the amounts of $1,029,322 on preferred stock and $2,495,871 on common stock. The current assets at that time were shown at $33,001,600, indicating Packard Motor Car Company's healthy financial condition.

All the personnel of Packard Motor Car Company were saddened on November 12, 1923, by the death of William Dowd Packard at his home in Warren, Ohio, one of the founding brothers of the Ohio Automobile Company, which later became the Packard Motor Car Company. He was 62 years old and had been in ill health for a number of years.

In 1924 it became stylish to copy the famous Packard radiator and hood shoulder line; Buick, Studebaker, and many others did it in a slightly modified form. On December 23, 1923, Packard announced the introduction of mechanical four-wheel brakes on the new Packard Single Six. The new models were exhibited at the National Automobile Show in New York in January 1924. The new car also featured demountable disc wheels and a stop lamp mounted in the rear energized by a switch actuated by the brake pedal. The Single Six wheelbase remained at 126 and 133 in. Eleven body styles were offered in the price range of $2,585 to $3,675.

The open models featured a striped belt molding originating at the radiator shell, continuing through the hood sides, cowl panels, doors, rear quarters, around the rear, and returning to the radiator shell on the opposite side. This design feature was subsequently copied by most competitive automobile manufacturers. Packard enjoyed a successful year in 1924, building and selling 13,382 cars domestically, and about 9,000 vehicles through Packard Export Division.

The Packard Single Six Series 226 and 233, as well as the Packard Single Eight Series 136 and 143, continued through 1924 unchanged, except the Packard Fuelizer was incorporated in the intake manifold to facilitate cold-weather starting. This device, in effect, was a glow plug to help vaporize the fuel-air charge before it entered the cylinders.

On March 31, 1924, the United States Court of Claims ordered the United States Government to pay Packard Motor Car Company the final $425,540 that it withheld in the settlement of the contract of July 27, 1917, for the purchase of 4,800 Packard Trucks for military use.

1924 Packard Single Six 126. (Source: P.M.C. Co.)

Former President William Taft and a Packard. (Source: P.M.C. Co.)

Packard paid cash dividends on common stock during 1924 amounting to $2,852,424, the equivalent of 12%, and in addition paid an extra cash dividend of 3%. The Packard financial report of November 1, 1924, showed the then-current assets of $31,348,572, of which $15,652,833 was in cash and United States Treasury Notes, indicating a very sound financial condition. Packard production and sales in 1924 amounted to 14,220 units domestically and about 7,000 cars through Packard Export.

On January 1, 1925, Packard reduced the prices of the Single Six closed models down to the price levels of the open models in the line. These reductions amounted to $640 to $840, and the price range became $2,585 for the five-passenger sedan model to $2,885 for the seven-passenger limousine.

The second series Packard Eight and third series Packard Six were introduced on February 2, 1925, with the Single designation omitted. While the appearance of both the Packard Six and Packard Eight were considerably improved by the use of disc wheels of a smaller diameter and balloon tires, making the cars noticeably lower, the important changes were in the chassis construction.

The front spring layout was completely changed; the springs were shackled at the front and anchored at the rear. This arrangement provided better geometry with the steering knuckle upward movement arc and the drag link arc. The balloon tire sizes were 33 x 5.77-in. on the Packard Six, and 33 x 6.75-in. on the Packard Eight. The Packard Eight engine dimensions remained the same, while the Packard Six engine cylinder bore was increased from 3-3/8 in. to 3-1/2-in. The crankshaft main bearing journals were increased to 2-3/8 in., and the crankpin diameter was increased to 2-1/8 in., making these dimensions the same as that of the Packard Eight engine.

The Skinner Oil Rectifying system (in effect, a miniature refinery) and the Bijur one-shot chassis lubrication system were provided as standard equipment on the Packard Eight models. The Skinner Oil Rectifier unit was located on the exhaust manifold, just aft of the carburetor and the fuelizer, and extended the oil change period two-fold.

On August 6, 1925, Packard Motor Car Company announced the addition of a Club Sedan in both the Six and the Eight lines. The Packard Six Club Sedan was mounted on a 133-in. wheelbase chassis, while the

Packard Eight Club Sedan was on a 143-in. wheelbase chassis. Both models featured a new one-piece windshield which also became a feature of other Packard models. The prices were $2,725 for the Six model and $4,890 for the Eight. At about the same time, what was perhaps the first of Packard's Custom models made its debut. It was a Packard two-passenger Coupe by custom coach builder Holbrook. This sporty coupe carried full equipment on a 136-in. wheelbase chassis and was priced at $5,775.

On September 1, 1925, Packard Motor Car Company announced the distribution of $7,282,440 to its preferred stockholders, to retire the entire issue of preferred stock certificates at $110 per share. After paying off the preferred stock, the company had on hand approximately $14,500,000 in cash and marketable securities. With the retirement of preferred stock, the company announced it had no obligations, except those incurred on a day-to-day operation basis, which were paid promptly to enjoy the cash discount from the suppliers.

More Aircraft/Marine Engines

The nation was shocked and bereaved upon hearing the news over radio that the twisting violent storm over central Ohio near Ada broke the airship "Shenandoah" into three pieces on the early morning of September 3, 1925. Lieutenant Commander Zachary Lansdowne, United States Navy, and 13 crewmembers lost their lives. Most of the crew perished when the control cabin (gondola) broke away and fell to the ground. The only survivors were those in the rear portion of the airship, as it broke away from the front portion and drifted to the ground.

Colonel C.G. Hall, a United States Army observer aboard the airship, stated after the disaster that, "had we had wireless communication, we undoubtedly would have been able to steer away from the storm entirely." As it was, the commander had to steer away from the approaching clouds without any guidance from the weather bureau authorities. In those years, air-shear phenomena had not been recognized.

One of the last communications from Commander Lansdowne stated that the six 300-hp Packard aircraft engines performed extremely well in all flights, giving no malfunction nor problems whatsoever. The subsequent disaster of the German dirigible Hindenburg, destroyed by fire upon mooring at the hangar in Lakehurst, New Jersey, on May 6, 1937, with the loss of 35 lives, and the loss of the airship Macon in the Pacific Ocean in February 1933, convinced the United States Government to stop building rigid dirigibles.

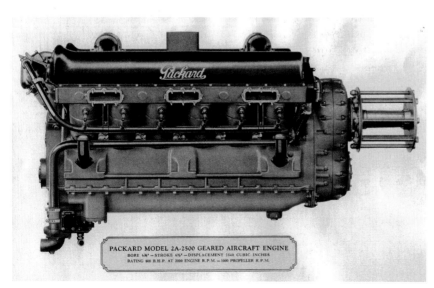

Packard 2A-2500 aircraft engine (with reduction gear). (Source: P.M.C. Co.)

Orlin Johnson and Gar Wood (right), mechanic and pilot of Miss America X. (Source: P.M.C. Co.)

Packard Motor Car Company, however, continued the development of Packard aircraft engines, departing from the original Liberty designs, to create a character all their own. During the 1920s Packard developed V-12 aircraft engines of 1,500- and 2,500-cu.-in. displacements identified as A-1500 and A-2500, respectively.

The United States Navy had a particular interest and confidence in Packard aircraft engines, and had many Navy aircraft designed with Packard aircraft engines in their planning. Gar Wood, famous powerboat racer, used Liberty and, later, converted Packard A-2500 aircraft engines for his Gold Cup and Harmsworth Trophy Miss Americas. His experience provided the United States Navy with valuable data for the PT boats.

In 1933 Gar Wood easily defeated Hubert Scott-Paine in the British Miss England, and later established a world record of 124.915 mph. Gar Wood never announced Miss America X's total horsepower of 8,000; in fact, he never told anyone exactly what the horsepower was, but let it be assumed that 6,400 was the peak. Gar Wood earned the title of "Father of the PTs." During this period Packard Motor Car Company

Unlimited Miss America X powered by four Packard engines. (Source: P.M.C. Co.)

experimented with many aircraft engine designs: V-, W-, and X-type, and a nine-cylinder diesel aircraft engine. Lionel M. Woolson was chief research engineer on these projects.

Continued Growth and Innovations

Because of the use of the finest materials and most skilled workmanship, Packard enjoyed passenger-car sales of 23,948 cars domestically and approximately 10,000 cars through the Export Division. Velvet mohair, wool broadcloth, or wool Bedford Cord upholstery with wool plush carpets were used on all closed models. Silver-plated hardware adorned the interiors of the closed models. Finest top-grain leather was used in the open car upholstery, and featured nickel-plated body hardware and windshield stanchions (pillars). Nitro-cellulose lacquers by Valentine, Mimax (PPG), DuPont, and Cook provided Packard with brilliant colors no competitor could equal.

The second series Packard Eight and third series Packard Six were continued through July 1926 with only minor changes in standard production models. In early 1926 Packard made available a number of Custom models "created by Master Designers" and built by custom body builders. These Custom models included a Sedan Cabriolet by Judkins, a Town Cabriolet by Derham, a Town Cabriolet by Fleetwood, a seven-passenger "Inside Drive" Limousine by Holbrook, and a Town Cabriolet, Convertible Coupe, and four-passenger Sedan by Dietrich. A special color brochure was printed to show these special Packard models.

Eleven body styles were available in the Packard Eight production models, on wheelbase lengths of 136 and 143 in. Twelve body styles were available in the Packard Six production models, having wheelbase lengths of 126 and 133 in. The price range was $3,875 to $4,950 for the Eight and $2,785 to $3,675 for the Six. Packard built and sold 29,588 cars domestically in 1926.

On August 1, 1926, Packard announced the new third series Packard Eight and the fourth series Packard Six. The Packard Eight engine bore was increased to 3-1/2 in., and with a 5-in. stroke provided a piston displacement of 384.8 cu. in., developing 109 hp. A new clutch manufactured by the Long Manufacturing Company provided two lining-faced driven discs. Packard introduced the "Angle-Set" differential (other manufacturers, who some 10 years later picked up this feature, called it "Hypoid"). This differential carrier had the centerline of the drive pinion below the centerline of the ring gear (driven gear), providing the combined advantages of a spiral bevel gear and a worm drive. In operation it gave more tooth contact in rolling action,

Gilda Gray and a 1927 Packard 336 phaeton. (Source: P.M.C. Co.)

and resulted in quieter operation. The driving pinion location allowed the propeller shaft to be lower and, in effect, permitted the lowering of the floor and a reduced overall height of the body.

Eight body styles were available in both the Packard Eight and Packard Six, spanning a price range of $2,585 to $5,100. During 1927 Packard Motor Car Company built and sold 31,355 cars domestically.

The 1928 Packard Six Series 526 and 533, and the Packard Eight Custom Eight Series 443, were introduced in late July 1927 and deliveries started in August. The 1928 Packard Eight was limited to one wheelbase length, 143 in., and introduced four hinged rectangular doors on each hood side panel for ventilation. The 1928 Packard Six had vertical louvres on the hood side panels. The Packard Eight engine dimensions remained the same as previous models; however, new "Invar strut" slotted skirt pistons were used to control piston expansion due to heat. New concentric dual-coil valve springs were used to eliminate valve float at high engine speeds. The forged crankshaft crankpins were drilled out, removing about 12 oz of steel from each crankpin. This, in effect, reduced the crankshaft weight by approximately 6 lb, thereby reducing the rotational weight without sacrificing crankshaft torsional strength.

The Packard Six engine remained the same having a 3-1/2-in. bore and a 5-in. stroke. The chassis remained the same; however, the tire size changed to 32 x 6.00-in. on the 126-in. wheelbase chassis, and 32 x 6.75-in. on the 133-in. wheelbase chassis. The Packard Eight Series 443 used 32 x 6.75-in. tires. The new tire and wheel sizes reduced the overall height of all models.

Both the Eight and the Six had a new external oil manifold along the left side of the cylinder block, with a drilled oil passage to each cylinder bore. During cold starting, when the choke was used, the choke operation opened the valve to the oil manifold. This allowed oil to be injected under pressure to each cylinder bore, replacing the oil that was generally washed off the cylinder wall during the cold start when using the choke.

The 1928 Packard body styling was fresh and pleasing, with a new cadet-type exterior windshield sun visor on the closed models. The sport phaeton and other open models sported an insert panel that was spear-pointed at each end, extending from the cowl panel to the rear quarter panel, just below the belt line. This styling feature was soon copied by competitors. The instruments were grouped under a single glass panel on

Faye Wray and a 1928 Packard Six 526. (Source: P.M.C. Co.)

the wood-grain instrument board. Wood-spoke, steel-disc, or wire-spoke wheels were available on all Packard Eight models, with the spare wheels and tires carried in wells in the front fenders.

On November 19, 1927, the right of Packard to patent the radiator shell (casing) design and radiator cap shape was upheld by the United States Patent Office. The decision was granted despite the appeal of Harry H. Bassett, who claimed that the radiator, because of its peculiar design, was not patentable. Buick in 1928 continued its radiator shell (similar to Packard), but discontinued the design shape in 1929. Studebaker discontinued the design shape in 1928.

New five-passenger phaeton and four-passenger runabout body styles were added to the Packard Six Series 533 and Packard Eight Series 443. Ten body styles were available in the Packard Six, and seven in the Packard Eight. A new four-passenger coupe was also available in the Packard Eight Series 443. The price range was from $2,275 for the Packard Six phaeton to $5,250 for the Packard Eight Sedan Limousine Series 443.

On March 1, 1928, the Packard Eight Standard Line was added. The Standard Line chassis, engine, and body styles were essentially identical to those of the Packard Custom Line, except for less-costly interior refinements, and a considerably lower price range. The Packard Standard Line was priced at $3,550 to $3,850 to fill the gap between the Packard Six and Packard Eight Custom Line prices. In addition, Packard offered 20 individual body styles by custom coach builders Derham, Dietrich, Fleetwood, Holbrook, Judkins, LeBaron, Murphy, and Rollston. During this period Packard used oil paintings by famous artists in their color advertising schemes. Apparently the Packard models pleased the buying public, as 42,961 cars were sold domestically in 1928, and export sales increased proportionately.

Packard Dies

On March 20, 1928, Packard Motor Car Company officials and all employees were saddened by the news of the passing of James Ward Packard, who died at the Cleveland Clinic Hospital at age 64, where he had been a patient since November 1926. He was treated for a malignant growth by all medical science available. He was survived by his widow, three sisters, and a nephew, Warren Packard of Detroit.

On May 24, 1928, Hugh J. Ferry succeeded Richard P. Joy as treasurer of Packard Motor Car Company. Ferry had been assistant treasurer since 1919.

1929 Packard sixth series sedan. (Source: P.M.C. Co.)

On July 9, 1928, Packard Motor Car Company reduced the prices on all its Packard Six models by $300. This move was in anticipation of a line of cars to be introduced in the fall. Prospective buyers were advised of the plan before the sale of the Packard Six. The Packard Six model was replaced by the new Packard Standard Eight when the 1929 sixth series Packard went into production in August 1928, and was publicly announced the week of September 8, 1928.

	Sixth Series (1929) Packard Standard Eight Series 626 and 633	Sixth Series (1929) Packard Custom Eight Series 640	Sixth Series (1929) Packard Deluxe Eight Series 645
Wheelbase (in.)	126.5/133.5	140.5	145
Price	$2,435	$3,850	$4,585-5,985
No. of Cylinders / Engine	IL-8	IL-8	IL-8
Bore x Stroke (in.)	3.18 × 5.00	3.50 × 5.00	3.50 × 5.00
Horsepower	90 adv, 32.5 SAE	105 adv, 39.2 SAE	105 adv, 39.2 SAE
Body Styles	10	9	9
Other Features	combination tail lamp, stop lamp, and back-up lamp laminated safety glass windshield		

The longer chassis of the Deluxe Eight Series 645 was used for mounting individual custom body styles, created by custom coach builders Dietrich, LeBaron, Rollston, and many others. The standard production Packard bodies were refined giving a more slender and agile look. The dimensions between the belt molding and window reveals on closed models were reduced giving a lower appearance. The belt molding on open models was carried along the top of the door outer panels and around the rear, just below the top rear curtain lower edge. The famous Packard radiator shell was more slender and now carried the Packard family badge on the forward upper part. The front of the radiator shell also housed the actuating mechanism for the automatic radiator shutters, powered by a group of aneroid discs within a chamber in the radiator upper tank.

The Moto-Meter was omitted, replaced by a temperature gauge in the instrument panel group. A chromium-plated wire bail replaced the Moto-Meter on the radiator cap. Hemisphere-shaped headlamp bodies, with an embossed Packard shoulder line along the upper portion, were used in place of the former drum shapes. Rectangular-shaped doors on the hood sides were adjustable to provide ventilation for the engine space. Disc wheels were standard, and wire-spoke wheels were an option, with the spares carried in the wells of the fully crowned front fenders. A folding trunk rack was provided in the rear of the models with fender wells. All bright metal exterior parts including the bumpers were chromium-plated.

The Standard Eight engine was, in fact, a scaled-down Custom Eight engine, including the nine-main-bearing crankshaft. The Standard Eight engine used a North East ignition system and Owen-Dyneto starting and charging systems. All chassis models featured a new kick shackle at the rear end of the left front spring to eliminate front wheel fight. The Watson Stabilizer friction-type shock absorbers were replaced by new double-acting rotary-vane-type shock absorbers mounted near each end of the front and rear axles. The shock absorber arms were connected by links to each corner of the frame. A very important feature that contributed greatly to Packard's famous riding characteristics was the individual fabric-covered coil springs in the seat cushion and seat back.

Packard Motor Car Company enjoyed a prosperous sales year in 1929, when a record of 44,634 cars were built and sold domestically, with a proportionate increase in export sales. This record would not be exceeded until 1936.

Black Thursday

The seventh series Packard was introduced on October 1, 1929. It consisted of the Standard Eight Line Series 726 and 733, the Custom Eight Line Series 740, and the Deluxe Eight Line Series 745. Fifteen "Individual Custom" designs were available by custom coachbuilders, including Brewster, Dietrich, LeBaron, and Rollston.

	Packard Standard Eight Series 726 and 733	Packard Custom Eight Series 740	Packard Deluxe Eight Series 745
Wheelbase (in.)	127.5/134.5	140.5	145.5
Price	$2,425-2,775	$3,190-3,885	$4,585-5,350
No. of Cylinders / Engine	IL-8	IL-8	IL-8
Bore x Stroke (in.)	3.18 × 5.00	3.50 × 5.00	3.50 × 5.00
Horsepower	90 adv, 32.5 SAE	106 adv, 39.2 SAE	106 adv, 39.2 SAE
Body Styles	11	11	11
Other Features			

While Packard cars were selling at a record pace, "Black Thursday," October 24, 1929, descended on the American economy and triggered the cruel Depression that was soon to follow. Despite the foreboding signs of business downturn, the builders of fine cars started the "multi-cylinder" engine race. While it had been rumored for several months, on January 1, 1930, Cadillac announced their new (165-185 hp) V-16-cylinder-powered cars. They were available on a 148-in. wheelbase chassis, offering 33 body styles priced from $5,350 to $9,700.

Packard, while having new engine and chassis designs on the drawing board, was not about to give up their sales advantage. In January 1930 at the National Auto Show in New York, Packard unveiled their new Speedster Series 734.

Packard on endurance test at proving grounds, 1930. (Source: P.M.C. Co.)

1930 Packard sport roadster. (Source: P.M.C. Co.)

1930 Packard Speedster Series 734

Wheelbase (in.)	134.5
Price	$5,200-6,000
No. of Cylinders / Engine	8
Bore x Stroke (in.)	3.50 × 5.00
Horsepower	125/145 adv, 39.2 SAE
Body Styles	two-passenger runabout with staggered seats, four-passenger phaeton, four-passenger Victoria, and four-passenger Sport sedan
Other Features	

The 384.8-cu.-in.-displacement, eight-cylinder engine developed 125 hp with the standard cylinder head and 145 hp with the optional (6.0 to 1 compression ratio) cylinder head with 18-mm spark plugs. The engine also had a two-barrel updraft carburetor of Packard design built by the Detroit Lubricator Company. Optional rear axle ratios of 3.33 to 1 and 4.33 to 1 were available, making a top speed of over 100 mph possible. The styling was distinctly Speedster motif, with shorter windshield pillars (stanchions) and door pillars, making the cars much lower with a long slender appearance. The runabout had a boat-tail rear deck which was the forerunner of many boat-tail speedsters to follow. While the original plans called for only 70 cars, it had been reported that because of the overwhelming demand, the production of this series exceeded 100 units. For the 1930 model year, 28,318 Packard cars were built and sold domestically.

1930 Packard 740 sport phaeton. (Source: P.M.C. Co.)

On January 26, 1930, Russell A. Alger, former director and vice president of Packard Motor Car Company, died in a New York hospital of pneumonia following an operation. Alger was one of the prominent businessmen who, with Henry B. Joy, established Packard Motor Car Company in Detroit. He was survived by his widow, Mrs. Marion J. Alger, two sisters, two daughters, a son, Russel A. Alger Jr., and a brother, Colonel Frederick M. Alger.

Diesel Aircraft Engine

On April 5, 1930, Packard Motor Car Company announced and made the first showing of its high-speed diesel aircraft engine. It was coincident with the opening of the All-American Aircraft Exposition at the Detroit City Airport. Basically the engine was a four-stroke, mixed-cycle, solid-injection, nine-cylinder, air-cooled, radial diesel engine weighing only 510 lb dry. The cylinder bore was 4-13/16 in. and the piston stroke was 6 in. The compression ratio was 16 to 1, and it developed 225 hp at 1900 rpm. The general appearance and installation size was comparable to other radial aircraft engines. It had an overall diameter of 45 in., and was provided with an eight-stud mounting ring of 22-in. diameter.

The outstanding feature of this engine was the crankcase design of a single-piece magnesium-alloy casting, weighing only 34 lb except for the load-carrying diaphragm and cover. The most unusual design was the method of mounting and securing the cylinder barrels to the crankcase. Instead of the usual mounting studs in the crankcase and mounting flange on the open end of the cylinder barrels, the Packard engine crankcase had no studs. The cylinder barrels were held in place by two circular steel-alloy "hoops" passing over the front and rear areas of the cylinder mounting flanges. Turnbuckles on the hoops provided the necessary compressive forces on the cylinder flanges.

Three Packard radial diesel engines on Ford Tri-motor aeroplane, 1930. (Source: P.M.C. Co.)

The most salient safety feature of all was the use of domestic fuel oil for combustion, with a flash point of 169°F. Because of the single valve arrangement and the mixed cycle, the exhaust pipe temperature never exceeded 200°F and the cylinder head temperatures never reached 350°F. With respect to altitude test flights in Stinson and Waco aeroplanes, altitudes of over 18,000 ft. had been reached.

Packard diesel aircraft engines had been successfully tested and used in the Bellanca Special, the Buhl Air Sedan, the Ford Tri-motor transport, the Ryan foursome, the Stewart twin transport, the Stinson Detroiter and Junior, the Towle Amphibian, the Verville Aircoach, the Waco straight wing, and the U.S. Army primary trainer. The Packard diesel aircraft engine received unqualified acceptance from aircraft builders and the United States Army Air Corps.

On April 15, 1930, Edward R. Macauley, assistant to H.W. Peters, vice president of distribution, was appointed manager of the aircraft and marine division of Packard Motor Car Company. Macauley was a son of Alvan Macauley and handled all sales of the aircraft engines to the users. Macauley remained in this capacity until 1940, when he was appointed manager of the passenger-car styling division. Captain Lionel M. Woolson was the chief engineer of the aircraft engine division and was responsible for the development of the Packard diesel aircraft engine. Felix A. Kummer was secretary and compiled and recorded all technical specifications of aircraft and marine engines. U.S. Government contracts were handled by George H. Brodie, vice president of Packard Motor Car Company. The diesel aircraft engines were designed and built in Packard's aircraft engine building constructed during World War I. Unfortunately, Captain Woolson was killed on April 26, 1930, when his diesel-engine-powered aeroplane crashed into the side of a hill during a blinding snowstorm near Attica, New York.

In May 1930, Clyde R. Paton joined Packard Motor Car Company. He was a member of the National Advisory Committee for Aeronautics from 1923 through 1925, serving as Assistant Mechanical Aeronautical engineer in the Langley Memorial Laboratories at Langley Field, Virginia. He joined the Studebaker Corporation as experimental engineer in 1925, serving in that capacity until May 1930, when he was appointed chief experimental engineer at Packard Motor Car Company.

Alvan and Edward Macauley in 1935. (Source: P.M.C. Co.)

Capt. Lionel M. Woolson and Col. Charles Lindbergh. (Source: P.M.C. Co.)

The Depression Years

On September 6, 1930, the Packard eighth series was announced, and production began on September 9th. The body lines remained practically identical to the seventh series. The name Custom Eight was dropped from the 140-in. wheelbase car, and both the 840 and 845 were known as the Packard Deluxe Eight.

	Packard Standard Eight Series 826	Packard Standard Eight Series 833	Packard Deluxe Eight Series 840	Packard Deluxe Eight Series 845
Wheelbase (in.)	127.5	134.5	140.5	145.5
Price	$2,385	$2,425-3,465	$3,795-3,950	$4,150-4,285
No. of Cylinders / Engine	IL-8	IL-8	IL-8	IL-8
Bore x Stroke (in.)	3.18 × 5.00	3.18 × 5.00	3.50 × 5.00	3.50 × 5.00
Horsepower	100 adv, 32.5 SAE	100 adv, 32.5 SAE	120 adv, 39.2 SAE	120 adv, 39.2 SAE
Body Styles	five-passenger sedan	11	9	seven-passenger sedan and limousine
Other Features		vibration damper and automatic chassis lubricator		

1931 Packard 845 sport sedan. (Source: P.M.C. Co.)

With the introduction of the eighth series Packard began operating its own Individual Custom body department, managed by R.B. Birge, former vice president and general manager of LeBaron Inc.

Packard experimental engineering garage in 1932. (Source: P.M.C. Co.)

Of major importance was the new vibration damper combining the advantages of vulcanized composition rubber discs and spring-controlled friction members. By means of this design an unusual "lost work" span was obtainable to cover all harmonic ranges that were to be expected in the operation of the engines. Another important feature was the Packard-Bijur completely automatic chassis lubricator. The lubricant served 36 different areas on the chassis and carried enough lubricant for about 3,000-3,500 miles of driving.

Despite the outstanding features of the 1930 and 1931 Packard cars, Packard sales dropped to 28,318 units in 1930 and 16,256 units domestically in 1931. The severity of the Depression was taking its toll.

On June 23, 1931, Packard announced their entirely new model lines for 1932, referred to as the ninth series and consisting of two lines.

	Packard Standard Eight Series 901/902	Packard Deluxe Eight Series 903/904
Wheelbase (in.)	129.5/136.5	142/147
Price	$2,485-3,445	$3,680-4,550
No. of Cylinders / Engine	IL-8	IL-8
Bore x Stroke (in.)	3.18 × 5.00	3.50 × 5.00
Horsepower	110 adv, 32.5 SAE	135 adv, 39.2 SAE
Body Styles	11	9
Other Features	ride control and harmonic front-end stabilizer	

Of the numerous chassis features the ninth series possessed, two of the most important were the ride control and the harmonic front-end stabilizer. The ride control consisted of a push-pull control knob located to the left of the steering column that changed the shock absorber adjustment to meet the varying road conditions at the driver's will. The function of the harmonic front-end stabilizer was to dampen out the torsional movements of the front-end structure caused by the uneven road surfaces. The harmonic stabilizer consisted of a lead weight at each end of the front bumper, molded around a ribbed cast-iron sleeve with an oil-less bushing, riding up and down on the central assembly bolt. The lead weight was positioned between two sets of calibrated concentric coil springs to control the reciprocating lead mass movement of definite frequency.

The in-line eight-cylinder engines, while basically the same as those of the previous models, had numerous improvements, among which were steel-backed main bearings, Perfect Circle #85 oil control piston rings, cylinder wall lubrication by metered oil spray orifices on the trailing side of the connecting rods, and the elimination of the external manifold for cylinder lubrication. A combination air cleaner and intake silencer was adopted.

The ninth series transmission was a four-speed synchromesh, silent in all forward speeds. Later in the model year the transmission was changed to a three-speed silent in all forward speeds. The chassis springs were metal-covered and permanently lubricated, while all other points on the chassis were lubricated by the Bijur automatic lubrication system. The clutch was a two-driven-disc type, with vacuum-power operation to allow gear shifting without depressing the pedal.

All chassis models had a wider tread and a new double-drop frame with a heavy-steel X-member in the center, and 1-3/4 in. lower than the previous models. Disc wheels were standard and wire-spoke wheels were optional. The tire size was 6.50 x 19-in. on the Packard Standard Eight and 7.00 x 19-in. on the Packard Deluxe Eight. Four-speed synchromesh transmissions and free-wheeling were available on all models.

The body styling had been considerably improved, making the ninth series Packard cars among the most beautiful cars ever built. The radiator shell and winterfront shutters were V-shaped, and the Packard family coat of arms badge was moved from the front of the radiator shell upper tank cover to a location over the starting crankhole cover. The radiator filler cap was larger and the wire bail lock was eliminated. The windshield was slanted 7 degrees to help eliminate light glare and to improve appearance. The graceful roof line had no external sun visor. The enclosed models had inside sun visors, hinged at the top, and all models had adjustable (front) seats and steering column.

The engine compartment ventilation was accomplished by adjustable rectangular doors on the hood sides of all models. The front fenders were long, sweeping, crowned, clamshell-type with wells for the spare wheels and tires. Folding trunk racks were fitted at the rear of the models with six wheels and fender wells. The lower body panels and door panel inside surfaces were covered with waterproof composition sheets to dampen body rumble frequencies.

Like all fine car manufacturers, Packard Motor Car Company placed all its chips on the (ninth series) 1932 models, producing some of the finest and most beautiful cars ever built.

On January 1, 1932, Packard Motor Car Company startled the automotive industry with the introduction of the new Packard Twelve and Packard Light Eight at the National Automobile Show in New York City. Thus, with the continuation of the Packard Standard Eight and the Packard Deluxe Eight, the Packard line for 1932 included four series of cars, with 41 production body types and several Individual Custom models by custom coach builders. In the interest of thousands of car buyers, in whom the desire to own a Packard had been assiduously cultivated over many years, Packard's well-chosen course was to meet the situation by offering a new and lighter Packard in a lower price class than the lines had previously occupied.

The new Packard Light Eight, which the automotive trade had long awaited with keenest anticipation, was characteristically Packard. Although it was radically new in a number of respects, for its size was surprisingly a big car, it appeared sleek and agile. The radiator shell retained the characteristic Packard lines, but it narrowed down and gracefully swept forward at the bottom to merge with the front fenders at the center, omitting the usual dust shield. The front fenders themselves gave a streamlined appearance to the car. They were made in a one-piece steel stamping of clamshell design, with front area and hood ledges acting as a cover to conceal the front of the frame, front axle, springs, and other chassis parts. The chassis of the

Jean Harlow and a 1932 Packard 905 sport phaeton. (Source: P.M.C. Co.)

Packard Light Eight was an entirely new design. The frame had a deep double drop, with 8-in. side rails, and a large box-girder X-member in the center.

	1932 Packard Light Eight	1932 Packard Twelve Series 905 and 906
Wheelbase (in.)	128	142/147
Price	$2,485	$3,790-4,090
No. of Cylinders / Engine	IL-8	V-12
Bore x Stroke (in.)	3.18 × 5.00	3.43 × 3.50
Horsepower	110 adv, 32.5 SAE	160 adv, 56.7 SAE
Body Styles	five-passenger sedan, two- to four-passenger coupe, five-passenger Victoria, and two- to four-passenger convertible coupe	
Other Features	finger control free-wheeling and vacuum clutch operation when changing gear speeds	

The new Packard Twelve Series 905 and 906 was originally called the Twin Six for the early 1932 models, but was later changed to Packard Twelve to obviate any thoughts that it might be a carryover of the 1916-23 Packard Twin Six. Having over a decade and a half of twelve-cylinder experience in aircraft and marine engines, Packard employed the newest technology in creating the new ninth series Packard Twelve. It was designed and developed by the most brilliant engineers and designers in the automotive industry, among whom were Jesse Vincent, Clyde Paton, Thomas Milton, and Cornelius Van Ranst.

The cylinder banks had a 67-degree included angle, and the L-head-type cylinder heads were of an aluminum alloy, removable for servicing. The valves of 1-1/2-in. diameter were placed at an angle of approximately 120 degrees in a lateral transverse position, actuated by rocker levers. The rocker lever clearance (lash) was controlled by hydraulic units at the lever pivot point. Once the hydraulic unit took up the valve clearance by hydraulic pressure, the unit remained stationary with no movement, and consequently no wear. The combustion chamber of the cylinder heads was of an unusual machined triangular shape, and with the gable-shaped piston heads, it reduced the heat rejection area providing more heat for useful power.

With the deepening of the economic depression, Packard sales for 1932 numbered only 11,058 units domestically, the lowest since 1922. Even though Packard outsold its nearest competitor in the fine-car field by two to one, 11,000 cars per year was not enough for Packard Motor Car Company. They had to seriously consider entering the medium-priced field in the $1,000 to $1,500 range.

On July 1, 1932, Clyde Paton, who had been in charge of all experimental and research operations, was appointed chief engineer, succeeding A. Moorehouse, who resigned to become a consulting engineer. Paton had wide and varied experience in automotive and aeronautical engineering. He served in the U.S. Army Air Corps during World War I and for a considerable time afterward. A number of patents are held in his name, many of which had to do with suspension systems, front-end stability, as well as the dampening of body panels to eliminate vibrations and road noises. Paton was responsible for the design and development of all the fine Packard cars until World War II.

As of July 1, 1932, the top-most executives of Packard Motor Car Company included: Alvan Macauley, president and general manager; M.A. Cudlip, vice president and secretary; Hugh J. Ferry, treasurer; H.W. Peters, vice president of distribution; Milton Tibbetts, vice president and patent counsel; R.F. Roberts, vice president of manufacturing; J.G. Vincent, vice president of engineering; C.R. Paton, chief engineer; J.R. Ferguson, assistant chief engineer; and J.H. Marks, purchasing manager.

1933 Packard 1006 Dietrich runabout. (Source: P.M.C. Co.)

On November 15, 1932, even though Alvan Macauley expressed optimism that the lowest point of the Depression had been reached, he had the distasteful responsibility of reporting to the board of directors that Packard Motor Car Company had experienced a net loss during the first nine months of 1932.

The Packard tenth series models were introduced at the National Auto Show in New York in January 1933. They offered three lines of cars: the Packard Eight Series 1001 and 1002, the Packard Super Eight Series 1003 and 1004, and the Packard Twelve Series 1005 and 1006. These Packard models were available on six chassis in five wheelbase lengths. Forty-one production body types were available in addition to 17 Individual Custom bodies by custom coach builders. The price range was $2,150 to $2,880 for the Eight, $2,750 to $3,590 for the Super Eight, and $3,720 to $4,650 for the Twelve.

The new Series 1001 and 1002 superseded the ninth series Packard Standard Eight and covered the market range occupied by both the Light Eight and the Standard Eight in 1932. Bodies in general carried forth on the Packard lines by redefining the streamlined motif introduced the year before. The interiors of the tenth series were based on those of the ninth series Packard Twelve, using luxurious, smooth, wool broadcloth, now in a variety of bright solid colors including maroon, beige, chestnut brown, green, and blue.

The front-end appearance was a distinctly Packard radiator shell, grille, and hood. The long sweeping fenders were skirted, following the tire outline to further conceal the chassis undercarriage. The windshield had a greater slope, and the front door windows were divided in half so that the forward half could be opened outwardly to ventilate the car without annoying drafts. The windshield and door window garnish moldings were of the screwless type. The bumpers of the Packard Eight and Packard Super Eight were the new one-piece chrome-plated face bar, while the Packard Twelve continued the two-piece face bar with the stabilizers on the outer ends.

All Packard frames were of the double-drop X-member type whose member legs were braced in the center and at the joint with the frame side rails by wide steel gusset plates. The Packard Eight was powered by a

1933 Packard Twelve 1006 sport sedan. (Source: P.M.C. Co.)

1933 Packard Eight models parked in company's Harper Garage in Detroit. (Source: P.M.C. Co.)

120-hp eight-in-line engine, the Packard Super Eight was powered by a 145-hp eight-in-line engine. All Packard eight-cylinder engine crankshafts were supported by nine main bearings. The increase in horsepower was due to the increase in compression ratio to 6 to 1.

The Packard Twelve engine size was increased to 445.5 cu. in. by increasing the stroke to 4 in., developing 160 hp with the standard 6 to 1 compression ratio head and 176 hp with the optional 6.7 to 1 cylinder head. The Stromberg EE-3 two-bore carburetor was recalibrated for the larger-displacement engine. The engine crankshaft was supported on four large main bearings. Auto-Lite double breaker distributor with dual coils and condensers was continued on the Packard Twelve, and a new two-coil Auto-Lite double breaker ignition system was incorporated in both Eights. The Dyneto generator output capacity had been increased, while the Dyneto starting motor remained the same. The battery was a 19-plate 6V type. The Solar-Ray headlight system used three lamp filaments with a provision for tilting the left lamp by means of a control switch when passing.

The transmissions of all Packard models were of the three-speed helical gear type (silent in all forward speeds), similar to the ninth series four-speed type except the first speed (low-low) was omitted, as it was unnecessary with the high engine torque output. The new leakproof universal joints on the propeller shaft permitted the elimination of the intermediate propeller shaft on the long-wheelbase models. The four-wheel service brakes were of the Bendix duo-servo type, with 14 x 2-1/4-in. centrifuge brake drums on the Eights, and 14 x 2-1/2-in. centrifuge brake drums on the Twelve. There was a provision for driver regulation of power brake booster assist by a regulator control lever. The clutch manufactured by Long Manufacturing Company was a single-disc type, with the clutch pedal geometry so designed to reduce pedal effort, eliminating the need for automatic clutch control.

In 1933, Packard realized that the economic conditions were not improving sufficiently, so they embarked on a sweeping cost-control program, while maintaining enough funds for product development. Dividends were discontinued and all unnecessary expenses were curtailed, as the Bank Holidays and tight capital compounded the problems. On April 1, 1933, Packard Motor Car Company reported a deficit of $6,824,312 after depreciation and the creation of a $1,000,000 reserve against possible losses on bank deposits. This compared with a net loss of $2,909,117 and a deficit of $9,654,770 for 1932 after dividends were paid. Packard Motor Car Company maintained a working capital of $18,000,000.

Arlene Judge and a 1934 Packard 1101 convertible roadster. (Source: P.M.C. Co.)

The 1934 eleventh series Packards started production in August 1933 and deliveries on September 1st. The wheelbase of the Packard Standard Eight was increased from 127-1/2 in. to 129-1/4 in. New chassis of 141-1/4- and 146-7/8-in. wheelbase lengths were added to Packard Standard Eight Series 1102 and Packard Super Eight Series 1105, respectively. Full-flow (100%) oil filters and water-cooled oil temperature regulators were incorporated on all Packard engines. The use of Gemmer-Marles steering gears improved steering control with less effort. The steering wheels were color-matched to the garnish molding finishes. The front fenders were carried farther down in front, and both front and rear fenders had a trough (gutter) beneath their edges to carry road water away.

The prices of all eleventh series Packard models were increased by $200 over the corresponding previous models, except the coupe roadster which was increased by $330. The Packard Twelve models were increased by $100. A new streamlined speedster Sport Coupe (Series 1106), with all-metal roof, was displayed as a prototype at the National Automobile Show in New York in January 1934.

Despite the depressed economic conditions and sales of only 9,081 cars domestically in 1933, Packard was able to show a net profit of $107,081 after depreciation and other charges. On May 26, 1934, Max M. Gilman was elected vice president and general manager of Packard Motor Car Company. He was previously vice president of distribution.

1934 Packard 1106 special sport coupe. (Source: P.M.C. Co.)

1934 Packard 1106 speedster. (Source: P.M.C. Co.)

Medium-Priced Car

During the first few months of 1934, Packard's engineering department underwent a massive realignment of personnel. Approximately 90 engineers were added to chief engineer Paton's staff to form the nucleus of a separate section of the engineering department. The responsibility of this section was to design and develop a smaller car that could be built at lower cost, retain the Packard identity, and be able to be sold at a reasonable medium price. Among the new incoming engineers were: Earl H. Smith, assistant chief engineer, formerly assistant chief engineer of Pontiac; Erwin A. Weiss, chief chassis engineer, formerly of Willys-Overland Company and Continental Motors Corporation engineering departments; and Erwin L. Bare, chief body engineer, formerly of Hupp Motor Car Company.

In the meantime the manufacturing and sales organizations for the medium-priced car were put into place. George T. Christopher was named assistant vice president in charge of manufacturing. Christopher had previously been factory manager of Buick Motor Division of General Motors Corporation. Lyman W. Slack was appointed sales promotion manager of Packard medium-priced car division. Slack held a similar position at the Pontiac Motor Division of General Motors.

With an outlay of over $6,200,000 to develop the new car, plus the cost of new manufacturing equipment and rearrangement of plant facilities, and with sales in 1932 falling to a record low of 6,552 domestically, it was anticipated that Packard Motor Car Company would sustain a great loss in 1934.

In January 1935 the long-heralded medium-priced Packard car, the 120, made its debut at the National Automobile Show in New York. The styling was distinctively Packard, up to and including the Packard creased shoulder line in the radiator shell and hood. Even though in the early prototypes conventional

F.D. Roosevelt Jr. and a 1934 Packard Twelve 1207 sport phaeton. (Source: P.M.C. Co.)

radiator shell and hood designs were considered, the recognizable Packard hood shoulder line was decided on. The first Packard 120 came off the production line on February 23, 1935, at which time Alvan Macauley and Frank Cousens (Mayor of Detroit) participated in nationwide radio broadcast of the ceremonies commemorating this great historic event.

1935 Packard 120

Wheelbase (in.)	120
Price	$990-1,395
No. of Cylinders / Engine	IL-8
Bore x Stroke (in.)	3.25 × 4.25
Horsepower	120 adv, 33.8 SAE
Body Styles	7
Other Features	exclusive "Angle Set" (hypoid) rear axle, sturdy X-member frame, Bendix duo-servo hydraulic four-wheel brakes, and low-pressure 7.00-in. tires on 16-in. wheels

The most outstanding mechanical feature of the Packard 120 was the "Safety-Flex" coil spring front suspension, which featured steel-cored rubber bushings at the lower support arm pivot points, and rubber-insulated longitudinal torque arms to absorb and control the braking forces. Both of these important features are being copied today. This suspension system contributed to the smoothest, easiest ride of any car on the road.

Amelia Earhart in New York City in 1935 with a Packard 1204 phaeton. (Source: P.M.C. Co.)

The engine employed aluminum-alloy pistons, 6.5 to 1 compression ratio cylinder head, exhaust valves of Austenitic steel alloy, and full-length water jackets on the cylinder block. The radiator was mounted in a yoke having a single mounting point on the frame bracket that provided a fulcrum for the front-end assembly, eliminating front-end sheet metal movement.

In April 1935, Macauley revealed the 1934 losses. The cost of developing and introducing the Packard 120 was approximately $3,541,500 and the twelfth series tooling $1,559,975. The factory and branch losses due to low volume amounted to $2,239,074. The net loss was $7,290,549 for 1934.

On April 11, 1935, Macauley entered his office to find it bedecked with flowers, commemorating the 25th anniversary of his joining Packard Motor Car Company in 1910. On August 1, 1935, Macauley announced that George T. Christopher was new vice president of manufacturing, succeeding R.F. Roberts, who retired after 32 years of service.

The Packard 120-B was introduced on October 1, 1935, along with the other fourteenth series Packard models. While the 120-B was basically the same as its predecessor, a number of detail improvements had been made, including: the timing chain width was increased to 1-1/4-in., the stroke of the engine was increased to 4-1/4-in., giving a piston displacement of 282 cu. in., and the resulting power output was 120 hp at 3,800 rpm. The price of the Packard 120-B sedan was $1,075.

Packard Motor Car Company built and sold 37,653 cars domestically in 1935, and earned approximately $3,000,000 after payment of dividends and paying the costs of the development and new model launch.

The fourteenth series Packard Eight, Super Eight, and Twelve remained fundamentally the same as the twelfth series, except the frontal appearance was modified; the radiator shell was narrower at the bottom and raked rearward at a greater angle. The price range remained essentially the same.

Mechanically, as a safeguard against annoying vapor lock, a metal shield was placed over and in back of the fuel pump to stimulate air circulation. The ignition systems were by Delco-Remy and were equipped with a

Packard 120 pace car at 1936 Indianapolis 500. (Source: Indianapolis Motor Speedway Corp.)

Jack Benny and a 1936 Packard 1407 formal sedan. (Source: P.M.C. Co.)

vacuum advance mechanism. In late 1935 Packard Motor Car Company installed a completely mechanized method of heat-treating, quenching, and drawing of steel parts in an atmosphere that prevented scaling. The process was fully automatic, requiring only a single operator to charge the hardening furnace.

In January 1936, William M. Packer was appointed general sales manager of Packard. Also named were: R.E. Chamberlain, assistant sales manager of the Eastern division, and Lyman W. Slack, assistant sales manager of the Western division. On February 1, 1936, Hugh J. Ferry was appointed secretary-treasurer of Packard.

Lowest-Priced Car

The 1937 Packard fifteenth series models were introduced on September 1, 1936, comprising four basic lines. Of utmost interest in the Packard introduction was the announcement of the addition of a new smaller car, identified as the Packard 115-C. It was the lowest-priced car sold under the Packard name in the history of the company.

1937 Packard 115-C

Wheelbase (in.)	115
Price	$795-1,295
No. of Cylinders / Engine	L-6
Bore x Stroke (in.)	3.43 × 4.25
Horsepower	100 adv, 29.4 SAE
Body Styles	8 including station wagon
Other Features	

The Packard 120-C continued unchanged except for minor details. The Deluxe model was identified as the Packard 120-CD, equipped with such luxury items as an electric clock, deluxe interior trim, and Marshal springs in the seat and back cushions. In mid-1937 the Packard 120-CD was offered with a 138-in. wheelbase chassis, available in a seven-passenger sedan and a seven-passenger limousine, priced at $1,900 and

$2,050, respectively. A total of nine body types were available in the Packard 120-C and 120-CD models, both powered by 120-hp, in-line eight-cylinder engines.

The major changes in the fifteenth series Packard Super Eight Series 1500, 1501, and 1502 were that they were now powered by the 320-cu.-in.-displacement engine previously used in the fourteenth series Packard Eight, available in 14 production body types, on three chassis of 127-, 134-, and 139-in. wheelbase lengths. The Packard Super Eight was priced at $2,335 to $3,350 for the production body types.

The fifteenth series Packard Twelve Series 1506, 1507, and 1508 remained generally the same as the fourteenth series models. The Super Eight and Twelve now featured the Packard Safe-T-Flex front suspension, with its components heavier and stronger than those of the Packard 120-C. Delco-Remy ignition, starting, and charging electrical systems were used on the 115-C, while Auto-Lite electrical equipment was used on the 120-C, 120-CD, Super Eight, and Twelve. The lighting circuits of all models were protected by thermostatic circuit breakers instead of by fuses. The spark plugs of all engines were 14 mm. The price range of the fifteenth series Packard models was $895 for the 115-C sedan to $4,285 for the Packard Twelve Series 1508 seven-passenger sedan, and $6,435 for the Packard LeBaron town car.

Joy Dies

On November 6, 1936, Henry Bourne Joy, former president of Packard Motor Car Company, died at his home in Detroit at the age of 72.

Emerging from the Depression

The automotive industry enjoyed a prosperous year in 1937; industry sales totals reached 3,483,752 units, just short of the record of 3,880,247 in 1929. Packard also prospered, with a boost in sales by the 115-C. Packard built and sold 95,455 cars domestically in 1937, a record that would not be surpassed until 1949. In April 1937 Packard Motor Car Company reported that first-quarter earnings were in the neighborhood of $2,500,000, approximately double the earnings in the corresponding quarter of 1936.

On April 16, 1937, Packard Motor Car Company held its first conference with representatives of the United Automobile Workers of America union. The meeting lasted only 7 minutes, but the list of demands (22) presented by the union was the longest list received by any company in the automobile industry. An election held by the National Labor Relations Board at the Packard plant on April 29, 1937, gave a landslide vote in favor of the union: 11,588 for the UAW to 2,655 against. Following the election, Alvan Macauley said, "Our employees have expressed their desire by their vote. We are pleased that the matter was determined peacefully and with apparent good will all around."

While Packard Motor Car Company enjoyed amiable relations with the UAW, the labor strikes in other automotive plants and in other industries during 1938 depressed the economic conditions nationwide. Only 1,891,021 cars were sold domestically industry-wide in 1938, about one-half of the 1937 sales.

Four lines of cars were offered by Packard Motor Car Company in 1938. The Packard Six (formerly 115) Series 1600, Packard Eight (formerly 120) Series 1601 and 1602, Packard Super Eight Series 1603, 1604, and 1605, and the Packard Twelve Series 1607 and 1608. The Six and Eight had longer wheelbase lengths: 122 and 127 in., respectively. The double-trussed X-member frame was continued with changes and improvements to accommodate the longer wheelbases. The frame X-member arrangement consisted of

I-beam rails of tapered section, deepest at the center, welded to reinforce the rivet fastening. From the terminals of the X-members at the front and rear, the frame side rails continued of box section form to the front and rear ends.

The rear semi-elliptical suspension springs incorporated controlled friction-dampening by the use of embedded buttons separating the spring leaves at their outer ends. Three types of buttons were used: hard rubber, graphite-impregnated bronze, and antimony-lead alloy. Completely isolated rubber shackles were used at both spring ends to prevent metal-to-metal contact. The rear suspension also incorporated a transverse lateral stabilizer, pivoted at one end at the frame and at the opposite end to the axle spring pad. Rubber isolation was used at the pivot mountings. The body mountings were also isolated at the body bolts.

The body styling was all-new with an all-steel roof and V-type two-piece windshields. The windshield defrosting was accomplished by defroster outlets in the garnish moldings that delivered heat from the car heater. All sedan models had built-in trunks at the rear, and legroom was increased by the longer wheelbase and longer floor pan. The Packard Eight models offered three choices of interior upholstery cloth: a new flat Bedford Cord, plain all-wool broadcloth, and patterned all-wool broadcloth.

The radiator and front fenders were mounted on a cushioned support on the frame front cross-member. In addition to the usual vent of horizontal hood side louvres, the under-hood ventilation was aided by built-in cooling tunnels in the front fender inside aprons.

The engines of the Packard Six and Packard Eight were essentially the same as those of the fifteenth series models. In the Packard Eight a 7.05 to 1 compression ratio aluminum-alloy cylinder head was available in addition to the standard 6.6 to 1 ratio aluminum cylinder head. The Packard Six engine output was rated at 100 hp, while the Packard Eight engine output was 120 hp. Five body styles were available on the Packard Six, priced at $878 to $1,023. The Packard Eight was available in seven body styles on the 127-in. wheelbase chassis, priced at $1,250 to $1,540. Also available on the 148-in. wheelbase chassis were the seven-passenger sedan and seven-passenger limousine priced at $1,955 and $2,110, respectively.

The Packard Super Eight Series 1603, 1604, and 1605, and the Packard Twelve Series 1607 and 1608 were considerably refined and improved. They were available on 127-, 134-, and 139-in. wheelbase chassis, in addition to the 148-in. wheelbase special chassis. The Super Eight offered eleven body types priced at $2,790 to $3,970. The Twelve was available with ten production body styles, priced at $4,135 to $5,390. In addition, individual Custom models were offered, built by custom coach builders. Despite the proliferation of models offered by Packard Motor Car Company, the domestic sales dropped to 49,163 cars for 1938, following the trend industry-wide.

1938 Packard Darrin convertible. (Source: P.M.C. Co.)

1939 Packard V-12 custom town car. (Source: P.M.C. Co.)

On October 14, 1938, the seventeenth series Packard 1939 models were introduced.

	1939 Packard Six Series 1700	1939 Packard Eight Series 1701/1702 (seventeenth series 120)	1939 Packard Super Eight Series 1703/1705	1939 Packard Twelve Series 1707/1708
Wheelbase (in.)	122	127/148	127/148	134/139
Price	$888-1,404	$1,099-1,636 / $1,703-1,856	$1,650-2,130 / $2,156-2,294	$4,140-6,730 / $4,485-6,880
No. of Cylinders / Engine	6	IL-8	IL-8	V-12
Bore x Stroke (in.)	3.50 × 4.25	3.25 × 4.25	3.18 × 5.00	3.43 × 4.25
Horsepower	100 adv, 29.4 SAE	120 adv, 33.8 SAE	130 adv, 32.5 SAE	175 adv, 56.7 SAE
Body Styles	6	7 / seven-passenger sedan and limousine	4 / seven-passenger sedan and limousine	8 / 3 plus Rollston All-Weather Town Car
Other Features		steering-column-mounted gearshift lever		

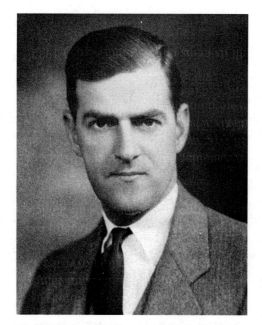

Hugh Hitchcock. (Source: P.M.C. Co.)

During 1939 Packard Motor Car Company offered unbodied long-wheelbase chassis up to 156 in. in the Packard Six, Packard Eight (120), and the Packard Super Eight models.

On April 17, 1939, Alvan Macauley was elected chairman of the board, and relinquished the presidency at his own request to Max M. Gilman, who was elected by the board of directors. Under Macauley's leadership Packard Motor Car Company's net worth had increased from $17,000,000 in 1916 to over $42,000,000 in 1939. On May 15, 1939, Packard Motor Car Company announced price reductions of $100 on the Packard Six and Packard 120 models, and up to $303 on the Packard Super Eight models; the Packard Twelve prices remained unchanged.

On May 15, 1939, William M. Packer, former general sales manager, was elected vice president of distribution of Packard. During 1939 Packard domestic sales amounted to 62,005 cars, of which less than 450 units were Packard Twelve models. That same month, Hugh W. Hitchcock, graduate of the University of

Michigan and West Point Academy, was appointed advertising manager after 17 years of advertising experience with Packard. He had been assistant advertising manager since 1933. It was during Hitchcock's management that Packard displayed the finest in good advertising, and was so recognized by many art and literary organizations.

The 1940 Packard eighteenth series models were introduced on September 1, 1939, comprising four lines. The Packard 160 Super Eight and the Packard 180 Custom Super Eight superseded the 1939 seventeenth series Packard Super Eight and the Packard Twelve.

	1940 Packard 110 Series 1800	1940 Packard 120 Series 1801	1940 Packard Super Eight 160 Series 1803/1804/1805	1940 Packard Custom Super Eight 180 Series 1806/1807/1808
Wheelbase (in.)	122	127	127/138/148	127/138/148
Price	$869-1,200	1,038-1,573	$1,614-2,179	$2,243-6,332
No. of Cylinders / Engine	L-6	IL-8	IL-8	IL-8
Bore x Stroke (in.)	3.50 × 4.25	3.25 × 4.25	3.50 × 4.62	3.50 × 4.62
Horsepower	100 adv, 29.4 SAE	120 adv, 33.8 SAE	160 adv, 39.2 SAE	160 adv, 39.2 SAE
Body Styles	6	8 including Darrin, convertible Victoria	9	9 including Darrin four-door convertible sedan and Rollston Town Car
Other Features	three-speed all-silent transmission with optional Packard Econo-Drive (overdrive) unit			

The Packard 160 and 180 engines incorporated barrel-type Wilcox-Rich (Eaton) hydraulic tappets to take up the valve clearance. The crankshaft was supported on nine main bearings in a crankcase cast integral with the cylinder block.

The front suspension was the Safe-T-Flex, the same as on previous models with the exception that the rear end of the torque arm was now attached to the frame by means of a steel-cored rubber bushing (Harris type), in place of the spherical rubber bearing.

Stylewise the eighteenth series Packard carried the traditional V-type Packard radiator shell and hood, but they were narrower and had vertical louvres. Chromium-plated die-cast grilles with vertical louvres were placed one on each side of the radiator shell. The bullet-shaped headlamps featured the new sealed-beam

1940 Packard Darrin.
(Source: P.M.C. Co.)

headlamp units. The hood sides had a longitudinal grille with chromium-plated die-cast louvres, adorned at the forward end by a Packard cloisonne with the model identification number.

First Air Conditioner

In November 1939 the first automobile mechanical refrigeration (air-conditioning) system was offered by Packard Motor Car Company as factory-installed extra-cost optional equipment. This was the forerunner of all automotive air-conditioning equipment used today.

Preparation for War

During the week of March 11, 1939, Packard Motor Car Company announced the acceptance of a $2,000,000 contract for the production of supermarine engines for the United States Navy. The engines at that time developed 1,200 hp each, and were to be used by the United States Navy to power the ultra-high-speed Motor Torpedo Boats, then under construction. The Packard 4M 2500 was a 2,498-cu.-in.-displacement, V-12-cylinder, water-cooled engine, constructed of aluminum-alloy crankcases and valve housings. The cylinders were of chrome-moly-steel alloy, with welded-on sheet-steel water jackets. The crankshaft was a chrome-moly-steel forging, precisely machined and carried on eight bronze-lead-tin bearings.

Suffering from the unceasing bombing by Nazi aircraft in early 1940, Great Britain realized they needed more aircraft to defend Britain and to ultimately win the war. Their greatest need was for aircraft engines. Naturally, they came to the United States for such help. On June 1, 1940, the British Government granted the United States the authority to build the Rolls-Royce twelve-cylinder aircraft engines. At the same time, British Government officials sought an appropriate United States automotive manufacturer to build these engines for the Royal Air Force aircraft and replacements. While they contacted all the prominent American automobile manufacturers with their request, they were referred to Packard Motor Car Company as the only manufacturer having the engineering talent and production skills to produce such a precise aircraft engine.

George Christopher. (Source: P.M.C. Co.)

A satisfactory arrangement was worked out between the Packard Motor Car Company and Rolls-Royce Limited. Vice president George H. Brodie negotiated the government contracts for Packard, while vice president Jesse G. Vincent was responsible for all engineering, modifications, and development aspects. Marsden Ware was named chief engineer of the marine engine division, and Clyde R. Paton became chief engineer of the aircraft engine division, with J.R. Ferguson as his assistant.

The first contract called for 3,000 Rolls-Royce aircraft engines for Great Britain and 6,000 for United States military aircraft. Ultimately, Packard Motor Car Company built in excess of 55,500 Rolls-Royce and Merlin aircraft engines.

In the late summer of 1940, Max M. Gilman resigned as president of Packard Motor Car Company because of ill health stemming from a serious automobile accident. George T. Christopher, vice president of manufacturing, was named acting general manager, and handled the administration in this capacity until he was elected president in 1942.

The 1941 Packard nineteenth series cars, introduced in September 1940, were in fact carryover eighteenth series models with many refinements. Noticeably, the sealed-beam headlamps fared into the front fenders, with parking lamps mounted above them. The hood side louvres were eliminated, with a short hood-release handle located in their place. A spear-shaped molding was placed below the hood shoulder line, with the barb at the radiator shell end.

The Packard nineteenth series models consisted of four lines: the Packard 110 Series 1900, the Packard 120 Series 1901, the Packard Super Eight 160 Series 1903, 1904, 1905, and the Packard Super Eight 180 Series 1906, 1907, and 1908. Eleven body types were available in the 110, including a wood-constructed station wagon model. Eight body styles were offered in the Packard 120, nine in the Packard 160, including a seven-passenger sedan and a seven-passenger limousine, and eleven in the Packard 180, including seven custom-built bodies by Darrin, LeBaron, and Rollston. Electro-hydraulic window regulators and partition window regulators were standard equipment, and air conditioning was available on the Packard 180. The price range of the nineteenth series models was from $927 for the Packard 110 business coupe to $5,595 for the Packard 180 seven-passenger limousine by LeBaron.

Clipper Series

In April 1941 the new Packard Clipper Series 1951 was introduced.

1941 Packard Clipper Series 1951

Wheelbase (in.)	127
Price	$1,420
No. of Cylinders / Engine	IL-8
Bore × Stroke (in.)	3.25 × 4.25
Horsepower	125 adv, 33.8 SAE
Body Styles	four-door sedan
Other Features	

1941 Packard Clipper 1951. (Source: P.M.C. Co.)

The Clipper was not a derivative of any other previous Packard model, but a completely new chassis, suspension, body, and all new styling. The new Packard Clipper was longer, wider, and lower. It had a long tapered hood with the exclusive Packard traditional shoulder line, blending gracefully into the narrow, horizontally louvred radiator grille. Horizontally louvred grilles were also located below the headlamps on each side. The sealed-beam headlamps were fared into the front fenders, and the rear edge of the fender faded gently into the front door outer panels. Concealed body sills below the doors replaced the former running boards, and the rear doors were hinged at the center (B) pillar.

The Clipper was on a new X-member-type frame with pressed-steel lower suspension arms supporting the coil springs. The double-acting opposed piston shock absorbers provided the upper suspension arms. Threaded, hardened, steel-alloy bushings and pins were the pivots for the suspension.

During 1941 Packard proliferated into 40 body styles, and domestic sales amounted to 69,653 cars, of which about 16,600 were Clipper models. The styling of the Packard Clipper was well-accepted, being declared by most art critics as "the most beautiful car of the year." The Clipper styling was extended to the balance of the car lines for the 1942 Packard twentieth series. Unfortunately, Pearl Harbor and the subsequent four years of World War II gave the competitors time to catch up; thus, Packard lost the competitive edge of the Packard Clipper styling.

The twentieth series Packard was announced on August 25, 1941. The 110 and 120 model designations were dropped, and were called the Packard Six and Packard Eight, respectively. Emphasis was placed on the Packard Clipper lines.

	1942 Packard Clipper Six Series 2000	1942 Packard Clipper Eight Series 2001	1942 Packard Clipper Super 160 Series 2003	1942 Packard Clipper Custom Super 180 Series 2006
Wheelbase (in.)	120	120	127	127
Price		$1,248 - 2,215		$4,595
No. of Cylinders / Engine	6	IL-8	IL-8	IL-8
Bore x Stroke (in.)	3.50 × 4.25	3.25 × 4.25	3.50 × 4.62	3.50 × 4.62
Horsepower	105 adv, 29.4 SAE	125 adv, 33.8 SAE	165 adv, 39.2 SAE	165 adv, 39.2 SAE
Body Styles		four-door sedan and two-door club sedan		Darrin convertible
Other Features				

The conventional (non-clipper) Packard models were carryovers from the nineteenth series. The convertible models were available in the Packard Series 2020, 2021, and 2023, powered by 105-, 125-, and 165-hp

1942 Packard Super Clipper 2003. (Source: P.M.C. Co.)

engines, respectively. The Packard (Six) taxicab model was continued. The Packard Super Eight 160 Series 2004 and 2005 were available in a six-passenger touring sedan, seven-passenger sedan, and seven-passenger limousine on wheelbase lengths of 138- and 148-in. The Packard Super Eight 180 Series 2006, 2007, and 2008 were available in a six-passenger Formal Sedan, touring sedan, and an All-Weather Cabriolet (by Rollston) on a 138-in. wheelbase chassis, a seven-passenger touring sedan, a seven-passenger limousine, a seven-passenger touring sedan by LeBaron, a seven-passenger limousine by LeBaron, and an All-Weather Town Car by Rollston on a 148-in. wheelbase chassis. A Special Darrin two-door convertible was built on a 127-in. wheelbase Series 2006 chassis.

The price range was from $1,283 for the Packard Clipper club sedan to $5,795 for the seven-passenger limousine by LeBaron. Only about 34,000 cars were sold by Packard Motor Car Company in 1942 because of the curtailment of all automobile production early in the year due to the war.

World War II

After Pearl Harbor, December 7, 1941, automotive production was systematically suspended and all effort by engineering and manufacturing was concentrated on the precise building of the Rolls-Royce and Packard-built Merlin aircraft engines, and the Packard 4M 2500 marine engines. Jesse Vincent remained as vice president of engineering, with Marsden Ware as chief engineer of the marine engine division. J.R. Ferguson was named chief engineer of the aircraft engine division in early 1942, succeeding Clyde R. Paton, who was on leave from Packard, carrying out an assignment for the United States Army Air Corps.

During World War II (1942-45), Packard Motor Car Company established records in providing quantity production of Rolls-Royce, Merlin, and Packard 4M-2500 marine engines. There was an early tie-up in starting, due to the United States Government's insistence of 10 years to amortize the $30,000,000 in plant expansion, while Packard wanted only five years. This was amicably resolved through the diplomatic

Assembly of Packard-built Merlin aircraft engines, 1945. (Source: P.M.C. Co.)

negotiations by George H. Brodie. Production of the Rolls-Royce aircraft engines started on August 1, 1941, and within five months, they were building 840 engines per month in a plant area of 1,200,000 sq. ft., employing about 14,000 skilled workers. By August 1, 1942, Packard Motor Car Company had 20,000 employees on defense work.

During these four years (1942-45) Packard engineers redesigned the Rolls-Royce aircraft engine, raising the horsepower output from 1,200 to 1,500, then to 1,800. This was made possible by redesigning the aluminum-alloy castings, relieving the concentrated high-stress areas by providing adequate fillets and reducing the number of bolts by 40%. This allowed an increase in supercharger boost and incorporation of a linter-cooler (heat exchanger) between the two-stage supercharger outlet and the intake manifold log. The latest aircraft engines with the greatest output, called the "Packard-Built Rolls-Royce Merlin," were used on the Curtis P-40D and the North American P-51 Mustang United States Air Force fighter aircraft. The Rolls-Royce aircraft engines built to British Royal Air Force specifications were used in the Hurricane and Spitfire fighters and the DeHavilland bombers, and for replacement engines for British aircraft.

The improvements in the Rolls-Royce and Merlin aircraft engines, and their outstanding performance, should be credited to the engineering talents of Col. Jesse Vincent, William H. Graves, J.R. Ferguson, Forest R. McFarland, and Herbert L. Misch. Ralph DePalma, Thomas W. Milton, and C.W. Van Ranst were consulting engineers. Dennis (Duke) Nalon, Russell Snowberger, and Richard Van Emerick were the field service engineers. The combined effort of all these engineers contributed to the reliability of the Rolls-Royce and Merlin aircraft engines, and the extended flying time between engine overhauls. Packard built 55,523 Rolls-Royce and Merlin aircraft engines from 1942 to 1945. After 1945 the Packard-built Rolls-Royce Merlin aircraft engine was the most sought-after engine to power the unlimited power boat racers.

The Packard 4M-2500 marine engines were built in the former aircraft engine building, using the skills of the former aircraft engine employees. Under the engineering guidance of Marsden Ware, the 4M-2500 horsepower was increased from 1,200 to 1,350, and then to 1,500. The Packard 4M-2500 marine engines gave a

The 50,000th Packard-built aircraft engine. (Source: P.M.C. Co.)

brilliant account of themselves in the rescue of General Douglas MacArthur in March 1942, and the subsequent United States Naval victories in the Pacific and European theatres of war. Packard built more than 35,000 4M-2500 marine engines during 1942-45.

At the same time Packard Motor Car Company provided the finest engine maintenance and reconditioning schools with the United States Navy and the United States Army Air Corps. The precise and skillful instruction was provided by Army Air Corps, Navy, and Packard instruction personnel. Felix A. Kummer was manager of the Packard Marine Engine School, and naturally so, because since the late-20s Kummer was secretary and assistant to the late Captain Lionel M. Woolson in the development of Packard aircraft engines.

Post-War Years

During this period the Packard design and styling personnel were not idle. Edward Macauley was negotiating government contracts under the guidance of vice president George Brodie. John Rinehart and the other stylists were working with the engineering department to process engineering drawings, photographs, technical data, and printed manuals.

In April 1945 Clyde Paton returned to Packard to reorganize the automotive engineering department, which was almost totally disbanded after Pearl Harbor. Upon completing this assignment, Paton left Packard Motor Car Company to join the central engineering staff of the Ford Motor Company. J.R. Ferguson was permanently appointed chief engineer of the automotive and aircraft engineering divisions. William H. Graves was appointed executive engineer reporting to Vincent. Milton Tibbetts remained vice president and patent counsel.

Administratively, Alvan Macauley was chairman of the board, George T. Christopher president and general manager, George H. Brodie assistant to the president, J.H. Marks executive vice president, Hugh J. Ferry vice president, secretary, and treasurer, E.C. Hoelzle vice president and comptroller, and W.B. Hoge assistant comptroller.

In the meantime the manufacturing and sales divisions were reorganized. George C. Reifel remained as vice president of manufacturing, with M.F. Macauley as manager of manufacturing control, and capable staff members to carry out their responsibilities. R.R. Rees was director of purchasing.

With William Packer leaving Packard Motor Car Company to establish his own dealership, Lyman Slack was elected vice president and general sales manager, with C.E. Briggs as assistant sales manager. Hugh Hitchcock was appointed director of advertising and public relations, and Karl Greiner was appointed parts and service manager.

In manufacturing, since the Briggs Manufacturing Company was tooled to build the 1941 Packard Clipper bodies, it was a wise decision to have them build all the Packard bodies. They had a relatively new plant, constructed for aircraft fuselage component manufacturing during World War II. This plant, located at Warren and Conner Avenues in Detroit, was devoted exclusively to building Packard bodies. Body manufacturing lines were laid out to be able to produce a maximum of 70 bodies per hour. These bodies were hauled, seven to a truck-trailer unit, to Packard's twin assembly lines at 1580 East Grand Boulevard, a distance of about 3 miles.

The new twin assembly lines were located to occupy a court between two rows of factory buildings, protected by a roof. This decision was a very good and important one because it placed the assembly lines in close proximity to the adjoining subassembly buildings. Pits were located under the conveyor lines providing adequate room to enable the workmen to assemble under-chassis parts as the cars moved down the line. This method was a new approach to manufacturing and was frowned upon by many in the automotive industry at the time. However, it proved to be very successful, with the conveying systems delivering the parts and subassemblies just in time from the surrounding three levels of buildings, coordinated by teletype.

The first twenty-first series Packard Clipper was built on October 19, 1945, and 2,721 more Packard Clippers were built by the end of 1945. But all was not rosy, as many logistics problems had to be solved before full production could be attained.

Prior to World War II, Packard purchased most of the steel for stamped parts and sheet steel from the smaller steel producers. The steel plants were bought up by steel fabricators during and after the war to provide the needed steel for their factories. This put Packard in an unfavorable position; while the demand for the 1946 Packard Clipper was at an all-time high, Packard with all its effort could obtain only enough steel to build approximately 4,000 cars per month. Thus, Packard had to take extraordinary steps to obtain the needed steel.

In early 1947 Packard Motor Car Company, in cooperation with other steel users, set up an arrangement to provide advance funds to the Detrola International Corporation, who had acquired the electric smelting furnaces previously installed and set up by the Andrews Steel Company in Newport, Kentucky. It was late 1947 before this plan was fully in operation. In the meantime, Packard had its dealers ship thousands of tons of scrap steel to Detroit, which was a necessary ingredient for steel production. Packard, in turn, forwarded this scrap steel to help its steel producers, as well as for its own foundry in Detroit.

The net result was that Packard production, limited by lack of basic materials and by labor strikes at supplier plants during 1946-47, was able to build only 36,435 cars in 1946, and about 48,000 cars in 1947. The low car production and resultant low sales caused Packard to sustain financial losses in both years.

Immediately after V-J Day, Packard stylists went full steam ahead in the design of the new post-war models. The styling that emerged was influenced by the many pre-war special exhibit models, including those of competitive manufacturers. Packard Motor Car Company also engaged consumer research firms to determine what styles the post-war buyers wanted in new cars.

Interior of 1947 Packard Custom sedan. (Source: P.M.C. Co.)

1947 Packard Clipper Deluxe. (Source: P.M.C. Co.)

Ironically, as Edward Macauley quoted, "The three things that all American males can do are play piano, sing bass, and style cars." The research showed that there was a wide spectrum of style favorites, ranging from weird, dynamic, locomotive dinosaurs to future spacecraft. Fortunately, Packard followed the designs utilizing the artistic talents of their own stylists. During 1946-47 Packard designed and built two convertible prototypes using the Packard Super Clipper and Packard Custom Super Clipper chassis and body lines, powered by eight-in-line engines of 150 and 165 hp, respectively.

In the mid-summer of 1947, Lyman Slack left Packard to establish his own dealership in the Pacific Northwest. He was succeeded by Karl Greiner as vice president and general sales manage, and E.D. Longenecker was appointed parts and service manager.

The twenty-second series Packard was introduced in August 1947 by the first showing of the new Packard Super Eight convertible. It was displayed in Packard showrooms with an atmosphere of a carnival or circus because of the sales promotion tactics favored by the newly established Sales Promotion Department.

	Packard Eight Series 2201	**Packard Super Eight Series 2202/2222**	**Packard Custom Eight Series 2206/2226**
Wheelbase (in.)	120	120/141	127/148
Price	$2,125-3,350	$2,690-3,800	$3,750-4,868
No. of Cylinders / Engine	IL-8	IL-8	IL-8
Bore x Stroke (in.)	3.50 × 3.75	3.50 × 4.25	3.50 × 4.62
Horsepower	135 adv, 33.8 SAE	150 adv, 39.2 SAE	165 adv, 39.2 SAE
Body Styles	3	convertible / seven-passenger sedan and limousine	/ seven-passenger sedan and limousine
Other Features		electro-hydraulic window regulators and top operating mechanism	engine had nine-main-bearing crankshaft and hydraulic tappets

The twenty-second series Packard was available in seventeen body styles, including a wood-paneled four-door station wagon. The six-cylinder engines were discontinued, except for taxicab models. Packard built and sold 77,843 cars domestically in 1948, and 21,055 through export. Another 46,543 were sold during 1949 before the introduction of the twenty-third series Packard. During 1948 Packard Motor Car Company made a profit in excess of $15,000,000, and dividend payments were restored in 1949. The design and building of Packard marine and industrial engines (a marine conversion of 327-cu.-in.- and 356-cu.-in.-displacement passenger-car engines) also added to Packard's profits.

1948 Packard Super convertible 2297. (Source: P.M.C. Co.)

The Packard twenty-third series was introduced in May 1949, celebrating the Golden Anniversary of Packard Motor Car Company. They were the finest Packard cars built up to that time.

	Twenty-Third Series Packard Eight	**Twenty-Third Series Packard Super Eight**	**Twenty-Third Series Packard Custom Eight**
Wheelbase (in.)	120	127	127
Price	$2,150-2,395	$2,385-3,350	$3,750-4,295
No. of Cylinders / Engine	IL-8	IL-8	IL-8
Bore x Stroke (in.)	3.50 × 3.75	3.50 × 4.25	3.50 × 4.62
Horsepower	135 adv, 39.2 SAE	150 adv, 39.2 SAE	165 adv, 39.2 SAE
Body Styles	four-door sedan and two-door club sedan, four-door station wagon	four-door sedan and two-door club sedan (convertible coupe in Super Deluxe model)	four-door sedan and convertible coupe
Other Features			sedan had all-wool broadcloth upholstery in a choice of solid colors; convertible was upholstered in top-grain leather and Bedford Cord material combinations

The Packard Ultramatic Drive transmission was standard on the Custom Eight, and was later available on other Packard models at extra cost. At the time of introduction, the Packard Ultramatic Drive was hailed as the finest automatic transmission up to that time. It consisted mainly of an aluminum-encased torque converter having two turbines, a reactor, and the converter driving pump, which was the converter outer housing. The torque converter multiplied the engine torque by 2.4 to 1, while the planetary gear train gave an additional 2.2 to 1 torque multiplication in low range. The torque converter also featured a direct-drive clutch with a lock-up feature, later copied by all competitors.

During 1949 Packard sales were 97,771 units domestically, and about 12,000 cars through export. For 1950 Packard built and sold 73,155 cars domestically and about 10,000 cars through export.

Instrument panel of 1949 Packard twenty-third series. Source: P.M.C. Co.)

In retrospect, it would appear that Packard unnecessarily spent a lot of money, time, and effort to design, style, and tool for the building of the twenty-second and twenty-third series cars. After World War II, the limiting factor was the availability of material and the ability to build enough cars to satisfy the potential buyers' demands. The Packard Clipper twenty-first series models might have satisfactorily filled the need through 1950.

The 1950s

After 1945 the automobile buyers' tastes were influenced by the appearance of foreign sports cars. By late 1948 came the introduction of the gaudy tail lamps, fins on the rear fenders, and slab-sided bodies that appeared on competitive makes of cars. Packard chose to follow the styling trend by lowering the hood and raising the front fender and belt line to just below the window lower edge, to give a sports-car look.

Indeed, Edward Macauley, John Rinehart, and Charles Phaneuf did a magnificent job of creating a new Packard style for 1951. The styling of the Packard 200, 300, and Patrician 400 won the acclaim of "The most beautiful car of the year" by the Society of Motion Picture Art Directors on October 12, 1950. At that time such an honor and recognition was highly prized, since it could not be bought with a four-page advertisement.

1951 Packard 300 sedan. (Source: P.M.C. Co.)

1951 Packard Patrician. (Source: P.M.C. Co.)

While the traditional Packard hood shoulder line was all but lost in the previous two series of cars, a concave flute along each side of the hood gave the twenty-fourth series Packard a new shoulder line effect. At the time of introduction, only a four-door sedan, a two-door club sedan, and a two-door business coupe were available in the Packard 200 series. The Packard 300 and Patrician 400 were available in four-door sedan models only. The Packard 250 two-door hardtop and two-door convertible were not available until March 1951. The main styling feature was the slab-sided body panels with a pressed outline for the rear fender. A curved one-piece windshield replaced the former, flat, two-piece V windshield. The Patrician 400 interior trim was of wool broadcloth in solid colors of maroon, blue, green, and tan.

Unfortunately for Packard, almost coincident with the introduction of the twenty-fourth series was the start of the Korean War. This resulted in a shortage of materials (especially nickel), and all automobile manufacturers had to abide by the United States Government allotment of critical materials. The net result was that Packard could obtain only enough material to build and sell 66,999 cars domestically. This was not enough to show a profit in 1951.

Meanwhile Packard enjoyed a splendid relationship with the United States Government, and was paid fairly for their service to the military. At their own expense they continued the development of the 3,330-cu.-in. 16-cylinder marine engine for the Motor Torpedo Boats, based on the 4M-2500 marine engine. Packard also designed and developed a 12-cylinder non-magnetic diesel marine engine for the United States Navy minesweepers. With the outbreak of hostilities, Packard was again asked to build a quantity of United States Air Force J-47 aircraft engines.

Because of the urgent request of the United States Air Force, Packard hurriedly purchased the Mahon Steel Company plant on Mount Elliott Avenue in Detroit for the purpose of forging the critical-alloy buckets for the turbine rotor of the jet aircraft engine. Concurrently Packard constructed a new modern plant on the real estate of the Packard Proving Grounds in Utica, Michigan. The plant, while built for the purpose of manufacture, assembly, and test of the J-47 jet aircraft engine, turned out to be a good investment because it later served as a manufacturing facility for automotive production after the Korean War had ended. However, the Mahon plant was not such a good idea, since it was old, inefficient, and obsolete. It, in effect, became a millstone around Packard's neck.

The twenty-fifth series Packard cars were introduced on November 14, 1951. They were carryovers of the twenty-fourth series with minor changes such as: stainless-steel-stamped louvres on the Patrician rear quarter, and the Cormorant on the Patrician and 300 had its wings folded rearward as if alighting. The chromium-plated parts had a reduced amount or no nickel under the chromium, in accordance with the United States restrictions on the use of nickel during the Korean War.

1952 Pan-American custom sports convertible (with the author driving). (Source: P.M.C. Co.)

The model line-up was essentially the same with the price range from $2,495 to $3,795. A Formal Sedan by Dietrich was made available on a Patrician chassis by special order. At the National Auto Show in New York in January 1952, a special Pan-American show model was displayed. It was, in fact, a special convertible model built with all the body panels 4 in. lower, and the windshield upper panel an additional 2 in. lower, bringing the overall height down to 56 in. The show model was a two-passenger vehicle with a long rear deck, designed by Richard Arbib and built by Henney Motor Company of Freeport, Illinois. The Pan-American was demonstrated at the Michigan Fairgrounds in Detroit in early September 1952.

The Era of Nance

By mid-1951 the impending unravelling of the fibers of the Packard Motor Car Company structure became visible, when the board of directors showed their displeasure at the company's post-war performance. Regardless of the reasons, or the impact of World War II and the Korean War, the board directed president Hugh Ferry to terminate the advertising contract with Young and Rubicam Incorporated, who had been Packard's advertising agency exclusively since 1932. This was accomplished by Ferry's letter dated July 3, 1951, the termination was effective October 3, 1951.

At the same time Ferry was given the task of locating his successor from outside the company. Ferry, upon accepting the presidency on January 1, 1950, did so with his insistence that it was a temporary measure until a successor was located. By February 1952, after reviewing the qualifications and performance records of a number of candidates, the board of directors and Ferry agreed on the one person to whom they would offer the Packard presidency. In May 1952 it was announced that Ferry would relinquish the presidency of Packard Motor Car Company to James J. Nance on June 30, 1952. Nance's previous performance was judged by his great sales and profit success as president and CEO of the Hot Point Division of General Electric Company after World War II, although he lacked automotive experience.

Nance did not make personnel changes immediately; however, Col. Jesse Vincent, executive vice president, retired in early 1952 and was succeeded by LeRoy Spencer. In mid-1952 Spencer resigned and returned to Earle C. Anthony Incorporated in San Francisco, California. Also in mid-1952, Hugh Hitchcock was

James Nance. (Source: P.M.C. Co.)

succeeded as director of advertising and public relations by Patrick Monahan. During Monahan's advertising administration, Nance would make spontaneous and final advertising decisions. Maxon Incorporated was the advertising agency in 1952, until they were replaced by Ruthrauff and Ryan Incorporated late in the year. Hitchcock was given the assignment of handling customer relations.

In early 1952, John Rinehart, chief stylist, was prevailed upon by William C. Ford to join the Ford Design Group in creating and designing the new Continental Mark II. He was offered complete autonomy in design, and a monetary offer he couldn't refuse. Shortly afterward, Charles Phaneuf and Robert Jones joined Rinehart's staff at Ford. Richard A. Teague succeeded John Rinehart as chief stylist at Packard. Subsequently, E.P.J. Cunningham and Franklin Hershey joined Packard's styling staff.

Nance's presence was felt in the changes of personnel in the sales and manufacturing divisions. However, there were no extensive changes in engineering. John Z. DeLorean joined Packard engineering in 1952. Herbert L. Misch returned to passenger-car engineering as assistant chief engineer.

The twenty-sixth series Packard was introduced in November 1952, with modest styling changes in ornamentation, mostly knobs and handles. Given the limited funds for model styling changes, Teague and Cunningham (interior designer) did an excellent job of providing modest styling improvements and quality interiors. One of Nance's mandates for Packard Motor Car Company was that there would be a Clipper line and a Packard line.

	1953 Clipper	1953 Packard Cavalier 300 Series 2602	1953 Packard Patrician 400 Series 2606	1953 Packard Patrician 400 Series 2631
Wheelbase (in.)	122	127	127	122
Price	$2,595	$3,245	$3,740	$3,280
No. of Cylinders / Engine	IL-8	IL-8	IL-8	IL-8
Bore x Stroke (in.)	3.50 × 3.75	3.50 × 4.25	3.50 × 4.25	3.50 × 4.25
Horsepower	150 adv, 39.2 SAE	160 adv, 39.2 SAE	180 adv, 39.2 SAE	160 adv, 39.2 SAE
Body Styles	5	4	4	two-door convertible and two-door hardtop
Other Features				

Author, Michael Kollins, and a 1953 Packard Clipper Deluxe.

The Packard Caribbean sport convertible was introduced at the National Auto Show in New York in January 1953. The Caribbean was, in fact, a production Packard Convertible model modified by the Mitchell-Bentley Corporation in Ionia, Michigan. It featured unique styling, some of which was borrowed from the Packard Pan-American show model of 1952, including the raised front scoop on the hood. It was equipped with 15-in. chromium-plated wire-spoke wheels, with the single spare carried in the rear (Continental style). During 1953 Packard was able to muster domestic sales of 71,079 units, and about 9,000 units through export. This was an upward spurt from about 67,000 units during 1951-52. As a result Packard was able to show a profit for 1953, the last in Packard history.

Not content with the traditional conservative philosophies of Packard, James Nance had his assistant for finance, W.R. Grant, institute sweeping spending on such limited products as seven-passenger sedans, seven-passenger limousines, Caribbean, ambulance and funeral cars, marine, industrial engines, and other "no-win" projects.

With his responsibilities assumed by Grant, E.C. Hoelzle, former vice president and comptroller, retired. W.B. Hoge, assistant comptroller, also retired promptly. In February 1953, William H. Klenke became assistant to E.H. Brodie, vice president for defense, with J.P. Ostrander named manager of marine and industrial sales. O.E. Rodgers was appointed chief engineer for the jet engine division.

With the winding down of the Korean War, Herbert Misch returned to passenger-car engineering as assistant chief engineer. In April 1953, Rodger Bremer succeeded Russell R. Rees as director of purchasing. John Raisbeck left Packard in August 1953 and joined the Kaiser-Frazer Corporation as assistant sales manager. Ray P. Powers, former manufacturing manager of Lincoln-Mercury Division, joined Packard Motor Car Company as vice president of manufacturing, replacing George Reifel who had retired in September 1953.

While Packard had planned to continue having the Briggs Manufacturing Company build Packard bodies, the purchase of twelve Briggs plants by the Chrysler Corporation in November 1953 presented another obstacle for Packard Motor Car Company. Nance had no great affection for L.L. Colbert, so he would not have the Chrysler Corporation building bodies for Packard and Clipper cars. Historically, this should not have been a problem. In the past, Fisher Body Corporation built closed bodies for Chrysler in 1924, and built the two-door coach for Dodge Brothers in 1925. The Seaman Body Corporation (Nash Motors Company) built custom bodies for Cadillac and many other fine cars. Packard built the boat-tail speedster for Cadillac and other fine-car manufacturers.

Thus, Nance, Grant, and Powers hastily negotiated the lease purchase of the Conner Avenue plant from the Chrysler Corporation, in order to make Packard and Clipper bodies. The terms of the contract were not made known, and were not necessarily favorable for Packard. The transfer of the plant took place in early 1954, and by late 1954 it was decided to convert and use the Conner plant for the entire car assembly, abandoning assembly at the East Grand Boulevard plant in Detroit.

In the meantime, in 1954, the design and development work for the 1955 Packard and Clipper was directed by Herbert Misch. Chief Research Engineer Forest McFarland and his engineers, John DeLorean and William Allison, developed the new V-8 engine, Twin Ultramatic Drive, and the Torsion Level suspension.

In late 1953, the Packard Panther, a special prototype of fiberglass construction, was built. During the February 1954 Speed Week at Daytona Beach, Florida, the Packard Panther made its debut with Dick Rathman driving it on the beach at 110.9 mph (officially). This car was powered by a supercharged 359-cu.-in. eight-in-line engine. Three more Packard Panthers were built in 1954.

Merger with Studebaker

On June 22, 1954, a merger agreement between the Studebaker Corporation and Packard Motor Car Company was signed by James J. Nance and Harold S. Vance. The merger created the Studebaker-Packard Corporation, incorporated in Michigan and Indiana on October 1, 1954. Nance was elected chairman of the board of directors. The merger progressed smoothly, with most of the officers retaining their positions in their respective divisions. Clarence E. Briggs remained vice-president and general sales manager, with Roy Abernathy named assistant sales manager, W.E. Macke merchandising manager, and Charles P. Noonan manager of marketing services.

Charles D. Scribner became vice president of industrial relations, succeeding Wayne Brownell. Neil S. Brown was appointed manufacturing manager after the retirement of M.F. Macauley. In November 1954, Roy Abernathy resigned, and later became vice president of sales for the American Motors Corporation. In December B.C. Budd (manager of Packard Export) retired and was succeeded by R.A. Hutchinson, export manager for the Studebaker-Packard Corporation. On January 1, 1955, Roy P. Powers was elected vice president of operations, with Neil Brown appointed general manufacturing manager of body manufacturing and car assembly. Earl M. Douglas was named manager of foundry, engine, and transmission manufacture.

By this time the entire car assembly was moved to the Conner Avenue plant. This became the first attempt at "just in time" material handling and manufacturing methods in the automotive industry, but with dire consequences. Every bit of plant area was used for body building and assembly.

Apparently, Nance was convinced by the vice president of manufacturing that the compact space and area was adequate, and would result in efficient plant operation. It might have been so, if every part and every assembly operation were perfect, but anyone familiar with manufacturing knows that would have been "Utopia." Consequently, the parts components and assembly operations were not perfect, and a repair area was needed, but was unavailable.

Hastily, the Packard management had to acquire repair space in the unused metal shed plant of the Edward G. Budd Manufacturing Company on Conner at Vernor Avenues, just south of the Packard assembly plant. While the Budd plant may have served well for wheel manufacturing or as a stamping plant, it was not ideal for car assembly or repair.

As the cars were built at the assembly plant (Conner at Warren Avenues), those needing repair were driven or towed to the repair plant where the repair operations were performed. Many of these repairs were crude, cobbled, and costly. It may have been that these "botched-up" cars disillusioned the Packard buyers, and the buying public lost confidence in Packard.

On February 1, 1955, Walter R. Grant was elevated to executive vice president. D.C. Gaskin was named president of Studebaker-Packard Corporation of Canada Ltd. On June 1, 1955, William M Schmidt was elected vice president of design and styling, succeeding Edward Macauley who was unceremoniously discharged. C.E. Briggs resigned as vice president of sales on July 1, 1955, and was succeeded by Daniel O'Madigan. On July 15th Briggs was elected vice president of sales of the Chrysler Division, Chrysler Corporation.

During July, Richard A. Teague was elevated to director of styling. In August 1955, George Brodie was elected vice president and general manager of government sales and industrial divisions of the Studebaker-Packard Corporation. Peter S. Barno was named director of industrial relations for the Studebaker division, and Dana Norton was appointed director of industrial relations for the Packard-Clipper division. On October

1956 Packard Predictor. (Source: Studebaker-Packard Corp.)

15th, Robert P. Laughna was elected vice president and general manager of the Packard-Clipper division, while Harold E. Churchill was elected vice president and general manager of the Studebaker division.

The new Packard and Clipper cars were announced on November 1, 1955. Under the direction of Richard Teague, the Packard Predictor (a show car) was built during 1956. Erwin A. Weiss retired as assistant chief engineer on July 15, 1956.

	1956 Packard	**1956 Clipper**
Wheelbase (in.)	127	122
Price	$4,160-5,595	$2,730-3,165
No. of Cylinders / Engine	V-8	V-8
Bore x Stroke (in.)	4.12 × 3.50	4.00 × 3.50
Horsepower	290 (310 Caribbean) adv, 54.3 SAE	240/275 adv, 51.2 SAE
Body Styles	4	3
Other Features	electromechanical gearshift with pushbutton control	

With dismal domestic sales of only 28,396 Packard and Clipper cars for 1956, the Studebaker-Packard Corporation negotiated a management arrangement with the Curtiss-Wright Corporation in July. In this move the Curtiss-Wright Corporation obtained control of the Studebaker-Packard Corporation, and was able to use it as a tax write-off. James Nance resigned on August 1, 1956. Packard and Clipper components and materials were sold off as surplus. Car production was confined to the South Bend, Indiana, plants.

1958 Packard Hawk. (Source: Studebaker-Packard Corp.)

For 1957 and 1958 the Packard line became a badge-change engineered car, built on a Studebaker President chassis and components. Although it was a fine car, it failed to interest potential Packard buyers due to its lack of Packard character. The year 1958 was the last for the Packard crest and nameplate, as the Packard name vanished from the automotive world.

Postscript

Roy Abernathy went on to become president of American Motors Corporation. C.E. Briggs became vice president of Chrysler Corporation. Herbert L. Misch, as engineering vice president of Ford Motor Company, guided Ford Engineering through challenging times. John DeLorean, after becoming vice president of General Motors Corporation as general manager of the Pontiac and Chevrolet divisions, went on to organize the DeLorean Motor Corporation. William H. Graves returned to the University of Michigan as an engineering professor. Fred W. Adams and Richard A. Teague became vice presidents of American Motors Corporation. And Forest McFarland became assistant chief engineer of the Buick Division of General Motors Corporation.

Source: Underwood & Underwood

Edwin Thomas

While the production of the Thomas car did not reach millions, nor even hundreds of thousands, Edwin Thomas built some of the finest engines, motorcycles, and the most powerful cars of the time. He gave generous financial support to fledgling companies that furnished the "roots" to subsequent great automotive corporations. And winning the "Race Around the World" was one of America's most spectacular achievements.

* * *

E.R. Thomas Motor Company

The E.R. Thomas Motor Company was organized on April 25, 1900, by Edwin Ross Thomas for the purpose of manufacturing motor bicycles, motor tricycles, and their components for other manufacturers of motorcycles and automobiles. In addition they offered their own models, the Autobi No. 1, Autobi No. 2, the three-wheeled Autotri roadster, and the Autotwo. All Thomas vehicles were powered by the famous Thomas air-cooled engines.

	Autobi No. 1	Autobi No. 2	Autotri	Autotwo
Hp	2.25	1.5	3	3
Weight (lb)	95	75	90	100
Price	$250	$200	$350	$400

In late 1902 the E.R. Thomas Motor Company announced their first automobile, the 1903 Thomas Models 17 and 18. The cooling radiator was located up front in a sloping position. The engine power was transmitted

1901 Thomas Autobi No. 2 tandem.

by a lined cone clutch to the three-speed sliding-gear transmission, to the bevel gears and chain drive, through the roller-bearing axle to the rear wheels. The chassis suspension was by four parallel fully elliptical springs.

1903 Thomas Models 17 and 18

Wheelbase (in.)	78
Price	$1,250-1,400
No. of Cylinders / Engine	1
Bore x Stroke (in.)	5.00 × 6.00
Hp	8 ALAM
Body Styles	tonneau and touring
Other Features	water-cooled engine

The 1904 Thomas Flyer Models 22 and 23 were announced in November 1903.

	1904 Thomas Flyer Model 22	1904 Thomas Flyer Model 23
Wheelbase (in.)	84	92
Price	$2,650	$3,000
No. of Cylinders / Engine	3	3
Bore x Stroke (in.)	4.50 × 5.50	4.50 × 5.50
Hp	24 ALAM	24 ALAM
Body Styles	tonneau	limousine
Other Features		

The engine's three cylinders were cast separately, mounted on a crankcase of cast aluminum. The crankpins were set 120 degrees apart. The crankshaft was of drop-forged steel, carried on four babbited main bearings. At fairly high prices, the Thomas car sales were very limited.

1902 Thomas.

The new 1905 Thomas Flyer models 25, 26, 27, 29, and 30, were announced in November 1904.

1905 Thomas Flyers

Wheelbase (in.)	106/110/114/124
Price	$3,000-7,000
No. of Cylinders / Engine	T-4 / T-6
Bore x Stroke (in.)	5.00 × 5.50 / 5.50 × 5.50
Hp	40/50/60 adv, 40/43.8 ALAM
Body Styles	6
Other Features	

While the 60-hp engine was offered in the Model 27 touring and the Model 30 limousine, the primary reason for the development of the 60-hp, six-cylinder engine was E.R. Thomas' personal interest in participating in racing events. The Thomas Flyer racing car ran successfully in the Vanderbilt Cup race, the French Grand Prix, and many other notable events, including the Fairmount Park Race in Philadelphia.

For 1906 the Thomas Flyer was confined to one line, with Models 31, 32, 33, and 34.

1906 Thomas Flyer

Wheelbase (in.)	118
Price	$3,500-4,600
No. of Cylinders / Engine	T-4
Bore x Stroke (in.)	5.50 × 5.50
Hp	50 adv, 43.8 ALAM
Body Styles	touring, seven-passenger limousine, seven-passenger landaulet, seven-passenger semi-limousine
Other Features	

Thomas stock car in Fairmount Park Race, October 10, 1909. Haupt, driver, and Tom Wilkie, mechanic.

The main new features of the 1906 Thomas Flyer were the new dry-disc clutch, replacing the former lined-cone type, and a new four-speed speed-change gearbox (transmission) with an interlocking mechanism to prevent the possibility of getting into two gears at the same time. The ignition was of the high-tension type, featuring a synchronized system, a high-tension timer combined with a low-tension interrupter. A single coil was used, the primary current from a storage battery carried on the outside step. The production and sales of the 1906 Thomas Flyer amounted to approximately 500 cars.

Thomas-Detroit Company

Another important event happened in 1906, when Roy D. Chapin and Howard E. Coffin went to the Pacific coast to try to obtain financial support to organize their next venture. Their search was fruitless, so they decided to return to Detroit by train. However, on April 17, 1906 (the day after the tragic San Francisco earthquake), in the dining car of the Union Pacific train, they met Edwin Thomas. After several hours of Chapin's salesmanship, they were able to convince Thomas to provide financial support to organize the Thomas-Detroit Company, to build a less-costly version of the Thomas Flyer. It was arranged that the Thomas-Detroit car would be merchandised by Thomas dealers.

Edwin R. Thomas was elected president of the Thomas-Detroit Company, and Roy Chapin and Howard Coffin vice presidents. The Thomas-Detroit Company located a factory building, installed the machinery and tooling, and was in production by the end of 1906. The Thomas-Detroit car was essentially a downsized Thomas Flyer.

Thomas-Detroit

Wheelbase (in.)	110
Price	$1,500
No. of Cylinders / Engine	T-4
Bore x Stroke (in.)	5.00 × 4.75
Hp	40 ALAM
Body Styles	touring and roadster
Other Features	

General view of the Thomas-Detroit assembling floor showing arrangement of chassis. (Source: NAHC)

The car was successful, as over 500 Thomas Detroit cars were built and sold during the first six months of 1907. However, the Thomas-Detroit was not exactly the car that Chapin and Coffin had in mind. They convinced Hugh Chalmers to buy Edwin Thomas' interest in the Thomas-Detroit Company. In July 1908, the Thomas-Detroit Company became the Chalmers-Detroit Company, and became the Chalmers Motor Company in 1910. Impatient as Chapin and Coffin were, they organized the Hudson Motor Car Company on February 24, 1909, with Hugh Chalmers' blessing. (See Chalmers' story in Volume 3, and the chapter on the Hudson Motor Car Company in this volume.)

The Taxicab

For 1907 the E.R. Thomas Motor Company introduced two separate model lines: the Thomas 40 and the Thomas Flyer.

	1907 Thomas 40	**1907 Thomas Flyer**
Wheelbase (in.)	112.5	118
Price	$2,750	$4,000-5,200
No. of Cylinders / Engine	T-4	T-4
Bore x Stroke (in.)	3.62 × 4.13	5.75 × 5.50
Hp	40 adv, 21 ALAM	60 adv, 53 ALAM
Body Styles	3	6
Other Features		

On May 8, 1907, Thomas introduced a new "taxicab" Model G, built on a 103-in. wheelbase chassis, powered by a new, smaller, four-cylinder, T-head engine, rated at 20 hp ALAM. A new shaft drive was introduced. The body was a semi-limousine designed by Gustave Chedru, head of a French designing company. The taxicab sales became a lucrative source of income for the E.R. Thomas Motor Company, and was administered as a separate company operation.

Due to the great demand for a high-powered Thomas Flyer runabout, Thomas introduced such a vehicle on May 22, 1907, built on a 118-in. wheelbase chassis, using the powerful, four-cylinder, T-head engine rated at 60 hp ALAM, and having a five-main-bearing crankshaft. The ignition was by two separate systems: one by Bosch magneto, and the other by Atwater-Kent battery-powered coil and distributor, firing two sets of spark plugs. Approximately 700 Thomas cars were built and sold in 1907.

Every season the E.R. Thomas Motor Company added something new, something progressive, something above the ordinary lines, and 1908 was no exception. They entered the new season with a proliferation of cars, ranging in wheelbase lengths from 103 in. on the Model G to 140 in. on the Model K. Engine horsepower ranged from 20 (Model G) up to 70 for the reinstated six-cylinder, T-head engine of the Model K. The body styles included runabouts, tourabouts, touring cars, cabriolets, landaulets, and limousines; a line-up to satisfy any requirement. One of the outstanding new models was the 4-20 Model G town car designed by Gustave Chedru of France.

The Model 4-60 Thomas Flyer was redesigned by Howard Coffin and Gustave Chedru, and while the specifications remained the same as the 1907 model, the Flyer was much faster and smoother, and it was lower and racier in appearance to meet potential buyer expectations.

The 1908 Thomas, priced from $3,000 to $6,000, offered 21 body styles: six in the Thomas Model G (including the taxicab), five in the Model DX, five in the Model F, and five in the Model K, powered by the 70-hp, six-cylinder, T-head engine. Over 800 Thomas cars were built and sold in 1908.

During 1908 Thomas Flyer racing cars, designed by George Salzman, chief engineer, participated in the French Grand Prix, the Vanderbilt Cup race, and many notable contests, the most spectacular of which was the "Race Around the World" in 1908.

Thomas Flyer 4-60 touring.
(Source: NAHC)

1907 Thomas Model 36 "New York to Paris" car. (Source: Smithsonian Institution)

The Trip Around the World

The automobile industry has always had a full appreciation of publicity. It was the golden profit to be reaped in publicity and was the prime mover for almost all factory participation in racing. Since the beginning of the automobile industry, there had been literally thousands of contests promoted with the single purpose of gaining newspaper notoriety. There had been transcontinental trips by Packards, Wintons, Reos, and others in the early days. There was a transcontinental trip by a dozen Premiers in a caravan, by Whites and many others at later dates. They all had one goal in mind and they achieved it in greater or modest measure, depending on the time, when the contest was promoted, and other conditions.

The most prominent contest of all was the "Race Around the World," scheduled to be run from New York to San Francisco (4,300 miles), then ship to Valdez, Alaska, where cars were to be driven 1,200 miles to Nome, Alaska. From Nome the cars were to be shipped to East Cape, Siberia, and from there they were to be driven cross-country to Petrograd, Russia, and thence through Berlin, Germany, to Paris, France. The total land mileage was 11,350 miles.

The start of the race was from Times Square, New York City, on February 12, 1908, and was witnessed by a quarter-million people. There were six participants:

Car	Driver(s)	Mechanic	Extra
DeDion Bouton (French)	Boursier St. Chaffray	Hans Hansen	Mons. Autran
Moto Bloc (French)	Charles Godard	Hue Aethur	Maurice Livier
Sizaire-Naudin (French)	August Pons	Lucien Deschamps	Maurice Berthe
Zust (Italian)	Antonio Scarfoglio	Emilie Sartori	Henri Haaga
Protos (German)	Lt. H. Koeppen	Hans Knape	Ernst Mass
Thomas (American)	George Schuster	George Miller	Montague Roberts

Crew of the Thomas Flyer: George Schuster, driver, George Miller, Montague Roberts. Correspondent George McAdam in back seat with checkered cap. (Source: Motor World)

The Thomas entry was definitely a production stock car, powered by a four-cylinder engine rated at 60 hp. The others were all more or less within the stock car classification, but had more special equipment and work than the American car. In the midst of the cruel northern winter, reports declared an even more severe season in the western mountains.

The start of the race was delayed by 1 hour and 15 minutes because Mayor McLellan, who was to fire the starting pistol, had not appeared at his special grandstand by 11 a.m. After waiting 15 minutes more, Colgate Hoyt, president of the Automobile Club of America, fired the starting pistol at 11:15 a.m. The contestants proceeded up crowd-lined Broadway and the race was on its way.

The Sizaire-Naudin had gone as far as it could at Red Hook, New York; Moto-Bloc failed in Cedar Rapids, Iowa; the Protos loaded their car on the Oregon Railway and Navigation's rail car at Pocatello, Idaho, instead of driving under its own power. But, the Thomas, Zust, and DeDion reached the Pacific Coast under their own power, with the Thomas covering the route in 42 days. This was 11 days fewer than the Zust, and 14 days fewer than the DeDion.

No automobile had ever made the short journey of 1,200 miles from Valdez to Nome, Alaska, where there were no roads and hardly any trails. In the condition of the terrain at the time of the race, Alaska was perhaps the most impossible area to traverse anywhere on earth. Anyone who had visited Alaska before that time would know that the trail between Valdez and Nome was impossible for automobiles. But that fact remained for the Thomas crew to discover after their car had been shipped to Valdez from the Pacific Coast. After making a thorough personal inspection of 10 miles of the proposed route to Nome, George Schuster wisely decided that neither the Thomas car nor any other car could possibly make that drive. The Alaska portion of the race had to be eliminated. The Thomas car was returned to Seattle, Washington, where it was learned that the other contestants had already sailed for Vladivostok, without waiting for the return of the Thomas group from Alaska.

Thomas team encounters blizzard during "Race Around the World." (Source: Motor World)

The Thomas group took a cargo ship from Seattle and landed at Yokohama, Japan, and toured 350 miles of the Japan country. From Japan the Thomas car was shipped to Vladivostok, and from there they crossed Siberia, encountering difficulty in obtaining enough gasoline. They crossed Siberia, Russia, Poland, Germany, and France to Paris, making the 8,280 miles from Vladivostok to Paris in 49 days running time. The Thomas car reached Paris on July 30, 1908, making the final sprint of the journey at 50 miles per hour. The Thomas car had covered the land distance of approximately 11,350 miles in 170 days elapsed time; the daily average run was 152 miles, and the longest day's run was 420 miles. The Thomas car had covered 2,385 land miles in excess of the mileage of any other car entered.

Considering the condition of the roads, weather, and difficulties of making such a trip at that period of automobile development, the feat stands out as the most brilliant and extraordinary achievement ever carried out in the automobile industry. The single fact that three automobiles were able to run under their own power from New York to San Francisco at high speeds shed a new light on the sturdiness and reliability of the American motor car, and gave it tremendous publicity in public newsprint. How the passengers managed to ride is a mystery when one thinks of the thousands of miles riding in a tonneau seat that was insecure, uncomfortable, and dangerous, considering the Siberian route.

The Spoils of Victory

In the fall of 1908, the E.R. Thomas Motor Company announced their 1909 models. The G, F, and K were carried over into 1909, with refinements and new added features. The Model DX was discontinued, but a new Model L was added.

Thomas team in Siberia in 1908. (Source: Motor World)

1909 Thomas Model L

Wheelbase (in.)	122
Price	$3,000-4,500
No. of Cylinders / Engine	T-6
Bore x Stroke (in.)	4.25 × 5.50
Hp	40 adv, 43.35 ALAM
Body Styles	touring, Flyabout, Tourabout, limousine
Other Features	

The full Thomas line, available in 15 body types and a taxicab, was priced from $3,000 to $7,500. Apparently, winning the "Race Around the World" gave E.R. Thomas Motor Company a boost in sales, because 1,036 Thomas cars were built and sold in 1909.

For 1910, the 1909 Thomas models were carried over, but given new model designations. The G became the Model R-4-28, the L became the Model M-6-40, the F became the F-4-60, and the K was called the K-6-70. The Thomas cars were priced at $3,500 to $7,500. Sales momentum continued as approximately 1,000 Thomas cars were built and sold in 1910.

Trouble Begins

While the economic depression of 1910 may have had some effect on automobile sales in general, perhaps the Thomas management "rested on their oars" and expected the sales momentum to carry them through 1911. It did not, and the Thomas sales fell to a dismal low, with expenditures as high as ever. Edwin R. Thomas found himself in a quandary. He resolved his personal problem, by selling his interest in the E.R. Thomas Motor Company to Eugene Mayer and Company, New York bankers. But, Thomas wisely held on to the taxicab entity. The E.R. Thomas Motor Company was reorganized and the name was changed to E.R. Thomas Motor Car Company. The model line-up for 1911 did not change, except the E-4-30 superseded the R-4-28, and the price of the K-6-70 landaulet was increased to $7,600.

Pioneers of the U.S. Automobile Industry, Volume 2: The Small Independents

1910 Thomas Tourabout 6-40.

1910 Thomas Model K-6-70. (Source: NAHC)

1910 Thomas Flyers at Auto Show. (Source: NAHC)

1911 Thomas exhibit at Auto Show. (Source: NAHC)

Shortly after the acquisition of E.R. Thomas' interests in February 1911, the Eugene Mayer and Company of New York proudly bragged that Mayer had induced several of the Packard Motor Car Company executives to cast their lot with the E.R. Thomas Motor Car Company. E.L. Chalfant was made president, F.R. Humpage vice president, W.L. Gleason factory manager, and J.J. Ramsey treasurer. At the time of reorganization, several E.R. Thomas Motor Company key executives were discharged. However, all of this came to naught because, by February 1912, Chalfant had resigned and was succeeded by Humpage as president, and on August 29, 1912, the E.R. Thomas Motor Car Company was placed in receivership.

The financial difficulties that plagued the E.R. Thomas Motor Car Company for more than a year came to a conclusion on August 29, 1912, when Judge Hazel in the Federal Court in Buffalo, New York, placed the concern in the hands of George G. Finley and Adolph Rebadow as receivers, under bonds of $50,000, with instructions to continue the business of the company. Within the previous few weeks, however, it was reported that the production of the plant had stopped. In August 1912 three of the executives who left the Packard Motor Car Company to join E.R. Thomas Motor Car Company when it was reorganized had retired. They were president Humpage, vice president Gleason, treasurer Ramsey; sales manager Fitzsimons also resigned. On September 2, 1912, the Judge ordered that the assets of the E.R. Thomas Motor Car Company be sold at auction.

The auction of the assets of the E.R. Thomas Motor Car Company opened on March 17, 1913, with about 400 bidders in attendance. Fifteen of the Thomas cars were sold at $1,900 each. These and most of the other items were bought by C.A. Finnegan, president of the Empire Smelting Company. Finnegan officially bid and paid $51,000 for these items as one lot of Thomas' material. He paid an additional $5,360 for machinery, tools, lathes, and other material not included in the first lot. The total realized by the auction of the Thomas' assets was $256,400, a mere pittance compared to the investment.

1911 Thomas 6-70 Model K limousine. (Source: NAHC)

1912 Thomas 6-70 seven-passenger touring. (Source: NAHC)

Finnegan also bought the victorious "Race Around the World" Thomas Flyer race car for $200, and the famous New York to Paris trophy for $300. Ironically, the trophy was sold for more than the racer that won it.

Source: R.E. Olds Museum

Ransom Olds

Ransom Olds not only established the first mass-production line method of automobile manufacturing, he left a legacy from which all of mankind will continue to benefit. To his faithful associates he was not only the employer, but their mentor and tutor. Many of his "graduates" became great automotive pioneers in their own right.

* * *

Roots in Engines

Ransom E. Olds was born on June 3, 1864, the youngest son of Pliny Fisk and Sarah (Whipple) Olds. His father was a blacksmith by trade, and when Ransom was six years old, the family moved to Cleveland, Ohio, where Pliny Olds became employed as superintendent of the Variety Iron Works of Cleveland. Subsequent moves and trading of homes and properties brought the Olds family to Lansing, Michigan. Selling the vacant lot next to their home provided Pliny Olds with enough funds by 1880 to start a small engine shop on River Street in Lansing, known as the P.F. Olds and Sons. Ransom and his brother Wallace learned the engine-building trade by working alongside their father, who was also their business partner. In 1885, by the time Ransom was 20 years old, he had saved enough money to buy Wallace's share of the business.

By 1887 Ransom Olds was prosperous enough to afford experimenting with steam-powered road vehicles. Between 1887 and 1890 the Olds business changed from steam-powered vehicles to gasoline-engine-powered vehicles. In 1890 Ransom Olds bought out his father's interest in the business, and started experimenting with internal-combustion engines. He soon developed reliable and popular gasoline-fueled engines for farms, mills, mines, and marine use. By the time he was 30 years of age, he had organized the

Pliny and Sarah Olds. (Source: R.E. Olds History Center)

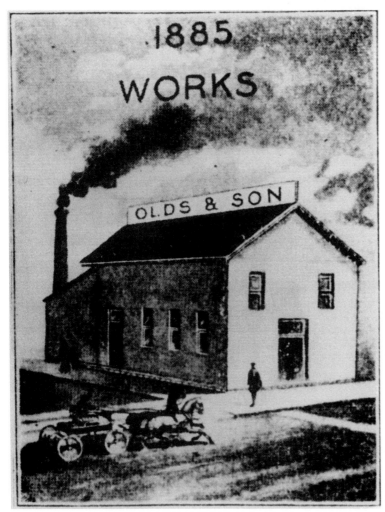

First shop of P.F. Olds and Sons. (Source: R.E. Olds History Center)

Olds gasoline engine advertisement. (Source: R.E. Olds History Center)

Olds Engine Works. The former, small, two-story engine shop was replaced by a modern, large factory with a foundry, pattern shop, and machine shop. The subsequent success of Olds transformed Lansing into the nation's leading gasoline-engine manufacturing center.

In 1895 Ransom Olds and his friend, Frank G. Clark, the son of a Lansing carriage builder, started construction of a gasoline-engine-powered vehicle, which they had intended to compete in the *Chicago Times-Herald* "Chicago to Evanston" race to be run in the fall of 1895. However, because of ensuing problems and delays, Olds and Clark did not complete their vehicle until a year later.

Stock of Olds gasoline engines. (Source: R.E. Olds History Center)

The Olds Motor Vehicle Company Incorporated was organized on August 21, 1897, with a capitalization of $50,000 and Ransom E. Olds as manager. By order of the board of directors, the goal of the company was to build a vehicle as nearly perfect as possible. Olds and his associates had successfully tested their first car prior to August 1897. Olds built four cars that year, of which three were sold. However, after that, Olds concentrated on building gasoline-fueled engines.

A Meeting With Destiny

The engine-building business was excellent, with sales outstripping production facilities. More capital was needed for expansion than could be raised in Lansing. Thus, Ransom E. Olds went to Newark, New Jersey, with the hope of raising the needed capital there. The Chamber of Commerce Group there, while trying to induce industry to locate in New Jersey, could not be convinced to invest in the automotive industry. Disappointed, dejected, and tired, Ransom Olds was waiting in the railroad station in Detroit to change trains for Lansing. A chance meeting in the station with an old friend and neighbor of his father, Samuel L. Smith, changed the course of events and the subsequent future of the automobile industry. Smith, who had previously been successful in the mining and lumbering business in Michigan, had retired and was now living in Detroit. He was understanding and listened to Olds as he told of his disappointment and failure to interest eastern bankers. Smith had previously invested in the Olds engine venture, and was pleased with how well it paid off.

That chance meeting in the Detroit railroad station subsequently shaped the destiny and future of Detroit and Michigan. Smith's confidence in the sincerity and sagacity of Ransom Olds was all that was needed to convince other financial investors that it would be wise to invest in a venture to be called the Olds Motor Works. However, one of the stipulations of the arrangement was that Smith's sons, Angus and Fred, would have important positions in the company. The Olds Motor Works was incorporated under the statutes of the laws of the state of Michigan on May 8, 1899, "for the purpose of manufacturing all kinds of machinery, engines, motors, carriages, and all kind of appliances therewith." The capitalization was $500,000 with

R.E. Olds (middle) in 1897.
(Source: R.E. Olds Museum)

20,000 shares of stock issued. Samuel Smith bought 19,960 shares, Fred Smith (Sam's son) 10 shares, Ransom E. Olds 10 shares, Edward Sparrow 10 shares, and James Seager 10 shares. Henry Russell later joined the Olds Motor Works.

Olds Motor Works

The Olds Motor Works obtained the assets of the Olds Engine Works of Lansing, and the Olds Motor Vehicle Company of Lansing which was dissolved in February 1900. Ransom Olds invested $400 cash into the venture. He received $50,000 in cash and $75,000 in Olds Motor Works stock for his Lansing businesses. Samuel Smith was elected president, Ransom Olds vice president and general manager. Olds also kept watch over the bookkeeping and expenditures, until Fred Smith was appointed secretary-treasurer.

Shortly after organization, construction began on the new plant at 1200 East Jefferson Avenue (corner of Helen Street) near the Belle Isle bridge. While most historical accounts of Olds' new factory in Detroit seem to indicate that the factory was designed and intended for the manufacture of automobiles, those accounts are not true; the plant was constructed primarily for the purpose of manufacturing engines more efficiently, and in greater numbers than could be done in Lansing.

In 1883, Charles Brady King, an automotive pioneer in his own right, designed and built a horizontal, two-cylinder, water-cooled, gasoline engine that was very efficient. Shortly afterward, King designed and built a cast-en-bloc four-cylinder engine that powered the first vehicle to be driven by King on the streets of Detroit on March 6, 1896. King joined the Olds Motor Works in 1900 after his release to inactive duty from the

United States Naval Reserve at the end of the Spanish-American War (1898-1900). King brought to Olds Motor Works his engineering skill, inventiveness, and some of his own engine designs, as well as the patent rights to the King Marine engine. Prior to his active duty in the United States Naval Reserve, King had been a well-established and prosperous engine builder in his own shop on St. Antoine Street in Detroit, and upon return to civilian life, King sold his patent rights on railway brakes and pneumatic tools to Chicago-based firms. (See the King chapter in Volume 4.)

Curved-Dash Runabout

Because of the widespread interest in horseless carriages, a few experimental vehicles were built at the factory. But when Samuel Smith learned that approximately $80,000 was spent in the development of one large gasoline-engine-powered automobile, he exercised his authority as president of the company and had the experiments stopped. Only by his sincere persuasion was Olds able to convince Smith to allow them to continue with the experiments and complete the small runabout, which Olds believed could be sold for $600. Thus, was the venture that led to the creation of the famous "curved-dash" Oldsmobile. Contrary to the many published automotive historical accounts, these experimental vehicles were only side projects in 1900 at the Olds Motor Works, which primarily built gasoline engines.

The Fire

The destiny of Olds Motor Works was decided on March 9, 1901. While most employees were out to lunch, disaster struck at the Olds Motor Works plant. An unattended tar pot ignited and started a blazing inferno that rapidly engulfed and destroyed the factory. By strange coincidence James J. Brady, the youthful timekeeper, was on the Jefferson Avenue streetcar near the Olds factory. Even though it was his day off, Brady, thinking only of the experimental cars in the corner of the factory, leapt off the street car and rushed into the blazing building, ignoring the orders of the police. By his intrepid courage and clear thinking, he was able to push one runabout to safety. His act of bravery saved one curved-dash runabout which was duplicated and became the most prized of youthful admirers. Even though all blueprints and drawings were destroyed, saving the one vehicle made it possible to duplicate the car. (In later years Brady organized the "Goodfellow Old Newsboys" so that no child would be without a gift for Christmas.)

Line-up of Olds prototypes in 1900, before the fire. (Source: R.E. Olds History Center)

Olds Motor Works plant fire, March 9, 1901. (Source: R.E. Olds History Center)

The First Subcontractor and Assembly Line

Ransom Olds was away from the plant at the time of the fire, and heard of the destruction later that day. Informed that the only property saved was the curved-dash runabout, Olds decided that the fastest way to get back into the automobile business was to make duplicates of the saved runabout. Through his sincerity and trustworthy reasoning, Olds was able to convince his financial supporters of his plans. To manufacture a car without a factory to manufacture the parts could be done only by improvising and innovating new methods of production. By subcontracting, Olds set the standard for the future of automobile production.

Since all drawings, specifications, and assembly lists were lost in the fire, the first step toward manufacturing was the complete disassembly of the runabout, even to the smallest detail parts. New drawings, prints, bills of material, and assembly instructions were prepared and quickly processed. Patterns, templates, and assembly jigs were promptly made. Engineering, purchasing, and production personnel then explored the city of Detroit for foundry, machine shop facilities, metal fabricators, and manufacturers of automotive parts, to build parts and components for the Olds Motor Works.

1901 Oldsmobile being driven up a steep grade by R.E. Olds (left) and Jonathan Maxwell. (Source: R.E. Olds Museum)

While one group of Olds personnel was seeking production capabilities another group was arranging an improvised method for car assembly. They swiftly cleared out the foundry building, which was not destroyed by the fire, and readied a small assembly line. The line consisted of wooden tables and stands mounted on wheeled dollies so they could be moved from one work station to the next as the parts and components brought in from outside suppliers were installed and assembled progressively in proper sequence to become a completed vehicle.

By Monday of the week following the fire, the work of making parts for the curved-dash runabout started in many of Detroit's small fabricating shops and factories. Many of the manufacturers and suppliers who participated in this innovative venture later became prominent automotive pioneers in their own right. Some of those who made the production of the curved-dash Oldsmobile runabout a reality were: Frederick O. Bezner (see the Chapin, *et al.* chapter in this volume), James J. Brady, Walter O. Briggs, Benjamin and Frank Briscoe (see Volume 3), David D. Buick (see Volume 4), Roy D. Chapin (see the Chapin, *et al.* chapter in this volume), Howard E. Coffin (see the Chapin, *et al.* chapter in this volume), Byron F. Everitt, Carl G. Fisher (see the Allison, *et al.* chapter in Volume 3), Fisher Brothers of Norwalk, Ohio, Charles D. Hastings (see the Hupp, *et al.* chapter in this volume), Clarence Hayes, George and Earl Holley, Robert C. Hupp (see the Hupp, *et al.* chapter in this volume), Roscoe B. Jackson (see the Chapin, *et al.* chapter in this volume), John Kelsey, Charles B. King (see Volume 4), Henry M. and Wilfred Leland (see Volume 3), Jonathan D. Maxwell (see Volume 4), Charles B. and David Wilson.

Lansing Comes Calling

In 1900 when Ransom Olds decided to move from Lansing and build his factory in Detroit, the first reaction of the Lansing citizens was one of indifference. However, after the Olds factory fire in Detroit and Olds' subsequent success with the curved-dash runabout, they had second thoughts. At an emergency meeting of the Lansing Businessmen's Association, a gentleman named Harris was selected to go to Detroit to try to persuade Ransom Olds to return to Lansing, Michigan.

While the improvised facilities were being arranged and the production process being worked out, R.E. Olds was visited by Harris representing the Lansing Businessmen's Association. Harris tried to convince Olds to move back to Lansing if an appropriate location could be found. Harris offered Olds (entirely on his own, but in the name of the Association) that it would buy and give to Olds the site formerly occupied by the Michigan State Fair, which had moved to Detroit for fiscal reasons.

Upon Harris' return to Lansing, his associates were pleased to learn that Olds would return to Lansing. However, they were astounded and shocked when the conditions were revealed, that Harris had offered and agreed to buy and give Olds the former site of the Michigan State Fair. After they recovered from the shock and composed themselves, the Association voted to approve Harris' action. They also agreed to contribute the money needed to buy the site from the bankers to whom the State Fair management was indebted. Hence, while the Olds managerial organization was taking place, a new factory was being erected for Olds car production in Lansing. Approximately 425 curved-dash Olds runabouts were built in 1901.

Roy Chapin

Roy Chapin, born in Lansing, Michigan, in 1880, was at this time in 1901 an engineering student at the University of Michigan. Chapin joined the Olds Motor Works during the reconstruction in Detroit after the fire. At $35 per month, Chapin worked as a gear filer under the supervision of Jonathan Maxwell, who helped the Apperson Brothers build the first car for Elwood Haynes in 1894.

Chapin, a 21-year-old with a much better than average education, was not dismayed by the dismal task of filing gears even though he was capable of writing good advertising copy and was a good photographer. By the fall of 1901, when Olds Motor Works was ready to display their car at the second annual National Automobile Show in New York, Chapin shocked R.E. Olds by offering to drive the curved-dash Olds runabout from Detroit to the New York Auto Show. After some thought Olds remarked, "Why not?" Chapin left Detroit on October 27th, arriving in New York on November 5, 1901. After the strenuous trip, and a bath for Chapin and the Olds runabout, it was admitted to the show.

Roy Chapin. (Source: R.E. Olds History Center)

Chapin's qualifications as an advertising writer and photographer were soon recognized and had a great influence on future automotive advertising in the United States. The first automotive ad to appear in a general weekly publication appeared in the February 15, 1902, issue of the *Saturday Evening Post*. The ad had Olds advertising copy and a car photograph by Roy Chapin, the work of a young man whose first camera was a simple black box with a pinhole instead of a lens. Chapin progressively worked his way up to become sales manager in 1906, before he left Olds Motor Works to organize the Thomas-Detroit Company.

Chapin's enthusiasm and ardent zeal for the automotive industry influenced his friends and classmates at the University of Michigan to join the Olds Motor Works. Howard Coffin joined Olds in 1902 as an engineer, and Roscoe Jackson as secretary. Through their influence they induced Frederick Bezner to leave the security of his executive position at the National Cash Register Company of Dayton, Ohio, to join them at Olds. This started the accelerated exodus of qualified, energetic, business executives from Dayton, Ohio, to the automotive industry in Detroit, Michigan, including Hugh Chalmers and Alvan Macauley.

Chapin's strenuous drive from Detroit to New York in November 1901 had a great impact on him; he struggled across almost roadless Ontario, Canada, to reach Niagara Falls, and then across New York State, where the driving conditions were even worse than those in Ontario. This arduous and tiring trip firmly implanted in Chapin's mind one serious fact that most automotive pioneers did not recognize, or chose to ignore: The lack of passable roads in our country would seriously limit or curtail the use of motor vehicles, and could, in effect, limit the sale of motor vehicles to urban areas only.

In 1901 all the roads in our nation, if laid end to end, would not cover the distance of Post Road from New York to Boston. It must be remembered that the Boston Post Road had been in existence since Benjamin Franklin laid it out, measured it, and set up mileposts when he was Royal Postmaster General of the American Colonies. This fact led Chapin to become one of the most ardent workers and campaigners for good roads in the United States.

Chapin contributed much to the cause of establishing the Bureau of Public Roads, a federal agency. While many politicians would have us believe that the fine network of interstate highways in the United States is totally of their doing, the fact is that businessmen, industrialists, and corporations were primarily responsible for the beginning of the enormous task of uniting our disunited states with adequate highway transport

1903 one-cylinder curved-dash Oldsmobile. (Source: Smithsonian Institution)

routes. Chapin's work for American highways was finally and officially recognized at the International Road Congress in 1930.

Olds Leaves Motor Works

Olds had a successful year in 1901, as 425 curved-dash runabouts were built and sold. In 1902 production and sales jumped to approximately 2,500 vehicles and to almost 4,000 in 1903. Even though the curved-dash runabout brought success to Olds Motor Works, and Olds' relationship with Samuel Smith always remained cordial and friendly, there were others in financial control (Smith's sons) who were enamored with big expensive cars. This led to contention, disagreements, and rift between them and R.E. Olds.

Ransom Olds left the Olds Motor Works at the end of 1903, after 20 years of hard work and "making his million." Samuel Smith retired after Olds left, and Henry Russell was elected president, Fred Smith was named vice president and general manager, and Angus Smith was appointed secretary-treasurer.

Olds exhibit at the 1903 New York Auto Show. (Source: Motor World)

In less than three years the Olds Motor Works venture gave a full return of all the investors' money through dividends, and the value of their stock holdings multiplied five-fold. Those who invested in Olds Motor Works stock and held it until 1908, shared in the distribution of $3,000,000 and General Motors stock when Olds Motor Works was purchased by General Motors in December 1908.

The departure of Ransom Olds dampened the spirits of the capable and skilled personnel left behind. While all the ardent, energetic young leaders such as Chapin, Coffin, Brady, Bezner, and Jackson worked as hard as ever, the future seemed to grow more dim. Roy Chapin was appointed sales manager in 1904, and for a short while their hopes and morale got a lift. Chapin criss-crossed the nation, signing up hardware merchants, implement stores, bicycle dealers, wagon and carriage dealers, saddlery and harness dealers, and sporting goods stores to sell Oldsmobile cars. Even though the sales of the curved-dash runabout kept climbing to a record 6,500 cars in 1905, and profits were increasing, the management unwittingly used most of the profits to overcome the ever-increasing losses from the manufacture and sale of the big cars. The curved-dash runabout was discontinued after the 1907 model year.

By 1906 the "four horsemen"—Chapin, Coffin, Brady, and Bezner—were tired of struggling to make their way in life, and the insecurity of Olds Motor Works induced them to form a pact of "all for one, and one for all," with their aim to become millionaires, which they accomplished in less than 10 years. They knew their first move had to be to leave Olds Motor Works. Chapin tendered his resignation on February 14, 1906, with the others following shortly after.

The obstinate belief that big profits were generated only by the manufacture and sale of big, expensive cars to wealthy and affluent people, had led many successful, thriving, small-car manufacturers down the road to bankruptcy and financial ruin. The massive Oldsmobile Model M touring car and the Model MR roadster were awesome to behold and exciting to drive, but were anything but profit makers. The prices spiraled from $650 for the curved-dash runabout in 1903, to $2,750 for the Model M touring and Model MR roadster, and $4,200 for the Model Z touring in July 1908. The stubborn insistence of Olds Motor Works financial backers and Samuel Smith's sons to build big cars nearly ruined Olds Motor Works, as sales dropped to 1,055 cars and profits eroded and slipped into losses. They were finally rescued by the newly formed General Motors Company in 1908.

Reo Motor Car Company

Meanwhile, in 1904, Ransom Olds had gone on a much-deserved vacation to the upper peninsula of Michigan. Upon his return to Lansing later in the year, much to his surprise he learned that the businessmen of Lansing had pooled their financial resources to organize a new company, and they asked Olds to be president. Their persuasive arguments convinced Olds to join them. The result was the R.E. Olds Company, organized and incorporated in Lansing, Michigan, in August 1904. The ground for the new plant was broken September 19th. On October 5th the name was changed to Reo Motor Car Company (representing Olds' initials) because the name Olds was already in use.

As soon as it was revealed that R.E. Olds would be heading up a new company, many of his former associates at Olds Motor Works sought to join him, including: Richard H. Scott, former superintendent of Olds Motor Works, who joined Reo as plant superintendent; Horace T. Thomas, graduate of Michigan State University and former engineer at Olds Motor Works, as chief engineer of Reo Motor Car Company, already working on the design of a new car; Raymond M. Owen, who with his brother Ralph had built a gasoline-engine-powered delivery vehicle for their dry-cleaning business in 1899, former secretary of Olds Motor Works, as general sales manager (see the Owen chapter in Volume 4).

1905 Reo. (Source: R.E. Olds Museum)

The first Reo car was completed around mid-October 1904, and tested for the balance of the year. It was introduced at the National Automobile Show in New York in January 1905.

1905 Reo

Wheelbase (in.)	72
Price	$1,250
No. of Cylinders / Engine	2
Bore x Stroke (in.)	4.50 × 6.00
Horsepower	16.2 ALAM
Body Styles	four-passenger detachable tonneau
Other Features	

The power was transmitted through a two-speed planetary transmission, a 1-1/4-in. by 5/8-in. Diamond roller chain, bevel-gear-type differential, and live rear axle shafts supported on Hyatt roller bearings. Reo at this time pioneered a new radiator of flattened tubes that were able to expand in case of freezing or overheating. It was a cross-flow radiator with the tubes in a transverse mounted position. Although Reo had almost 1,000 cars ordered, they could build only 864 cars during the first year. Olds also realized the need for a lower-priced runabout.

In September 1905 a runabout weighing about 900 lb was added to the 1906 Reo car line.

1906 Reo

Wheelbase (in.)	78
Price	$650
No. of Cylinders / Engine	1
Bore x Stroke (in.)	4.75 × 6.00
Horsepower	8 adv, 9.02 ALAM
Body Styles	runabout
Other Features	folding rear seat was $25 extra

Reo undercarriage. (Source: R.E. Olds Museum)

The engine was supported on two frame cross-members, and water-cooled by a tubular radiator in front of the hood. Having a tread of 55 in., the suspension was by three-quarter elliptical springs in front and semi-elliptical springs in the rear. The live rear axle had roller bearings within the axle tubes. The tubular front axle was fitted with Elliot-type steering knuckles and wheel spindles for wood-spoke artillery wheels.

On Tuesday, August 8, 1905, Percy F. Megargle of Rochester, New York, and D.F. Fassett started on a round-trip across the United States in a Reo car. During the run, the northern route was used going west from New York through Pennsylvania, Ohio, Indiana, Illinois, Iowa, Nebraska, Wyoming, Idaho, and Oregon to Portland, then south to San Francisco. The return eastward was through Nevada, Utah, Colorado, Kansas, Missouri, Kentucky, West Virginia, Maryland, Delaware, and New Jersey.

The car used was a 16-hp Reo tonneau model named "Reo Mountaineer." The object of the trip was not to establish a speed record, but rather to demonstrate Reo's dependability. The Reo completed the 11,742-mile journey experiencing only tire failures, getting bogged down in mud, and exterior scrapes and dents.

For 1906, in addition to the runabout and tonneau, two body types were added: the "Physician's vehicle" and the four-passenger depot wagon. These models were powered by the two-cylinder, 18-hp engine, and priced at $1,195 and $1,250, respectively. The 1906 models were carried over into the 1907 model year, except the two-cylinder-engine-powered models had a longer wheelbase of 94 in. to accommodate the depot wagon model, now called a limousine, extended to seat seven passengers. The limousine was priced at $2,500. Increased sales moved the Reo Motor Car Company into third place in 1907, behind Ford and Buick.

The 1907 models were continued during the 1908 calendar year with just a few refinements. The limousine model was discontinued, and a "gentleman's" was added to the line. Powered by the two-cylinder, 18-hp engine, the roadster was priced at $1,000. Despite 1907 being the year of national financial panic, Reo sales were good in 1908.

1908 Reo Sportsman's roadster. (Source: R.E. Olds Museum)

An important event took place in 1908: On May 16th, William C. Durant met with Benjamin Briscoe to discuss the possibility of forming a large automotive combine of about 20 companies. A meeting was held in Detroit late in May, with Briscoe, Durant, Henry Ford, and Ransom Olds in attendance. The meeting was held in a well-organized manner and ended cordially, but with no conclusion. However, at the next meeting in Detroit, attended by Durant, Briscoe, Ford, Olds, and a representative of J.P. Morgan and Company, Ford stated he would join the combine only on the basis of cash payment for his interests. Olds declared that he

1908 Reo touring. (Source: R.E. Olds Museum)

1909 Reo Gentleman's roadster. (Source: R.E. Olds Museum)

too would demand cash payment for the Reo interests. Because cash funds were not available, the discussion ended and the meeting was adjourned. Durant decided to go it alone. Thus, on September 16, 1908, Durant organized and incorporated the General Motors Company of New Jersey (who bought Olds Motor Works that year).

The 1909 Reo model line-up was the same as that for 1908, except the prices were reduced to $500 for the one-cylinder models and $1,000 for the two-cylinder models. The roadster was offered as a stripped-down model called a "semi-racer."

Owen as Merchandiser

In May 1909, Olds and Raymond Owen entered into a contract sales agreement whereby the R.M. Owen and Company would accept the entire production output of Reo Motor Car Company, and handle the wholesale distribution of Reo cars in the United States. It was reported that the sales agreement contract dollar value was approximately $50,000,000. Raymond Owen had previously relinquished the title of sales manager to R.C. Rueschaw. Apparently the buying public liked what Reo was doing, as production and sales of Reo cars progressively increased from 864 cars in 1905 to 6,592 in 1909.

1910 Reo Model R touring car. (Source: Reo Catalog)

*1910 Reo Model G runabout.
(Source: Reo Catalog)*

As the demand for more powerful four-cylinder engines increased, Reo obliged in August 1909 by introducing the new 1910 Reo powered by a new F-head four-cylinder engine of 4-in. bore and 4-1/2-in. stroke developing 30 hp. The engine located under the hood up front provided power through a multiple-disc dry clutch by shaft drive and a selective sliding-gear three-speed transmission with Reo's own shift pattern. A sand pan under the engine and transmission was held by six spring locks. The single-cylinder and two-cylinder engines were continued for 1910.

L.L. Whitman who had made it his business to break transcontinental records for the previous eight years, again distinguished himself by clipping 4 days, 10 hours, and 59 minutes off the previous record. He left New York at 12:01 a.m. on August 8, 1910, in a Reo car powered by a four-cylinder, 30-hp engine, and reached San Francisco on August 18, 1910, after covering the 3,557 miles in 10 days, 15 hours, and 12 minutes. This record helped Reo sales, but they still had difficulty producing enough Reo cars to satisfy the demand. The single-cylinder- and the two-cylinder-engine-powered cars were discontinued after 1910.

Reo Motor Truck Company

On October 8, 1910, it was announced that R.E. Olds had organized a separate company to build trucks, called the Reo Motor Truck Company. Incorporated under the laws of the state of Michigan, it was capitalized at $1,000,000, of which R.E. Olds held 49,998 shares of stock with a par value of $10 each. A former stove manufacturing plant was purchased, refurbished, and equipped to house the truck production. The truck chassis featured the Reo F-head engine and many Reo passenger-car components. The merchandising of the Reo trucks was also through the R.M. Owen and Company.

Owen Motor Car Company

Meanwhile in January 1910, Ralph Owen (Raymond's brother), with Angus Smith and Frank Robson, organized the Owen Motor Car Company with a capitalization of about $500,000. They erected a plant at 1620 East Grand Boulevard in Detroit. Their product was the Owen car powered by a 425-cu.-in.-displacement,

L.L. Whitman and Eugene Hammond on their August 1910 New York to San Francisco trip in a Reo. (Source: R.E. Olds Museum)

L-head, four-cylinder engine developing 50 hp. The 120-in. wheelbase chassis featured a double drop frame to lower the entire car, providing excellent road clearance through the use of 42-in. wheels. It was available in four body types using a common chassis: a runabout, close-coupled model, touring, and Berline. The Berline was priced at $5,400 and the other models at $4,000. Olds looked upon this venture favorably, and, in fact, allowed the Owen car to be merchandised through the Reo selling organization.

Needless to say, Owen car sales were not overwhelming. By November 1910 the Owen Motor Car Company was absorbed by the Reo Motor Car Company. The remainder of the Owen components and parts were shipped to the Reo truck plant in Lansing to be built into the completed car called the "R-O." The Owen plant in Detroit was sold to the K-R-I-T Motor Car Company. Ralph Owen rejoined the Reo sales organization. By the end of 1910 the exclusive sales contract between Reo Motor Car company and the R.M. Owen and Company was terminated. However, R.M. Owen and Company continued as a Reo distributor in New York City.

During 1911 Reo Motor Car Company built and sold only the cars powered by the four-cylinder engines.

	1911 Reo Model K	1911 Reo Models R and S
Wheelbase (in.)	98	108
Price	$850	$1,250
No. of Cylinders / Engine	4	4
Bore x Stroke (in.)	3.75 × 4.25	4.00 × 4.50
Horsepower	22.5	30
Body Styles	runabout	five-passenger touring, four-passenger roadster, four-passenger demi-tonneau
Other Features		

Reo the Fifth

In December 1911, what had been announced as the greatest achievement of Ransom E. Olds was brought forward under the name of "Reo the Fifth." What was so startling about the announcement was that Olds called it his "Farewell Car." Such a comment coming from one of the most dynamic pioneers was difficult to understand. Reo the Fifth was essentially a refined version of the 1910 Reo, powered by the F-head four-cylinder engine. However, the one important improvement was the relocation of the gearshift lever to the center of the front compartment floor. The wheelbase was increased to 112 in. Reo the Fifth in 1912 was available in three body styles: a touring, baby tonneau, and two-passenger roadster. Except for increasing wheelbase length and pricing fluctuations, the Reo the Fifth remained basically unchanged through the 1919 model year. A three-passenger coupe priced at $1,575 was added for 1915 only. A five-passenger touring and a roadster, both priced at $875, were the only body styles available in 1916.

In 1914 Reo Motor Car Company introduced two truck chassis models.

1914 Reo Truck Chassis

Wheelbase (in.)	130/146
Price	$1,800
No. of Cylinders / Engine	F-4
Bore x Stroke (in.)	4.00 × 4.50
Horsepower	25.6 ALAM
Body Styles	
Other Features	

In November 1914 Reo Motor Car Company announced a new Reo Model M.

1915 Reo Model M

Wheelbase (in.)	126
Price	$1,250
No. of Cylinders / Engine	F-6
Bore x Stroke (in.)	3.56 × 5.12
Horsepower	45 adv, 30.4 ALAM
Body Styles	seven-passenger touring
Other Features	called "Reo the Six"

The Reo the Six touring styling was unique in that it had a rolled top of the rear doors and the intermediate panel between the front and rear doors that extended across and around the front seat back, creating a dual cowl effect. It also had cantilever rear springs and a helical bevel-gear drive. However, the Reo dealers didn't receive the shipments of Reo the Six until 1916.

Reo Motor Car Company built and sold 13,516 cars and about 1,600 trucks in 1914, and earned a profit of $2,539,187 for the 1914 fiscal year. Reo Motor Car Company built and sold 14,693 cars domestically in 1915. Domestic sales were 23,814 cars in 1916 and 25,577 cars in 1917. The Reo the Fifth and Reo the Six Model M continued through 1919 with only minor specification and price changes.

Reo Speed Wagon

In August 1915 Reo Motor Truck Company surprised the industry by offering two new, lighter, 3/4-ton delivery trucks, called the "Reo Speed Wagon," a name that remained synonymous with Reo. It was available with either full panel sides or express body with screen sides.

1915 Reo Speed Wagon

Wheelbase (in.)	120
Price	$1,075 ($1,000 for chassis alone)
No. of Cylinders / Engine	F-4
Bore x Stroke (in.)	4.00 × 4.50
Horsepower	30 adv, 25.6 ALAM
Body Styles	
Other Features	express body, driver's seat, windshield, and canopy top

A four-passenger roadster and a seven-passenger Springfield-type sedan were added to the 1917 Reo the Six model line-up. Announced in August 1916, the prices were quoted as $1,150 for the roadster, $1,225 for the touring, and $1,750 for the Springfield seven-passenger sedan. During 1917 the Reo Motor Truck Company and the Reo Motor Car Company were consolidated into a single company for operating efficiency.

World War I

The declaration of war against Germany in April 1917 had a profound effect on Reo's performance during 1917-18. In late December 1917, Robert C. Rueschaw resigned as sales manager of Reo Motor Car Company, but eventually returned in December 1923.

In early 1918 the United States Government negotiated a contract with Reo Motor Car Company for the building of tractors with caterpillar driving treads. The United States Army Quartermaster Corps also requisitioned several thousand Reo Speed Wagons, as they were a favorite of the American doughboys in France. By the end of 1918, all passenger-car production was curtailed, and over 35% of Reo's production

1919 Reo coupe. (Source: R.E. Olds Museum)

1919 Reo sedan. (Source: R.E. Olds Museum)

were trucks and military vehicles. Reo Motor Car Company made a modest profit on the production of war materiel in 1918.

In reconverting the factories for civilian car production, Reo Motor Car Company decided to offer only the Reo the Fifth, discontinuing the Reo the Six for one year. Four body types—a five-passenger touring, three-passenger roadster, four-passenger coupe, and a five-passenger sedan—were offered with little or no change in technical specifications. Because the prices escalated from $875 in 1916 to a range of $1,375 to $2,175 in 1919, passenger car sales dropped to a dismal 7,307 cars, while truck sales increased to 9,185 units. Reo's sales performance was considered satisfactory.

Prosperity Through the Post-War Years

For 1920 Reo Motor Car Company discontinued the four-cylinder engines, replacing them with a refined F-head six-cylinder engine rated at 50 hp. Using the same frame and wheelbase dimensions of the former

1924 Reo touring. (Source: R.E. Olds Museum)

Reo the Fifth, the new model was identified as the T-6, which would remain basically unchanged through 1926. Four body types—touring, roadster, coupe, and sedan—were available, priced at $1,650 to $2,400.

At the stockholders' meeting on December 4, 1920, R.E. Olds reported that the Reo Motor Car Company surplus funds had increased from $6,390,333 to $9,747,309 for the fiscal year ending August 31, 1920. Because of R.E. Olds' conservative management, low volume production, and the faithful Reo Speed Wagon, Reo Motor Car Company felt the effects of the 1920-21 depression only slightly, earning just more than $1,000,000 in 1921. Many other automobile manufacturing companies went into bankruptcy during this difficult period. The Reo T-6 line-up continued through 1926, with a taxicab model being offered in 1922, and balloon tires becoming available in 1925. The prices fluctuated from $1,650 for the five-passenger touring in 1920 to $1,395 for the same model in 1926.

Olds Retires

On November 26, 1923, at the board of directors meeting, R.E. Olds declared that the net profits for 1923 were $5,603,478 compared to $3,140,524 in 1922. At this same meeting Olds announced his intention to retire as Reo president on December 24, 1923. The board of directors promptly elected him chairman of the board, and Richard H. Scott was elected president and general manager. Horace T. Thomas was elected vice president of engineering, and H.C. Rueschaw was re-elected general sales manager. Clarence Triphagen presented an optimistic sales plan, reporting that the Reo Motor Car Company had 800 dealers, 146 distributors, and 6 factory branches. He also stated that the Canadian Reo sales organization had 43 direct agencies in various parts of the world.

In December 1923 Reo Motor Car Company negotiated the purchase of the Duplex Truck Company plant in Lansing, Michigan, to provide additional facilities for truck and bus manufacturing. Reo took possession of the plant on January 1, 1924. The 20th annual stockholders report of December 3, 1924, showed that Reo Motor Car Company earned $3,412,041 for the fiscal year ending August 31, 1924.

During 1925 three important events occurred: the announcement of the Reo Heavy Duty 2-ton truck Model G; the hiring of Thomas T. O'Brien as assistant sales manager to specialize in the sale of Reo trucks and buses; and Fabio Sergardi (former chief engineer of Oldsmobile) becoming chief engineer of Reo Motor Car Company. The Reo Heavy Duty truck Model G with stake cargo area, on a 156-in. wheelbase chassis, was powered by the Reo T-6, six-cylinder, F-head engine of 50 hp. Priced at $1,985, the addition of this new

1925 Reo roadster. (Source: R.E. Olds Museum)

truck gave Reo a line-up of two truck models, three bus models, a taxicab, and six passenger-car models. The passenger-car sales amounted to approximately 13,700 cars domestically on a calendar year production of 16,035 units. Truck production and sales were steadily increasing. Reo Motor Car Company realized a net profit of over $4,000,000 for fiscal 1925. In October 1925 Robert Rueschaw retired for health reasons, and was succeeded by Clarence Triphagen as sales manager.

The Reo T-6 was continued through December 1926 with minor changes and refinements. The cars built between August 1 and December 31, 1926, were referred to as 1927 Reo T-6 models. The 1926 Reo passenger-car sales amounted to 10,255 cars, a decrease of about 25% from 1925. Even with the price reduction to a range of $1,395 to $1,765, the loss of sales was due to the Reo T-6 being out of date. This condition was corrected in January 1927. Despite the loss of passenger-car sales, truck sales made up for it, allowing Reo Motor Car Company to realize a net profit of over $4,250,000 for fiscal 1926.

When R.E. Olds relinquished the presidency of Reo in December 1923, he presumed that Richard Scott would carry out plans in the same way that he would have. However, Scott had plans all his own, which were not all bad. Scott knew Reo Motor Car Company could not carry on indefinitely with the refined T-6. Thus, he made plans for a totally new car, and assigned Fabio Sergardi and his team to design it.

Reo Flying Cloud

The new Reo Flying Cloud made its debut at the National Automobile Show in New York in January 1927. It was named after one of the fastest, most graceful, and famous of the great American Clipper sailing vessels. It bore no resemblance to the Reo T-6, which it replaced, except for the same price range.

The 1927 Reo Flying Cloud was an entirely new car, featuring a handsome new body with heavily rounded, rear-quarter upper panels joining the new roof with curved sides. The roof extended gently ahead of the windshield, forming a built-in sun visor. The new nickel-plated radiator shell was rounded gracefully. The longer hood sides carried the lower belt molding, while the doors, quarter, and rear panels had a double belt molding.

1927 Reo Flying Cloud

Wheelbase (in.)	121
Price	$1,595-1,845
No. of Cylinders / Engine	L-6
Bore x Stroke (in.)	3.25 × 5.00
Horsepower	73 adv, 25.35 SAE
Body Styles	roadster, brougham, four-passenger coupe, Victoria, five-passenger sedan
Other Features	Lockheed internal-expanding-shoe hydraulic four-wheel brakes

The window recesses were finished in body color, contrasting with the upper structure color, giving a two-tone effect. The roadster was upholstered in genuine leather, while the closed bodies were trimmed in mohair velvet or wool broadcloth. The engine had a seven-main-bearing crankshaft, and the cylinders and crankcase were cast en-bloc with a removable cylinder head. The ignition distributor was mounted overhead, with the lower shaft end engaged in the oil pump drive gear, driven by the camshaft. The single-plate dry clutch and transmission were mounted as a unit at the rear of the engine.

Perhaps the most innovative feature of the Reo Flying Cloud chassis was the new Lockheed internal-expanding-shoe hydraulic four-wheel brakes, acknowledged as the first in the American automotive industry.

1927 Reo Flying Cloud sedan. (Source: R.E. Olds Museum)

1927 Reo Flying Cloud roadster. (Source: R.E. Olds Museum)

The front axle and steering knuckles were of the reverse Elliot type. The wheel rim diameter was 18 in., mounted with 6.00-in. or 6.20-in. balloon cord tires. The front and rear suspension was by semi-elliptical springs, controlled by Lovejoy hydraulic shock absorbers, perhaps another industry first.

An invasion of the lower-priced field was made by Reo Motor Car Company on May 1, 1927, with the introduction of the Reo Wolverine. However, because the crowded production schedules of the Flying Cloud, the Reo Speed Wagons, buses, and trucks took up the entire production capacity of Reo manufacturing plants, a departure from previous Reo procedure was inaugurated whereby Reo purchased major components from reliable outside manufacturing sources. The Continental engine, Borg and Beck clutch, Warner transmission, Salisbury axles, and Hayes-Ionia bodies furnished the major units of the Reo Wolverine. These highly respected manufacturers supplied some of the finest components available to the automotive industry at that time.

1927 Reo Wolverine

Wheelbase (in.)	114
Price	$1,195
No. of Cylinders / Engine	L-6
Bore x Stroke (in.)	3.25 × 4.00
Horsepower	50 adv, 25.5 SAE
Body Styles	five-passenger brougham
Other Features	internal-expanding-shoe Lockheed hydraulic four-wheel brakes, semi-elliptical spring suspension, and Lovejoy shock absorbers

*1928 Reo Wolverine coupe.
(Source: R.E. Olds Museum)*

The Reo Wolverine was handsomely styled by Charles F. Magoffin, strongly resembling the Flying Cloud, though on a smaller scale. Perhaps the most innovative feature of the Reo Wolverine engine was the method of inducting the carburetor inlet primary inlet air through the valve spring chamber, providing a full crankcase ventilating system.

Apparently Scott's planning was correct, as the Reo Flying Cloud and the Reo Wolverine were enthusiastically accepted by the car-buying public. Reo domestic car sales jumped to 19,394 cars on a calendar year production of 28,765 units, almost double that of 1926. Reo Speed Wagons, trucks, and buses were selling favorably, especially in the export trade market.

Announced in August 1927, the Reo Flying Cloud and Reo Wolverine models for 1928 were basically carryovers of the 1927 models, except for the addition of other body styles. A new two- to four-passenger rumble seat sport coupe, priced at $1,625, was added to the Reo Flying Cloud line. The Reo Wolverine added a two- to four-passenger rumble seat cabriolet priced at $1,195 and a five-passenger four-door sedan priced at $1,295.

R.E. Olds Farm Company

After relinquishing the Reo presidency to Scott in December 1923, Ransom Olds spent a great deal of time traveling, commuting between Lansing and Daytona, Florida. He invested in Florida real estate during the Florida Land Boom in 1924-25. His first intent was to establish a rubber plantation on his Florida property. However, he found that the chances of success were meager, so he converted his land to fruit and agricultural production, called the R.E. Olds Farm Company. In 1928 Olds sold much of his Reo Motor Car Company stock, when it was at what appeared to be a record high. Somehow he must have sensed what would happen in late 1929.

With a booming economy and public acceptance of the Reo Flying Cloud and Reo Wolverine, Reo Motor Car Company enjoyed a very prosperous 1928. Domestic car sales of the Reo Flying Cloud and Wolverine amounted to 21,374 cars, setting a sales record that would never be matched. Truck sales were almost an equal amount, and Reo Motor Car Company earned a net profit of over $5,000,000 in 1928. In August 1928 Clarence Triphagen resigned as sales manager and was succeeded by C.E. Eldridge, former assistant sales manager.

Richard Scott did not want to lose the selling momentum gained in 1927-28, so he had the stylishly new 1929 Reo Flying Cloud Master introduced early, on March 2, 1928. The new Flying Cloud Master had all new body panels, accentuated by a single belt line molding, a higher chromium-plated radiator shell, a new higher hood, and bullet-shaped headlamps. The running boards were shorter, due to the longer, sweeping, one-piece front fenders, with deep crown but with smooth surfaces without the former creased terrace lines.

It was during 1929 that Reo Motor Car Company innovated a new merchandising drive called the "Car of the Month." A specially built Reo Flying Cloud Master sport sedan was allotted to each dealer, one per month. This particular car was priced at $1,970, but was finished with the most attractive colors and upholstered with the most luxurious materials available, in harmonizing tones.

The 1929 Reo Flying Cloud Master specifications were generally the same as those for the 1928 models, except that the engine cylinder bore size was increased to 3-3/8 in., increasing the power output to 80 hp. The tire cross-section was increased to 6.50 in., and the suspension springs used the silent block rubber mounting in place of the conventional bolt and shackle arrangement. The instrument cluster included an oil level gauge, actuated by a small pushbutton switch.

On December 1, 1928, the Reo Wolverine was superseded by the new Reo Flying Cloud Mate. While most of the technical specifications were the same as those for the 1928 Reo Wolverine, the Flying Cloud Mate enjoyed many of the refinements found in the Master, including the silent block rubber mountings for the suspension springs and Lovejoy hydraulic shock absorbers. The engine cylinder bore was increased to 3-3/8 in., providing increased power output to 50 hp.

The Reo Flying Cloud Mate had styling all its own. The belt line molding was omitted, and instead, the step line originating at the radiator shell swooped down to the body sill at the front door area. The outlined area was finished in a contrasting color, and had vertical hood side louvres of diminishing lengths.

In August 1929 the Reo Flying Cloud Master innovated a new constant-mesh, silent-second, synchronized transmission utilizing the herringbone double-helix gear set. Acknowledged as being far superior to the

1929 Reo Flying Cloud Master coupe. (Source: R.E. Olds Museum)

1929 Reo Flying Cloud Mate sedan. (Source: R.E. Olds Museum)

single helical gears, competitive car manufacturers did not use the herringbone gear setup because of the greater cost. The 1929 Reo Flying Cloud Master and Mate continued unchanged through December 31, 1929. However, those cars built between August 1 and December 31, 1929, were considered 1930 models.

Surviving the Depression

While Reo car and truck sales were satisfactory in early 1929, the sales broke off abruptly after "Black Thursday," October 24, 1929. Thus, domestic car sales dropped to 17,319 units in 1929, with reduced truck sales as well.

The 1930 Reo Flying Cloud models announced on January 1, 1930, were essentially carryover 1929 models, with new model designations. The model numbers represented the last two digits of the chassis wheelbase lengths; i.e., Models 15, 20, and 25. Model 15 was powered by the 60-hp, six-cylinder engine, and Models 20 and 25 were powered by the 80-hp, six-cylinder engines. Model 15 had a Budd all-steel body, while the 20 and 25 were Murray-manufactured composite wood and steel bodies.

Regardless of Reo's stylish and well-engineered cars, quality manufacturing, and good management, domestic car sales fell to approximately 11,450 cars, and Reo Motor Car Company sustained a loss of almost $2,000,000 in 1930. What went wrong? Nearly everyone in the automobile industry based their forecasts on expected continued prosperity. The stock market crash and the subsequent economic depression caused the great losses and the bankruptcy of many fine automobile manufacturers. This sudden downturn in business could not have come at a worse time for Reo. Richard Scott had great plans for Reo in 1931. He was perhaps overoptimistic about the future, and had extended the resources of Reo Motor Car Company too far.

Accepting the responsibility for the business reverses, Scott stepped down as general manager in February 1930, but retained the presidency of Reo Motor Car Company. William R. Wilson, a person with extensive automotive and financial experience, was elected vice president and general manager of Reo Motor Car Company. Elijah G. Poxson was appointed sales manager and Harold A. Roland was named Reo truck sales manager in September 1930, with Clarence Eldridge as assistant sales manager.

Reo Royale

Scott's original plans for the 1931 Reos were carried out on schedule, introducing the most magnificent Reo of all time, the "Reo Royale," on September 26, 1930. The companion Reo Flying Cloud Six and Flying Cloud Eight were also announced at that time. The 1931 Reo Royale was unmistakably new in every respect, down to each sheet metal part, body panel, and entirely new chassis.

1931 Reo Royale

Wheelbase (in.)	135
Price	$2,745
No. of Cylinders / Engine	IL-8
Bore x Stroke (in.)	3.37 × 5.00
Horsepower	125 adv, 36.34 SAE
Body Styles	two- to four-passenger coupe, four-passenger Victoria, five-passenger four-door sedan
Other Features	first American production car to have been aerodynamically tested in an aircraft testing wind tunnel

The new Royale design was the culmination of Amos Northup's brilliant designing skill and thousands of hours of relentless, dedicated effort to create a masterpiece of art. The V-shaped radiator shell, finished in body-color lacquer, housed the radiator and supported the thermostatically controlled vertical shutter blades. The front fenders were of the long, sweeping, clamshell type, providing wells for the spare tires, with all edges turned under. The shatterproof windshield was in a fixed, gently rearward sloping position, omitting the exterior sun visors. The body corners were rounded and the sloping windshield A-pillar extended downward to the running board splash shields. The hood sides had adjustable ventilating doors and the cowl had four ventilators. Inside-adjustable sun visors were provided. The rear body panel of compound curved shape extended rearward to conceal the fuel tank beaver-tail style.

The chassis was supported on metal-covered semi-elliptical springs and 6.50-in. balloon cord tires on 18-in. wheel rims. The L-head engine had a counterbalanced crankshaft supported on nine main bearings and thermostatically controlled aluminum-alloy pistons. The power was transmitted through a three-speed herringbone silent-second transmission and exposed propeller shaft through a semi-floating rear axle. (After the Reo Royale was in production, Amos Northup designed the famous Graham Blue Streak, which closely resembled the Reo Royale.)

Richard Scott (middle) with 1931 Reo Royale. (Source: J.A. Conde)

Herringbone silent-second transmission.
(Source: R.E. Olds Museum)

The 1931 Reo Flying Cloud models, also introduced on September 26, 1930, were redesigned by Fabio Sergardi, somewhat resembling the Reo Royale in some respects. They had a flat front radiator and vertical hood side panel louvres.

	1931 Reo Flying Cloud Six	**1931 Reo Flying Cloud Eight**
Wheelbase (in.)	125	130
Price	$1,695	$1,995
No. of Cylinders / Engine	L-6	IL-8
Bore x Stroke (in.)	3.37 × 5.00	3.37 × 5.00
Horsepower	80 adv, 27.25 SAE	125 adv, 36.34 SAE
Body Styles	two- to four-passenger coupe, four-passenger Victoria, five-passenger four-door sedan	
Other Features		

On April 15, 1931, Reo Motor Car Company announced a new, shorter Reo Royale Model 31, built on a 131-in. wheelbase chassis. Having the same engine, styling, and all other features of the longer Reo Royale, it offered all three body types at $2,145.

On June 1, 1931, Reo extended the commercial vehicle line by adding a 1-1/2-ton, four-cylinder-engine-powered Reo Speed Wagon. Reo Motor Car Company also introduced a 4-ton truck model, extending the Reo line of commercial vehicles into the higher-capacity range. The two new 1-1/2-ton models placed Reo in competition for truck buyers in the low-priced field, formerly not represented by Reo trucks.

	1931 Reo Speed Wagon Models 1-A, C	**1931 Reo Speed Wagon Models 1-B, D**	**1931 Reo Speed Wagon Models H, J, K**
Wheelbase (in.)	136/160	136/160	170/190
Price	$625	$725	
No. of Cylinders / Engine	4	6	6
Bore x Stroke (in.)			
Horsepower	57 adv	61 adv	101 adv
Body Styles			
Other Features	1-1/2-ton	1-1/2-ton	4-ton

On May 15, 1931, Reo Motor Car Company announced the new second series Flying Cloud models: 6-21, 8-21, and 8-25.

	1931 Reo Flying Cloud Model 6-21	1931 Reo Flying Cloud Models 8-21, 8-25
Wheelbase (in.)	121	121/125
Price	$995-1,110	$1,195-1,310 / $1,565-1,650
No. of Cylinders / Engine	L-6	IL-8
Bore x Stroke (in.)	3.37 × 5.00	3.00 × 4.75
Horsepower	80 adv, 27.25 SAE	90 adv, 28.8 SAE
Body Styles	6	6
Other Features	aircraft-type round dials with pointer instruments on the instrument panel	

These second series 1931 Reo Flying Cloud models closely resembled the Reo Royale, including the V-type radiator shell, sloping windshield, clamshell fenders, and Royale body panels. The most noticeable differences were that the Flying Cloud models had the outside cadet-type sun visor and vertical hood side panel louvres. The second series 1931 Reo Flying Cloud models were carried over through March 31, 1932. Those cars built between August 1, 1931, and March 31, 1932, were considered 1932 models.

In an attempt to attract car buyers during 1931, Reo proliferated in extending the choices of many chassis models and available body types. Even with a price of only $1,295 to $1,745 for the fine Flying Cloud models, and $2,145 to $2,745 for the prestigious Reo Royale, domestic car sales dropped to a disappointing 6,762 units. Truck sales also dropped to 5,166 units, resulting in a net loss of approximately $2,000,000 for Reo Motor Car Company in 1931. William Wilson resigned in September 1931, with Richard Scott reassuming the responsibility of general manager.

By August 3, 1931, a new style of finish, known as a pearlescent shade, had been applied to a number of Reo Royale models displayed in the New York and Detroit Reo salesrooms and at the factory in Lansing. The pigment of the finish was mainly of ground mother of pearl, carefully scaled from the inside of oyster shells. This mother of pearl lining retained the iridescent beauty of mother of pearl. While Reo pioneered this type of finish, most automotive paint suppliers were able to substitute finely ground aluminum for the mother of pearl. The car-buying public liked this style of finish, and as the result, metallic or polychromatic finishes are more popular than ever today.

1932 Reo Flying Cloud Eight sedan. (Source: R.E. Olds Museum)

On August 1, 1931, Reo did not make any significant changes for the 1932 models except to reduce prices on most models. However, Reo gained the distinction of having the longest-wheelbase production seven-passenger sedan in America. The 1932 Reo Royale seven-passenger sedan, and seven-passenger sedan with partition window, were carried on a 152-in. wheelbase chassis powered by the Reo Royale eight-in-line engine of 125 hp. The prices were $3,695 for the seven-passenger sedan and $3,895 for the seven-passenger sedan with partition window. A five-passenger sedan carried on the 152-in. wheelbase was added to the Reo Royale in April 1932, priced at $2,545.

Beginning in January 1932, vacuum-powered semi-automatic clutch control was standard equipment on the Reo Royale Models 8-31, 8-35, and 8-52, and the Reo Flying Cloud Model 8-25. It was available on other Flying Cloud models at extra cost. The operation of the semi-automatic clutch was controlled by a pushbutton on the toe board to the left of the pedal.

The aerodynamic styling introduced in September 1930 on the Reo Royale was continued through 1932 and extended to the other Reo models. Because of the deepening of the economic depression, and the resulting sagging sales, Reo Motor Car Company announced a new lower-priced Reo Flying Cloud Model S-6 on April 1, 1932. The styling characteristics of the Reo Royale were employed, including the V-shaped radiator shell and grille, which were given a greater slope and flared at the bottom to blend with the clamshell front fenders.

1932 Reo Flying Cloud Model S-6

Wheelbase (in.)	117
Price	$995-1,070
No. of Cylinders / Engine	L-6
Bore x Stroke (in.)	3.37 × 5.00
Horsepower	80 adv, 27.25 SAE
Body Styles	two- to four-passenger coupe, two- to four-passenger convertible coupe, five-passenger sedan
Other Features	

The 1932 Reo Royale Models 8-31, 8-35, and 8-52, priced at $1,785 to $3,395, were continued unchanged through December 31, 1932. On January 1, 1933, the Reo Flying Cloud models were confined to one chassis line on a 117-in. wheelbase, designated as Model S-2, priced at $795 to $920. The 1933 Reo Royale was reduced to two chassis model lines, the Royale models priced at $1,745 to $1,845 on a 131-in. wheelbase

1933 Reo Royale sedan.
(Source: R.E. Olds Museum)

chassis, and the Royale Custom models priced at $2,445 on the 135-in. wheelbase chassis. The Reo Royale 8-52 models were discontinued. All Reo Royales were powered by the famous L-head eight-in-line engine of 125 hp. The Flying Cloud models were powered by the 85-hp, L-head, six-cylinder engine. The Reo Flying Cloud and Reo Royale models had skirted fenders, whereby the wheel opening followed the tire outline and concealed the chassis components.

Reo Self Shifter

Perhaps the most innovative feature of the 1933 models was the Reo Self Shifter two-speed automatic transmission, introduced in April 1933. This device consisted of a two-speed automatic gear set, and an auxiliary gear train for manually selecting reverse, enclosed in a single transmission case mounted at the rear of the engine. The clutch pedal was entirely eliminated and gear selection was by a control on the instrument panel, within easy reach of the driver. The Self Shifter was designed and developed by chief engineer Horace T. Thomas, with the assistance of John Bethune and Albert B. Hayes. Reo Motor Car Company was granted a patent by the United States Patent Office in the names of Thomas, Hayes, and Bethune.

The Reo Flying Cloud Series S-2 was continued through July 31, 1933, at which time the Series S-3 superseded the S-2 and continued through November 30, 1933. The new 1934 Reo Flying Cloud Series 4-S began production on December 1, 1933. It featured skirted fenders, horizontal hood side louvres, and draftless window ventilation. Startix automatic starting was employed. The radiator filler cap was located under the hood, and the chassis spring rebound was controlled by aircraft-type telescoping shock absorbers. Reo also built the Flying Cloud chassis (less engine) and Hayes Ionia bodies for the Franklin Olympic series.

1934 Reo Flying Cloud Series 4-S

Wheelbase (in.)	118
Price	$795-920
No. of Cylinders / Engine	L-6
Bore x Stroke (in.)	3.37 × 5.00
Horsepower	85 adv, 27.25 SAE
Body Styles	7
Other Features	Stellite exhaust valve seats
	Self Shifter optional at $75

The 1934 Reo Royale Series N1 and N2 were announced on August 1, 1933, and consisted of the Standard line and the Elite line on a 131-in. wheelbase chassis, and the Custom line on a 135-in. wheelbase chassis. Each of the Reo Royale models was powered by the famous Royale eight-in-line engine of 125 hp. Three body types were available in each line: a two- to four-passenger coupe, a four-passenger Victoria, and a five-passenger sedan. The prices were $1,500 for the Standard, $1,600 for the Elite, and $1,720 for the Custom.

Olds is Back

To add to the problem of Reo's poor sales, two opposing groups were seeking control of Reo Motor Car Company: the Independent Stockholders' Committee headed by Richard Scott and the New Management Committee led by Ransom E. Olds. The turmoil began in early January 1934 when Scott relinquished his position as general manager, at which time Olds became chairman of the newly formed Executive Committee, which held the responsibility for the active direction of the Reo Motor Car Company.

Harry Teel (Scott's brother-in-law) resigned as works manager on February 17, 1934, and was succeeded by Ray A. DeVlieg. Subsequently, Horace Thomas was appointed chief research engineer, and R.J. Fitness was named chief engineer, succeeding Thomas. The struggle for control through stockholders' proxies continued through February, March, and into April 1934. On April 17, 1934, Scott resigned as president of Reo Motor Car Company, and Donald E. Bates was elected president the next day. Thus, the full control of Reo management was in the hands of R.E. Olds again.

In July 1934 Elijah G. Poxson was named the president of the newly formed Reo Sales Corporation, and Clarence A. Triphagen succeeded Poxson as sales manager of Reo Motor Car Company. R.A. Weinhardt joined Reo in July 1934 as assistant chief engineer for both the car and the truck divisions.

Prompted by continued satisfactory truck sales of 5,035 in 1934, two new 1935 Reo trucks were added to the line-up: a new 1/2-ton commercial unit of high-performance characteristics and a refined 1-1/2-ton Speed Wagon chassis.

	1935 Reo Truck Model S4P	**1935 Reo Truck Model D**
Wheelbase (in.)	118	139
Price	$495 chassis / $695 panel delivery wagon	$595 chassis
No. of Cylinders / Engine	L-6	L-6
Bore x Stroke (in.)	3.37 × 5.00	3.37 × 5.00
Horsepower	80 adv, 27.25 SAE	80 adv, 27.25 SAE
Body Styles		
Other Features	1/2-ton	1-1/2-ton

The 1935 Reo Flying Cloud series was announced in July 1934, consisting of seven newly styled body types, priced at $795 to $945, with the Self Shifter transmission optional at $75 extra. The 118-in. wheelbase chassis was powered by a 85-hp, L-head, six-cylinder engine. On January 1, 1935, a new lower-priced Reo Flying Cloud was added to the line.

	1935 Reo Flying Cloud Series 6A
Wheelbase (in.)	115
Price	$845-985
No. of Cylinders / Engine	L-6
Bore x Stroke (in.)	3.37 × 4.25
Horsepower	75 adv, 27.25 SAE
Body Styles	two-door and four-door sedans
Other Features	Midland Steeldraulic four-wheel brakes

While the conventional Reo three-speed helical gear transmission was standard, an entirely new semi-automatic self-shifter transmission was offered as an option at modest extra cost. New pressed-steel road wheels of 16-in. diameter were mounted with 6.25-in. balloon cord tires. The narrower V-shaped radiator shell and grille were inclined at approximately 15 degrees from vertical, and the dirigible-shaped headlamps were mounted on streamlined brackets on the radiator shell. The body panels were flatter, following the

*1935 Reo Flying Cloud.
(Source: R.E. Olds Museum)*

trend of the automobile industry. The body panel, doors, and other sheet metal part designs were also shared by Graham cars.

On January 1, 1935, insult was added to injury when they used the Royale name for the Reo Model 7S, desecrating the name of the former, luxurious, powerful Reo Royale. The Model 7S was an upgraded version of the 1935 Reo Flying Cloud Model 6A, priced at $985 and $1,030. Needless to say, the 1935 Reo passenger-car sales did not improve, amounting to only 3,894 units domestically. Fortunately, the Reo truck sales were increasing; the contracts with the United States Government and the contract to build delivery models for Mack Trucks added approximately $3,500,000 to the truck sales dollar volume. Domestic truck sales were 5,101.

Focusing on Trucks

Realizing that Reo's survival depended on truck manufacture and sales, Donald E. Bates and R.E. Olds agreed that Reo would concentrate on truck manufacturing for 1936. Reluctantly, the 1936 Reo Flying Cloud Model 6D was announced on November 1, 1935. Following the lead of many manufacturers, Reo limited the model lines to one chassis. Available as a standard line and a deluxe line in two-door coach and four-door sedan body types, it gave Reo a total of four models. With refined styling, the bodies were carried on a chassis of 115-in. wheelbase, powered by a six-cylinder engine, priced at $795 to $845. The engine cylinder block was cast of chrome-nickel-alloy steel, fitted with T-slot cam ground pistons. The engine output was increased to 90 hp despite the reduced piston displacement through the use of a high-compression aluminum-alloy cylinder head. The main feature of the new 1936 bodies was the pressed-steel roof panel insert that was insulated and also wired to act as the radio antenna.

With passenger-car sales down to a dismal 3,146 units domestically, Reo discontinued the production of passenger cars after the 1936 model. Truck sales amounted to 4,227 units. Certainly, 1936 was not a profitable year for Reo Motor Car Company.

In an effort to bolster Reo truck sales by expanding the truck lines, Reo Motor Car Company developed a new series of 1/2- and 3/4-ton truck models, and introduced them on November 1, 1936. The base 1/2-ton model, known as the Reo Speed Delivery, was available with an economy Reo Silver Crown engine or a more powerful optional Reo Silver Crown six-cylinder engine. The economical Reo Speed Delivery, equipped with an all-steel cab with safety glass and a pickup body, was priced at $555, and the six-cylinder-engine-powered 3/4-ton model was priced slightly higher.

Olds Retires to the Farm

In January 1937 R.E. Olds retired from the Reo Motor Car Company, cutting all connections with the company. Olds spent all his time in the vocation which had previously been his avocation. During the 1920s Olds had purchased real estate in Florida with the intent of growing rubber trees and producing latex. Since this venture was not successful he diverted the land to agricultural and fruit production, known as the R.E. Olds Farm Company. Through his effort and financial support, the Reo Ideal Power Lawn Mower was developed in Lansing, Michigan. This mower was perhaps the forerunner of the smaller personal lawn mowers for urban and suburban lawns of today. As chairman of the boards of the R.E. Olds Farm Company, Reo Ideal Lawn Mower Company, and the First Bond and Mortgage Company of Lansing, he was commuting between Lansing, Michigan, and Daytona, Florida, frequently. This kept R.E. Olds busy for the rest of his life. R.E. Olds passed away on August 26, 1950, at age 86.

R.E. Olds in 1946. (Source: Automotive News)

As for the Reo Motor Car Company, it suffered burdensome and crippling labor strikes, as did other automobile manufacturers. However, Reo never fully recovered from the after-effects of the cruel Depression. The labor strikes of 1937, the recession of 1938, and the power struggle for management control pushed the Reo Motor Car Company into receivership on December 17, 1938. At that time A.J. Brandt was appointed temporary trustee by United States District Judge Arthur F. Lederle.

The receivership proceedings filed in the United States District Court in Lansing were dismissed on February 28, 1939, by voluntary action of the plaintiffs. On March 6, 1939, Col. Frederick Glover was appointed general manager of the Reo Motor Car Company by Federal Trustee Theodore Fry. Col. Glover had previously served temporarily as president of Reo for about two months in the summer of 1938. The number of employees at Reo Motor Car Company numbered about 800 in March 1939.

After a proxy battle that delayed the Reo Motor Car Company stockholders meeting for eight days until April 26, 1939, a syndicate headed by Horace Thomas emerged victorious and named five new directors to replace the others. Named to the board of directors were: Col. E.J. Hall, William B. Mayo, E.J. Connolly, J.W. Robb, and F.G. Alborn. The hearing on reorganization was adjourned until July 10, 1939. While the request for a loan of $1,500,000 to $2,000,000 from the Reconstruction Finance Corporation to provide operating capital for Reo had been submitted several weeks before, it had not been acted upon by mid-August, as they were awaiting acceptable reorganization plans.

Reo Motors Incorporated

It wasn't until December 18, 1939, that the reorganization plans were confirmed by Judge Arthur F. Lederle. He ordered that the operation of Reo be placed under the control of the successor company, the Reo Motors

Incorporated. On January 1, 1940, the court decision released $2,000,000 of working capital to resume production of trucks and buses under the name of the new company. Reo Motors Incorporated was managed by a three-man trusteeship during the term of the six-year $2,000,000 Reconstruction Finance Corporation loan.

In early January 1940, Col. Fred Glover (acting general manager) was elected president and general manager of the newly formed Reo Motors Incorporated. Theodore I. Fry was elected vice president and Walter O. Wood was named secretary-treasurer. At the same time Reo Motors Incorporated asked for court approval to dispose of approximately $900,000 worth of surplus machinery to pay creditors. Col. Glover also stated that the production of 5,000 heavy-duty trucks would permit Reo Motors to break even and the production of 6,500 trucks would allow a profit. However, truck sales were only 625 in 1940, 1,543 in 1941, and 156 in 1942. During World War II (1941-45) most of the Reo trucks were of the heavy-duty type to fill the orders for military vehicles and to provide civilian versions of the trucks to manufacturers of war materiel. The idled Reo passenger-car plant in Lansing was leased to the Nash-Kelvinator Corporation for the manufacture of aircraft propellers. While the profits from military production were quite small, they nevertheless managed to keep Reo Motors Incorporated in business until the end of the war. Reo's reputation for performance and reliability made it a favorite of military personnel.

During 1946 and 1947, Reo built the panel Speed Wagon delivery models and heavy-duty trucks using the 1941 designs, but with restyled front ends on the Speed Wagon models. By 1948 Reo Motors Incorporated offered 30 different truck models, ranging from 2- to 4-ton capacities. Reo trucks were redesigned, providing maximum cargo capacity but with reduced wheelbase lengths. The wheelbases ranged from 125 to 165 in., powered by Reo-built six-cylinder engines from 89 to 101 hp, and Continental six-cylinder engines of 153- to 202-hp output. Reo emphasized the shorter wheelbase, handling ease, driver comfort, and the ease of servicing.

Reo Motors Incorporated was purchased by White Motor Company in 1957, becoming a subsidiary called the Lansing Division. White Motor Company purchased the Diamond T Motor Car Company of Chicago in 1958. For about nine years Reo and Diamond T operated as separate entities. In May 1967 Diamond T was

1948 Reo Heavy Duty tractor. (Source: Commercial Car Journal)

Line-up of Diamond-T trucks. (Source: Commercial Car Journal)

merged with the Reo Motors to become the Diamond-Reo Division of White Motor Company. The new Diamond-Reo trucks were a blend of the finest features of both the Diamond-T and the Reo heavy-duty trucks. They were powered by diesel engines with choices of 195 hp to 335 hp, and provided transportation companies with some of the finest trucks on the highways with numerous available options.

Through a series of acquisitions and mergers, the White Motor Company became a division of the Volvo Heavy Duty Truck Corporation, and the Diamond-Reo truck was discontinued.

1932 Peerless V-16. (Source: Crawford Museum)

Peerless

"All that the name implies," was the advertising slogan for the Peerless cars. Truly Peerless had no peers. But the revolving door in its upper management offices, and the Great Depression, eventually drove the company to drink.

* * *

Company Background

The Peerless Wringer Company was established in Cincinnati, Ohio, in 1874 to manufacture clothes wringers, clothespins, and washboards. Having earned worldwide acclaim for quality products, in order to expand their business the Peerless Wringer Company moved to Cleveland, Ohio, in 1889, because of Cleveland's favorable location and rail and water transportation facilities. Peerless Wringer Company merged with the Mercantile Manufacturing Company to form the Peerless Manufacturing Company.

Peerless continued building wringers and other laundering equipment. However, the bicycle fad of the 1890s lured Peerless into building bicycles in 1891. The bicycle business was very lucrative during the last decade of the 19th century. But, by the time Albert A. Pope gained control of most of the bicycle manufacturers, Peerless management looked to the horseless vehicle as its main product to sustain company income. They also looked for someone to operate this new horseless vehicle business.

Kittredge and Mooers

Lewis H. Kittredge was born in Harrisville, New Hampshire, in 1871. He graduated from the New Hampshire State College in 1896, and after one year of employment in New Jersey, he joined Peerless in 1897. Kittredge became general manager and secretary by 1899 and was named treasurer in 1901. He was largely responsible for convincing Peerless management to engage in horseless vehicle manufacturing, after the bicycle boom collapsed in 1900. He was named vice president and then elected president in 1906.

> **.....TRIANGLE**
>
> HIGH-GRADE LIGHT ROADSTER, 33 POUNDS. Price, $150.00.
>
> Only Frame Made that is Mechanically Correct, Securing Lightness Without Sacrifice of Strength.
>
> Observe the BEAUTIFUL LINES. No other wheel so SYMMETRICAL, so GRACEFUL.
>
> THE VERY BEST MATERIAL THROUGHOUT. They talk for themselves and sell strictly on their merits.
>
> **THE PEERLESS MANUFACTURING COMPANY,**
> CLEVELAND, OHIO.

Advertisement in The Referee, *May 1983.*

Louis P. Mooers. (Source: Crawford Museum)

Louis P. Mooers was born in Watertown, Massachusetts, in 1873. In July 1897, while he was superintendent of a bicycle manufacturer in New Haven, Connecticut, he started to build a car of his own design and completed it in Watertown, Massachusetts. The success of his car prompted him to look for someone to finance its production. When it became known that Peerless was looking for a chief engineer, Mooers applied and was hired in 1900.

The Early Automobiles

To engage in automobile manufacture quickly, Peerless was licensed by DeDion-Bouton to manufacture cars of their design. The vehicles built by Peerless, based on DeDion-Bouton designs, were at first tricycles and then four-wheeled vehicles called motorettes, powered by DeDion engines. The two-passenger Model B had a 2.75-hp, single-cylinder engine, while the Model C seated four and had a 3°-hp engine. Steering was by tiller, bevel-gear shaft drive, and semi-elliptical spring suspension.

In 1901 Mooers and his engineering staff began work on his own design. The prototype racing car powered by a two-cylinder engine of 4°-in. bore and 5°-in. stroke made its debut in August 1902 at the Brighton Beach Speedway, performing splendidly. The car was also successfully raced at the Glenville, Ohio, Speedway on September 16, 1902. This same design with minor changes later became the passenger car with tonneau-style body, employing the vertical two-cylinder engine of the same type.

The 1903 Peerless tonneau passenger car model was introduced at the National Auto Show in New York in October 1902. At that time Peerless Manufacturing Company was incorporated with a capital stock of $300,000 and the name was changed to Peerless Motor Car Company. Peerless Motor Car Company

Mooers in the Peerless at Glenville, September 16, 1902.

franchised the Banker Brothers as distributors in New York, Philadelphia, and Pittsburgh, Pennsylvania. A.C. Banker was franchised distributor in Chicago in January 1903.

The 1903 Peerless models were innovators of two very important features. The fly-ball governor interconnected to the accelerator to maintain uniform car speed; this was truly the forerunner of today's cruise control. They also had an upward-tilting steering column (offered on today's cars as a tilting steering column). All 1903 Peerless cars had shaft and bevel gear drive and semi-elliptical spring suspension.

1903 Peerless, showing the tilting steering column. (Source: Crawford Museum)

Peerless 24-hp racing car. (Source: Crawford Museum)

The first of the 1904 Peerless touring cars having a new 24-hp, four-cylinder, T-head engine were completed in late November 1903. The formal introduction took place at the New York National Auto Show. The new touring car models were built on a 92-in. wheelbase, while the limousine model had a 104-in. wheelbase and featured side entrance rear doors. Both models had a 56-in. tread and 34-in. tires on artillery-type wheels.

Success in Racing

In the meantime Mooers continued to build race cars, some of which he drove himself, while others were driven by Charles Wridgeway and Joseph Tracey. These daring drivers added to Peerless recognition and prominence by winning many races and establishing a world speed record at Ormond Beach, Florida. In June 1904, Mooers and Peerless management were able to lure Eli (Barney) Oldfield away from Winton Motor Carriage Company, where he had established himself as a consistent winning race driver. By the end of July 1904, Mooers had a new 60-hp race car called the "Green Dragon" ready for Oldfield.

Barney Oldfield in 1934. (Source: Indianapolis Motor Speedway Corp.)

By year's end the Peerless Green Dragon, driven by Oldfield, had established many new records, from 1 to 50 miles on circular tracks. Tragically, in the race at the St. Louis World's Fair on August 28, 1904, Oldfield, blinded by dust from the car ahead, crashed through the track fence into the spectators, taking the lives of two people. Oldfield was injured in the crash, but recovered promptly to continue his tight schedule setting new records at all major race tracks in the United States. The effective Peerless sales promotion was based on the advertisement that the production Peerless passenger cars were the offspring of the Green Dragon race cars.

Barney Oldfield remained the Green Dragon driving ace through 1904 and halfway through 1905, leaving Peerless

1904 Peerless Green Dragon. (Source: Crawford Museum)

shortly after Mooers resigned in April 1905. Mooers was soon afterward named chief engineer and designer for the Moon Motor Car Company of St. Louis, Missouri, where he developed many of his innovative designs. At Peerless, Mooers was succeeded by Charles Schmidt, a former chief engineer and race driver for the Packard Motor Car Company of Detroit.

On May 5-6, 1905, Wridgeway and his mechanic drove a 24-hp Peerless passenger car on a 1,000-mile endurance run at the Brighton Beach race track in 25 hours and 50 seconds, bettering by 4 hours, 3 minutes, and 36 seconds the record established by Charles Schmidt in a Packard at Grosse Pointe, Michigan, the year before.

Passenger Cars

For 1905 the three new Peerless models, priced at $3,200 to $6,250, all had four-cylinder engines and four-speed transmissions. Model 9 had a 24-hp engine, Model 11 had a 35-hp engine, and Model 12 had a 60-hp engine, which Peerless advertised was a duplicate of the Peerless Green Dragon racing engine. The touring car, tonneau, Victoria, and the limousine all had side doors to the rear compartment.

In the initial Glidden Tour from New York City to Bretton Woods, New Hampshire, in 1905, the two Peerless entries gave a good account of themselves. The following year, again in the Glidden Tour, a Peerless Model 14 attained a perfect score in the run from Buffalo, New York, to Bretton Woods.

The 1906 Peerless line consisted of two models: Model 14 having a 30-hp engine and a 107-in. wheelbase cassis, and Model 15 having a 45-hp engine and a 114-in. wheelbase chassis. The 1906 models were the last designed by Mooers, and had a price range of $3,900 to $5,250.

The 1907 line of Peerless cars continued the Model 15 and added a new, smaller, four-cylinder Model 16. However, the price range still ranged upward from $4,000. It could be said that the 1907 models were a fill-in until the new 1908 models were introduced. Despite the business downturn and the financial panic of 1907, Peerless did not have difficulty selling their cars because of the splendid reputation Peerless had earned. Starting in 1907, Peerless adopted the unforgettable slogan, "All That the Name Implies." In the

1908 Peerless. (Source: Crawford Museum)

meantime Schmidt was busy developing a new six-cylinder engine. The new Peerless 57-hp, six-cylinder engine with a bore of 4-7/8-in. and a stroke of 5-1/2-in. was primarily based on the earlier design of Mooers' four-cylinder, 30-hp engine. During 1907 wet storage batteries were added to the dry-cell ignition circuits, and in 1908 the Eisemann low-tension magneto was employed, giving Peerless a dual-ignition system.

The 1908 models introduced in late 1907 included the new Model 20 on a 133-in. wheelbase, featuring the new 57-hp, six-cylinder engine. The 30-hp engine was continued on the line designated as Model 18, having a 118-in. wheelbase. In mid-1908 a new close-coupled touring model was introduced. The seating arrangement placed the passengers between the front and rear axles, giving greater riding comfort for the rear-seat passengers. This model was the forerunner of the new 1909 models.

For 1909 the Peerless models had basically the same features as the 1908 models, except for longer wheelbases and a higher price. The Model 19 which succeeded the Model 18 had a four-cylinder engine, a 122-in. wheelbase, and a price range of $4,300 to $5,800. The Model 25, successor to Model 20, had a six-cylinder engine, a 136-in. wheelbase, and a price range of $6,000 to $7,300, attaining the snobbish distinction of being America's most expensive car. The assumed impression that a majority of American car buyers preferred European-built cars because of their snob appeal was a fallacy. During 1909 the total import of all makes of European cars was 1,624 units valued at approximately $2,900,000. This was less than the production and sales of Peerless alone. By 1914 the total of imported European cars dropped to about 300 cars valued at $620,000. In April, Peerless capitalization was increased from $600,000 to $3,000,000, indicating Peerless' sound financial condition.

The 1910 Peerless models were changed but very little, except a new, smaller, 20-hp, four-cylinder-engined Model 26 was added to the line. Model 19 became Model 27 and the six-cylinder Model 25 became Model 28. Both the Models 27 and 28 offered the close-coupled touring car. The price range remained at $4,300 to $7,000. During 1910 Peerless built and sold approximately 1,500 cars.

1908 Peerless. (Source: Crawford Museum)

In late 1910 Peerless introduced a line of heavy-duty trucks, featuring a chain-drive rear axle. Three model lines of Peerless cars were offered in 1911. Models 29 and 31 had four-cylinder engines, while Model 32 had a 45-hp, six-cylinder engine and a 136-in. wheelbase. The three model lines offered 19 body types, including the new roadsters and torpedo phaetons. The price range was from $4,300 to $7,000.

The 1912 Peerless models were introduced in May 1911.

	1912 Peerless Model D	**1912 Peerless Models J, H**	**1912 Peerless Model K**	**1912 Peerless Model L**
Wheelbase (in.)	113	124.5/125	137	140
Price	$4,200-4,300	$4,000-5,400	$5,000-6,200	$6,000-7,000
No. of Cylinders / Engine	4	L-6	L-6	L-6
Bore x Stroke (in.)	4.00 × 4.62	4.00 × 5.50	4.50 × 6.00	5.00 × 7.00
Horsepower	25.4 NACC	38.4 NACC	48.6 NACC	60 NACC
Body Styles	2	6	5	6
Other Features	Gray and Davis charging and lighting electrical systems			

1912 Peerless. (Source: Peerless Catalogue)

During 1912 Peerless engineers working with Gray and Davis developed an electrical starting system. While some Peerless cars were built with electric starting motors, electric starters became standard equipment on all 1913 models.

The Revolving Door Starts Turning

Peerless had a profitable year in 1912, selling approximately 1,700 cars. The motor truck end of Peerless business was no less profitable. American Express Company of New York purchased eleven additional trucks to supplement the five they had in operation. On April 20, 1912, Peerless announced that arrangements had been made to sell $1,100,000 in 6% bonds, and that Cleveland banks had already underwritten the entire issue. The company also had issued additional stock in the amount of $300,000. The additional financing was arranged to provide funds to pay off some of the floating indebtedness, and to further develop the commercial vehicle end of the company's business. This additional financing may have precipitated the future financial roller-coaster ride and management revolving-door process that plagued Peerless in later years.

Due to a successful business operation with substantial profits during 1912, Peerless declared a sizeable stock dividend in January 1913. Along with the announcement came the news of a change in financial control of Peerless. A large block of Peerless stock was acquired by a group of Cleveland financiers associated with the National Electric Company. Although Lewis Kittredge remained as president, Theodore Frech of National Electric was named general manager. E.H. Parkhurst, Peerless vice president, and F.I. Harding, Peerless treasurer, both resigned. Authorized capitalization of Peerless was again increased in 1913, this time to $7,000,000 to provide the company with paid-in capital stock of approximately $4,000,000 and assets of over $5,000,000. Business conditions were good for Peerless, and the profits continued to rise due to the increased demand for Peerless cars and commercial vehicles.

The 1913 designs and model line-up were essentially the same as for 1912, except for the electric starting, charging, and lighting systems. On April 25, 1913, Peerless announced their new 1914 models. The body designs were the same as the 1913 models, except the bright metal parts were nickel-plated instead of brass finish.

	1914 Peerless Model 38-6	1914 Peerless Model 48-6	1914 Peerless Model 60 Six
Wheelbase (in.)	125	137	140
Price	$4,300-5,500	$5,000-6,200	$7,000-7,200
No. of Cylinders / Engine	L-6	L-6	L-6
Bore x Stroke (in.)	4.00 × 5.50	4.50 × 6.00	5.00 × 7.00
Horsepower	38.4 NACC	48.6 NACC	60 NACC
Body Styles	7	5	5
Other Features			

Passenger car sales dropped in 1914, but with the outbreak of World War I in Europe came the increased demand for American-built motor vehicles. Britain, France, and Russia placed large orders for Peerless cars, trucks, and special military vehicles. Peerless, along with several American car manufacturers, supplied large quantities of military vehicles to the Allied forces. In addition to vehicles of their own design, Peerless also built four-wheel-drive trucks under license by the FWD Company. Peerless profits for the truck division for 1915 were over $2,500,000. While the truck business was good in 1914 and 1915, Peerless passenger car sales dipped dismally to below 650 units in 1914. Peerless suspended dividend payments in April 1914, even though the passenger cars distinguished themselves with many luxurious features, such as silver-plated interior hardware and the use of Swiss railway glass.

Peerless management must have realized that the car pricing structure was too high, and noted the success of Hudson, Chalmers, Buick, and Studebaker in the low-medium price range. Peerless for 1915 continued the 48-hp, six-cylinder-engine-powered model of the previous series, with the price range remaining at $4,900 to $6,200, but added two new "All Purpose Lines."

	1915 Peerless Model 54	1915 Peerless Model 55
Wheelbase (in.)	113	121
Price	$2,000-3,500	
No. of Cylinders / Engine	4	6
Bore x Stroke (in.)	3.75 × 5.00	3.50 × 5.00
Horsepower	22.5 NACC	29.4 NACC
Body Styles	5	5
Other Features		

The management decision must have been a good one, because Peerless built and sold approximately 7,400 units in 1915, including cars, trucks, and military vehicles. Peerless resumed dividend payments on July 1, 1915.

During 1914 and 1915 the Peerless engineers worked feverishly in the development of a new V-8 engine. Charles Schmidt left Peerless and returned to France in 1915, perhaps because of World War I. George Wadsworth temporarily became chief engineer, but was soon afterward succeeded by William R. Strickland. F.W. Slack, who joined Peerless in 1908, was named assistant chief engineer.

Peerless Truck and Motor Corporation

The second phase of the financial roller-coaster ride and the second orbit of the management revolving door occurred in November 1915, when the financial control of Peerless passed to a New York financial syndicate headed by Harrison Williams, who was identified with the Cleveland Electric Illuminating Company. On November 3, 1915, the Peerless Motor Car Company and the General Vehicle Company of Long Island City, New York, were merged to form the Peerless Truck and Motor Corporation, incorporated under the laws of Virginia, capitalized at $20,000,000. Harrison Williams was elected chairman of the board of directors. The Peerless Motor Car Company was represented on the board of Peerless Truck and Motor Corporation by L.H. Kittredge president, T.W. Frech vice president, B.J. Tremaine and F.S. Terry, identified with the National Electric Lamp Works of Cleveland. The management of the subsidiary General Vehicle Company who were on the board of directors of the Peerless Truck and Motor Corporation were P.D. Wagoner president, A.W. Burchard vice president, M.F. Westover secretary, and E.A. Carolan, A.K. Baylor, W.B. Potter, and F.C. Pratt, all identified with the General Electric Company. General Electric's interest in the Peerless Truck and Motor Corporation may have been with the relation to propulsion power and power transmission. As the result of the merger, the Peerless Truck and Motor Corporation in effect showed a production of approximately 7,400 units, and a net profit of $2,555,773 in 1915, equivalent to 19% of the new stock.

The New V-8

Regardless of the upheaval in change of management, Strickland and Slack, in collaboration with the Herschell-Spillman Engine Company of North Tonawanda, New York, were able to continue with the development of the new V-8 engine.

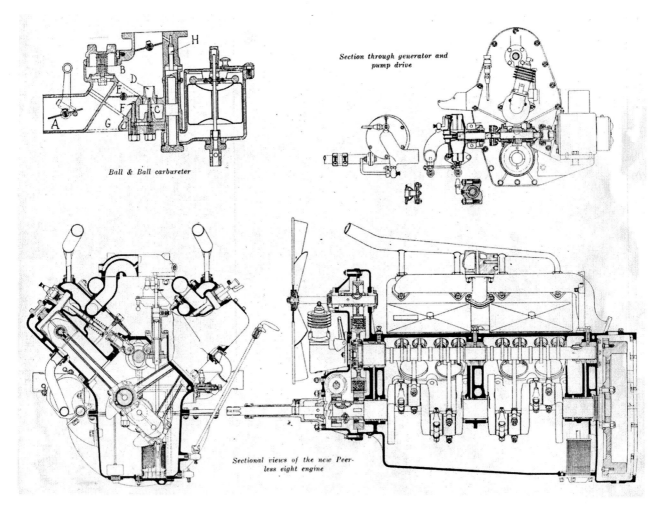

Sectional views of Peerless V-8 engine (bottom); Ball & Ball carburetor (top left); section through generator and pump drive (top right). (Source: Automotive Industries)

The new 1916 Peerless Model 56 was announced in December 1915, featuring the new Peerless designed and manufactured V-8 engine with cylinders staggered in the block, allowing side-by-side connecting rods. The valves were actuated by cam and rocker arm arrangement. The four- and six-cylinder engines were discontinued.

1916 Peerless Model 56

Wheelbase (in.)	125
Price	$1,890-3,060
No. of Cylinders / Engine	V-8
Bore x Stroke (in.)	3.25 × 5.00
Horsepower	80 adv, 33.8 NACC
Body Styles	seven-passenger touring, three-passenger roadster, seven-passenger limousine
Other Features	

Peerless enjoyed a production of approximately 9,000 units in 1916, and realized a net profit of $1,358,810, amounting to about 13.5% on the stock.

The 1917 Peerless models had only slight changes. Two new body styles were added: a coupe and a sedan model. The price range was increased to $2,340 to $3,690. While part of the Peerless production was for military vehicles, the combined total for 1917 was again approximately 9,000 units. Peerless earned a net profit of $1,065,869 for 1917, amounting to about 10.6% on the stock.

In early 1917 Peerless added a group of three new factory buildings, placing Peerless in a favorable position to obtain war materiel contracts. In the meantime Fred Slack redesigned the V-8 engine, placing the camshaft action directly against the roller tappets actuating the valves, eliminating the rocker arms. A new intake manifold and dual venturi carburetor, designed by Frederick Ball and Thomas Ball (father and son), called the "Ball and Ball," was manufactured by the Penberthy Injector Company of Detroit, Michigan, and was introduced on the 1918 Peerless cars. The carburetor was of a 1-1/8-in. double-venturi type, with two throttle valves operated progressively by the throttle linkage. At part throttle (one throttle valve open) it gave the engine smooth, economical performance. As the accelerator pedal was depressed further, the linkage opened the secondary throttle valve, giving the engine full-power "two-stage performance." Thus, Peerless introduced the "Two-Power-Range-Eight," and publicized it as a car of dual personality.

(It is noteworthy that this carburetor design was later used on the Chrysler 70, 72, 75, and the Imperial 80 models. The progressive throttle linkage design made four-barrel carburetors possible, and multiple-carburetor arrangements practical on high-performance engines in recent years. It is befitting to give Thomas Ball recognition for the enormous amount of work and experimentation he performed in pioneering and developing the "Electronic Fuel Injection" system as was used on the Chrysler 300D in 1958. Three decades later EFI had worldwide acceptance and usage. Thomas Ball designed the many Ball and Ball carburetors for Chrysler Corporation, manufactured by the Carter Carburetor Division of ACF.)

Another Turn

Shortly after the Armistice on November 11, 1918, those associated with General Electric Company gracefully withdrew their financial interest by selling their controlling shares of Peerless stock to a Cleveland-based financial group headed by B.G. Tremaine and F.S. Terry. This resulted in another orbit of the Peerless management revolving door. Tremaine was elected president and general manager of Peerless Truck and Motor Corporation. Kittredge remained as president of the subsidiary Peerless Motor Car Company.

The year 1919 brought about readjustment to peacetime civilian vehicle production and sales. The business recession of 1919-20 reduced the 1919 profits to $670,628 and $1,063,306 for 1920. The production for 1920 was approximately 1,500 vehicles. While Peerless produced more vehicles in 1921, the company was faced with a net loss of $103,665, reflecting the extreme inflation in operating costs and purchased material. Increasing the sales price of Peerless vehicles to $3,200 to $4,400 only meant more loss of business. The Peerless model line-up remained basically the same through 1924. A five-passenger sedan was added in 1922, and a four-passenger coupe and four-passenger phaeton were introduced in 1924.

Richard Collins

In late September 1921, the negotiations for the acquisition of control of the Peerless Truck and Motor Corporation by Richard H. Collins, former president of the Cadillac Motor Car Company, and his associates were in progress. B.G. Tremaine, president and general manager of Peerless, and George B. Siddall, secretary and attorney for Peerless, represented Peerless in the negotiations.

1919 Peerless. (Source: Crawford Museum)

The information with regard to the terms of the negotiations was included in circular letters sent to the stockholders. According to the circulars, Collins and his associates were to purchase at least 50,000 and not more than 80,000 shares of the 200,000 shares outstanding. He was to pay $50 per share, of which not less than $10 per share would be paid in cash, and the balance in six equal installments to be paid semi-annually over a period of three years, with interest at 6%. Stockholders who desired to sell their stock on the terms given were asked to deposit their shares with the Cleveland Trust Company.

On October 3, 1921, at the board of directors meeting of the Peerless Truck and Motor Corporation, the negotiations were consummated and the contract was signed after the initial payment of 20% of the approximately $4,500,000 involved in the deal. In Collins' contract with Peerless, at an annual salary of $150,000, it was also stipulated that when the new six-cylinder car (designated the Collins Six) made a profit of $1,000,000 for Peerless, Collins would receive a bonus of $65 per car sold.

Following the signing of the contract, Collins was elected president and general manager of the corporation, and the wheels of the management revolving door started to turn again. At the board meeting B.J. Tremaine, F.S. Terry, Lewis Kittredge, Theodore Frech, and Harrison Williams resigned as directors. They were succeeded by Richard Collins, Wilbur M. Collins (Richard's son), and F.A. Trester. H.A. Tremaine, Roland T. Meacham, and George B. Siddal remained on the board.

At that time Collins was recognized as one of the most versatile men in the automotive industry. His credentials as to his experience and success proved it. Collins entered the automotive industry as general manager of the Kansas City Office of Buick Motor Company in the early days of Buick. He directed the Buick sales and distribution in this vast and important territory with such definite success, that three-and-a-half years later he was called to Flint, Michigan, to assume the greater responsibilities of general sales manager of the Buick Motor Company. He was also elected a director of General Motors Company.

His achievements in the rehabilitation of Buick Motor Company, by improving the car, and strengthening the factory, distribution, and selling organizations, not only multiplied Buick sales and profits many times over, it also put Flint on the industrial map. His spectacular performance had already been recorded among the most sensational accomplishments in automobile history. In 1916 he was appointed assistant to William C. Durant, then president of General Motors Corporation, maintaining offices in New York and Detroit for the management of the most important affairs of the corporation. On July 16, 1917, R.H. Collins was elected president and general manager of the Cadillac Motor Car Company.

During his tenure as president of Cadillac, Collins doubled the company's sales volume and earnings, opened a factory branch office in Chicago, planned and finished a new sales and service building on Cass Avenue in Detroit, and had a new Cadillac manufacturing plant designed and erected on Clark Avenue in Detroit. This plant was considered the most modern and efficient at the time.

On November 21, 1921, reorganization of the personnel of Peerless Truck and Motor Corporation under Collins' management was started with the appointment of Benjamin J. Anibal as chief engineer to succeed W.R. Strickland, whose resignation was accepted. Fred W. Slack stayed on as assistant chief engineer. Anibal had been Cadillac's chief engineer for several years. C.R. Cunliffe was appointed general sales manager, succeeding Robert J. Schmunk, who also resigned after 14 years with Peerless. Cunliffe was previously the general manager of the Chicago Cadillac factory branch.

During Collins' management the image and prestige of Peerless soared, the profitability of the corporation went from a net loss of $103,665 in 1921 to a net profit of $1,005,112 in 1922, and the net profit after taxes for the first six months of 1923 was $955,995. The V-8 engine that had powered Peerless exclusively since 1916 underwent an extensive redesign, and the six-cylinder-powered car (originally planned to be the Collins Six) was being developed and readied for introduction. For 1923 and 1924 Peerless continued its V-8, now designated Model 66.

In August 1923, a disgruntled stockholder brought two class suits against Collins, under the guise of protecting the corporate funds and stockholder interests, claiming that the domination of the board of directors rested absolutely with Richard H. Collins. It turned out this masquerading minority stockholder was in fact a stock broker and had a private practice in which he collected $75,000 in lawyers' fees for his services in the settlement of the Peerless-Collins suit.

On November 15, 1923, the litigation was settled in the Common Pleas Court in Cleveland, Ohio, agreeing with Collins' original previous offer. Collins restored $150,000 to the Peerless treasury, and took a cut in his salary to $75,000 annually. He also agreed to forego the $65 bonus clause in his contract. It was Collins' fervent hope that the settlement would satisfy anyone who may have had grounds for criticism, and would allow the officers of the corporation to be free to accomplish all the plans for Peerless' success and well-being. Unfortunately, the damage was already done. Hanging the corporate "dirty laundry" in public view brought Peerless' image to its lowest level.

Tired of what had practically amounted to nothing less than persecution, Collins resigned on November 1, 1923, which was accepted at the board of directors meeting on December 26th. His son, Wilbur Collins, vice president and director, also tendered his resignation. After the meeting, the two left for California, where the senior Collins had purchased a home in Pasadena.

D.A. Burke, who had been vice president and director of sales under Collins, was appointed general manager at this board meeting. After the readjustment of personnel, the officers were as follows: George H. Laying vice president in charge of manufacturing, E.B. Wilson assistant sales director, F.A. Trester secretary, and John F. Porter treasurer. Burke, the new general manager, had long been a familiar figure in the automotive industry. For a long time he was in the sales division of Buick Motor Company. In 1920 he became the president and general manager of Sheridan Motor Car Company of Muncie, Indiana. In May 1921 Burke purchased the Sheridan Motor Car Company, and in August 1921 unloaded it to William C. Durant. In January 1922 Burke became one of Collins' associates, and was appointed general sales manager of Peerless, and in July 1923 he was elected to the board of directors of the company.

When Collins left, the Peerless Truck and Motor Corporation was in splendid physical and financial condition. It had $1,300,000 in the banks, and had just previously declared a regular quarterly dividend of $1 per share. Total registrations in the United States were 4,775 for 1923 against 4,200 in 1922. Unknown to the Peerless board at the time, Collins was being prevailed upon by Walter P. Chrysler, just prior to the introduction of the new Chrysler in January 1924. In early 1924 the R.H. Collins Automobile Company of Chicago was formed and Collins became the Chrysler distributor for all of northern Illinois and northwest Indiana. Collins was so successful in this merchandising and distributing enterprise that, in envy, the Chrysler Sales Division of the Chrysler Corporation bought the interests of the R.H. Collins Automobile Company in December 1927 and established it as the Chrysler Illinois Company. This was the first stage for Chrysler in establishing regional and zone offices in principal cities. After the sale to Chrysler, R.H. Collins retired to his estate in California.

D.A. Burke accepted the position of general manager, on the condition that he would stay on only until a permanent successor could be named. Prior to Collins' departure, Benjamin Anibal, chief engineer, resigned, and Fred Slack was appointed chief engineer in May 1923.

In early January 1924, Earl B. Wilson was appointed general sales manager, filling the position previously held by Burke. Through no fault of Burke, but as the result of unfavorable publicity, the buying public lost confidence and Peerless representation dropped. Peerless sales suffered accordingly, dropping to a low of 3,936 units for 1924, even while 1924 was a record year for most competing automobile companies.

New Designs

Three new prototypes of the new Peerless Six, developed from Collins' designs, were shown at the National Automobile Show in New York during January 5-12, 1924. This new Peerless Six, identified as the Model 6-70, was shown in three body types: a five-passenger phaeton, a two-passenger roadster, and a five-passenger sedan. This new car featured a new six-cylinder engine having a seven-main-bearing crankshaft with the cylinder bore of 3-1/2 in. and 5-in. stroke. However, there was difficulty in starting production of this car, and it finally emerged in mid-1924 as a 1925 model. No records could be located that would indicate that this car was manufactured on a production basis and sold as a 1924 model.

The 1923 and 1924 V-8 models were known as the Model 8-66. On November 20, 1924, Peerless introduced the new 1925 model known as the "Equipoised Eight" Model 8-67. While the 1923 and 1924 models had a refined Peerless radiator shell and hood styling design, the appearance of the new 1925 Peerless had been radically altered and improved by a complete radiator and hood shoulder redesign. While it retained the Peerless character and identity, the new hood and radiator shell lines introduced a new styling trend, which would be copied by competition in subsequent years.

An inherently balanced V-type engine, and an entirely new line of seven body models, were the outstanding features of the new 1925 Peerless Equipoised Eight. A heavier, counterweighted, "two-plane" crankshaft was responsible for the smoothness of the inherent balance of the new V-8 engine. In this engine, the crankshaft had the throws #2 and #3 in a plane at right angles to that of crank throws #1 and #4. Formerly, all four throws were in the same plane (a flat crank). The general layout was the same as for the crankshaft of a four-cylinder engine.

1924 Peerless Model 8-67. (Source: Crawford Museum)

In the Peerless crankshaft design, there were four forged-steel counterweights attached to the crankshaft by means of nickel-steel bolts. These counterweights were of such size and weight, and were placed in such location on the shaft, that they set up forces that were equal to and in opposite direction to those created by the inertia of the reciprocating engine parts. Consequently, the two sets of forces neutralized each other.

External, contracting, Lockheed hydraulic four-wheel brakes and 33 x 6.60-in. balloon tires were standard equipment on the Peerless Equipoised V-8 Model 8-67. A particularly important feature from an engine maintenance viewpoint was the adoption of the removable cylinder heads. In most other respects the chassis was, substantially, the same as the former Model 8-66.

Burke inherited a tremendous task of reconstructing the Peerless organization in 1924 after the adverse publicity in 1923. Needless to say, 1924 did not generate a profit, and the Peerless management was about to take off on another spin of the revolving door in February 1925.

While the Model 6-70 Collins Six was late in getting into production in mid-1924, approximately 1,700 of them were built and sold by the time the Peerless Six was refined and emerged as Model 6-72. The Peerless Model 6-72 had the same chassis components as those of the previous Model 6-70, except for the incorporation of external, contracting, four-wheel hydraulic brakes and balloon tires. The styling was improved considerably with the newly designed hood and radiator shell, almost identical to the Equipoised V-8 Model 8-67. The 6-72 was available in two wheelbase lengths of 126- and 133-in., offering seven body types, in the price range of $1,995 to $2,695.

Edward VerLinden

On February 18, 1925, the Peerless board of directors accepted Burke's resignation, and announced that Edward VerLinden was chosen as president and general manager of Peerless Truck and Motor Corporation. VerLinden came to Peerless with credentials of the highest caliber. After his formal education, VerLinden's career as an automotive executive began in 1906, when he organized the Michigan Auto Co. to manufacture

Edward VerLinden. (Source: R.E. Olds History Center)

components and parts for the automotive industry. In 1910 his company was purchased and absorbed into the General Motors Company. VerLinden went to Flint, Michigan, where he was appointed Master Mechanic at the Buick Motor Company. Within a year he became factory manager of Buick at Flint.

In 1912 he became works manager of the Olds Motor Works in Lansing, Michigan. On August 15, 1916, he was elected president and general manager of the Olds Motor Works, and at the same time was elected vice president of the General Motors Company, and a member of the board of directors. His management of Olds for nine years stood out as one of the most successful and prosperous periods in the history of that organization.

He resigned from Olds Motor Works and General Motors Corporation in August 1921 to join W.C. Durant as president of Durant Motors of Michigan. During his tenure at Durant he was responsible for the building of the large and efficient Durant manufacturing plant in Lansing, Michigan. VerLinden discontinued his association with Durant in late 1923, and purchased the Ryan-Bohn Foundry in Lansing. He then took his family to Europe for an extended vacation, the first real vacation he had ever indulged in.

On April 22, 1925, Edward VerLinden was formally elected president and general manager of Peerless at a board of directors meeting held at the Peerless plant. Leon R. German, who was a former vice president and comptroller at Olds Motor Works during VerLinden's tenure, and who was later the general sales manager of the Durant Motors Corporation of Michigan, was elected vice president of Peerless Truck and Motor Corporation. F.K. Trester, secretary, and John F. Porter, treasurer, were re-elected to their respective offices.

One of VerLinden's first observations was that the Peerless 6-72 was not selling as well as expected, and that Peerless needed a smaller, less-expensive, six-cylinder-engined car in the line-up. Therefore, plans for the design and development of such a vehicle were given the go-ahead at full throttle. In the meantime VerLinden reviewed the price structure of the Equipoised V-8, and decided to lower the prices. On March 11, 1925, substantial price reductions of $340 to $580 on all Peerless Equipoised Eight models were announced. This meant that the price range of the seven body types was $2,945 to $4,195 against the previous range of $3,285 to $4,450. The 1925 Peerless cars were promoted by aggressive and consistent advertising.

Further price reductions of $100 to $300, announced on August 13, 1925, on all Peerless models, helped to improve Peerless sales. Sales for the 1925 model year were 4,755 vehicles, and Peerless realized a modest profit of $126,804.

During the week of November 16, 1925, Peerless announced the 1926 models, and entered the medium-price market with the new Model 6-80.

Peerless

1926 Peerless. (Source: Peerless Sales Catalogue)

1926 Peerless Model 6-80

Wheelbase (in.)	116
Price	$1,595 (sedan)
No. of Cylinders / Engine	L-6
Bore x Stroke (in.)	3.25 × 4.62
Horsepower	63 adv, 25.3 NACC
Body Styles	6
Other Features	roadster model featured "boat-tail" rear deck styling

The L-head engine, designed by Peerless engineers but built by the Continental Motors Corporation, had a displacement of 230.2-cu.-in. and featured a seven-main-bearing crankshaft (similar to the 6-72), but the cast-iron cylinder block and cast-iron crankcase were cast together as a unit. This design offered a cost reduction.

Peerless Motor Car Corporation added a third series of six-cylinder-engine-powered models to the existing line of four chassis during the week of October 5, 1926. This new model, identified as the 6-90, was developed to compete in the $1,800 price class, midway between the 6-72 and 6-80 models.

1926 Peerless Model 6-90

Wheelbase (in.)	120
Price	$1,800
No. of Cylinders / Engine	L-6
Bore x Stroke (in.)	3.50 × 5.00
Horsepower	70 adv, 29.4 NACC
Body Styles	
Other Features	

VerLinden's decision in early 1925 to expand into six-cylinder-engined cars must have had merit, because it paid off handsomely. The all-time record of 10,430 vehicle sales, for a gross sales total of $19,301,301, resulted in a net profit of $919,883 for 1926. The December 31, 1926, report showed Peerless was in the healthiest financial condition of its history. Assets were $5,994,138 and the liabilities were only $546,838.

The 1927 model year started off with four car lines: the 6-72, 6-80, 6-90, and the Equipoised Eight Model 8-69, which were improved over the 1926 models. During the week of April 25, 1927, Peerless started shipments of the new Peerless Model 6-60.

1927 Peerless Model 6-60

Wheelbase (in.)	116
Price	$1,295-1,345
No. of Cylinders / Engine	6
Bore x Stroke (in.)	3.25 × 4.00
Horsepower	70 adv, 29.4 NACC
Body Styles	7
Other Features	external, contracting, Lockheed hydraulic four-wheel brakes

The car featured a Continental-built, seven-main-bearing engine, and body styling similar to that of the Model 6-80. Thus, during the 1927 model year, Peerless proliferated into five model lines, four wheelbase lengths, four different engines, and 26 body types. The overall price range was from $1,295 to $3,795. In 1927 the sales for the entire automobile industry dropped by approximately 600,000 units, reflecting a decline of about 19% industry-wide. While Peerless sales dropped slightly to 9,872 units, 1927 was still a profitable year for Peerless.

The last two weeks of July 1927 were spent in preparing the new model line-up for 1928. Models 6-60, 6-72, and 6-80 were refined and continued with minor changes. Model 6-90 became Model 6-91 in January 1928. The Equipoised Eight Model 8-69 and the Model 6-72 were continued for one more model year through December 31, 1928. The price range was from $1,295 to $3,295, reflecting lower prices in the more expensive models. While there was much optimism for Peerless personnel at year end, Peerless was about to embark on another orbit of the management revolving door.

1927 Peerless four-passenger coupe. (Source: Crawford Museum)

1928 Peerless. (Source: Crawford Museum)

Another Spin of the Door

On January 1, 1928, Edward VerLinden resigned as president and general manager of Peerless Motor Car Corporation. It seems that the financial backers just couldn't "stay out of the kitchen and let the chef do the cooking." The sudden departure of VerLinden left the board of directors unprepared, even though he left the Peerless Motor Car Corporation in a favorable sales position and a strong financial condition. VerLinden took his family on a two-month vacation, and returned to Cleveland on March 5, 1928, as director and chairman of the executive committee of the Jordan Motor Car Company. In the meantime the directorate and the management of Peerless remained intact, and the business was conducted by an executive committee composed of Robert M. Chalfee, Leon R. German, F.A. Trester, and Charles A. Tucker, with vice president German as acting general manager. On March 16, 1928, a Detroit investment banker announced that a group of Detroit and Cleveland businessmen had purchased a substantial interest in Peerless; however, the financial control remained with Cleveland businessmen.

A board of directors meeting was held on April 3, 1928, in Richmond, Virginia, to elect board members of the corporation. The directorate was composed of Chalfee, German, secretary-treasurer Trester, and general sales manager Tucker. The rest of the board was filled with prominent Detroit and Cleveland financial leaders. A special board meeting was held in Cleveland on April 19, 1928, to elect officers of the corporation. Robert Chalfee was elected chairman of the board, Leon German was elected president, A.F. Misch vice president of manufacturing, C.A. Tucker vice president in charge of sales, F.W. Slack vice president of engineering, and F.A Trester secretary-treasurer.

In August 1928 the first of the Peerless 1929 models was introduced, the refined Model 6-81 to replace the Model 6-80. The new body lines followed closely to that of the Model 6-91. The 6-81 had an increased cylinder bore from 3-1/4 in. to 3-3/8 in., increasing the displacement to 248 cu. in., and the Lockheed hydraulic four-wheel brakes were changed to internal expanding shoes. Numerous minor refinements included a larger flush-type radiator filler cap, and chromium plating for all external bright metal parts. Standard equipment now included bumpers front and rear, Lovejoy shock absorbers, automatic windshield

1929 Peerless Model 125 eight-cylinder engine. (Source: Crawford Museum)

wiper, etc., but all this came with a price increase of $200 to $245. Deluxe equipment of six wire-spoke wheels or six disc wheels, with a trunk rack at the rear, was available at $130 extra for disc or $180 for wire-spoke wheels.

On January 1, 1929, the new Model 6-61 was introduced with styling and other improvements similar to the 6-81, including the new internal expanding shoe Lockheed hydraulic four-wheel brakes. But, sadly, the splendidly crafted Peerless 6-72 engine and the magnificent Equipoised Eight V-8 engines were gone due to cost considerations. The 8-69 V-8 engine was replaced by a Continental-built in-line-eight (straight-eight) L-head engine of 3-3/8-in. bore and a 4-1/2-in. stroke, having five main bearings. The model number for the eight-cylinder-engine-powered car became Model 125. While the in-line engine developed adequate horsepower, it did not have the character of the 8-69. All 1929 models had improved styling, and the Model 125 had six wire-spoke wheels, fender wells (to carry the spares), and folding trunk rack at the rear as standard equipment. The price range was $995 to $2,295. Peerless had the hope of gaining a larger share of the market with lower prices, and indeed they did somewhat, with 8,318 registrations for 1929.

John Bohannon

Lo and behold, the management revolving door made another orbit in 1929. On July 1st, Leon R. German was replaced by John A. Bohannon as president of Peerless Motor Car Corporation. Bohannon was born in Knoxville, Tennessee, in 1895. In college he studied law and engineering. He entered the automobile industry as purchasing agent for a Detroit automobile manufacturer. After World War I he joined Nordyke and Marmon Company in the purchasing department, becoming a vice president in the mid-1920s. He resigned from Marmon in June 1929, and in July he joined Peerless Motor Car Corporation as president and general manager. Don P. Smith joined Peerless shortly after Bohannon. Smith had financial expertise, and was elected vice president and director of Peerless, assigned as financial advisor to Bohannon.

The Depression Hits

Alas, two ominous events took place in 1929 that would determine Peerless' destiny. The first was "Black Thursday," October 24, 1929, the stock market crash and the cruel business depression that followed. Then, on December 19, 1929, Fred Slack resigned as vice president and chief engineer of Peerless to become associated with the Overseas Division of the General Motors Export Corporation.

The 1929 Models 6-61A, 6-81, and 125 were carried over until December 31, 1929, as "first series" 1930 models. It was Bohannon's strong feeling that Peerless could not compete in the low- and medium-price field. Hence, he went all-out for the fine-car market. On December 28, 1929, the new (second series) 1930 Peerless Master Eight and Peerless Custom Eight were introduced.

	1930 Peerless Master Eight	1930 Peerless Custom Eight
Wheelbase (in.)	125/138	125/138
Price	$1,995-2,095	$2,795-3,145
No. of Cylinders / Engine	IL-8	IL-8
Bore x Stroke (in.)	3.37 × 4.50	3.37 × 4.50
Horsepower	120 adv, 36.34 SAE	120 adv, 36.34 SAE
Body Styles	5	6 including seven-passenger limousine
Other Features	Warner four-speed transmission	Warner four-speed transmission

The brakes were the new Bendix two-shoe internal expanding cable-operated four-wheel brakes. The superb Peerless body exterior styling and exquisite interior design was created by Alexis de Sakhnoffsky (a world-renowned Russian designer) and his design staff. The impressive exterior could only be described as a "majestic beauty," reflecting dignity, culture, and heritage of the highest order. The new body contours, the newly shaped radiator shell with automatic shutters, long hood with side ventilating doors, and the clamshell-type long front fenders were all designed to complement each other in the creation of this luxurious beauty. Six wire-spoke wheels were standard equipment, with the spares carried in the front fender wells.

In the spring of 1930 the Peerless Master Eight and Custom Eight were joined by a third Peerless Eight called the Standard Eight.

	1930 Peerless Standard Eight
Wheelbase (in.)	118
Price	$1,495-1,595
No. of Cylinders / Engine	IL-8
Bore x Stroke (in.)	2.87 × 4.37
Horsepower	85 adv, 26.35 SAE
Body Styles	5
Other Features	

The Standard Eight was similar in appearance to the Master Eight. The transmission was a three-speed, and the four-wheel mechanical brakes were cable operated. Certainly Peerless spanned the fine-car field completely, with an all eight-cylinder-engine-powered line-up.

Unfortunately the business depression took its toll during 1930. On March 24, 1930, Bohannon announced that a collaboration of Cleveland banks and eastern financiers had underwritten the proposed offering of Peerless stock of up to 125,000 shares, valued at $8 per share. During their tenure, Smith and Bohannon were able to cut operating expenses through consolidation of manufacturing operations. However, in April 1930, Smith submitted his resignation.

During 1930 the total sales of all makes of cars plummeted to about 60% of their 1929 level, and Peerless sales dropped to approximately 4,021 units. Despite the loss of sales, on January 5, 1931, Bohannon was able to announce that Peerless showed a net profit of $73,326 for the fiscal year ending September 30, 1930. This was due to close financial control.

But all was not as well as it seemed with Peerless. On June 22, 1931, the General Parts Corporation purchased the entire parts inventory of Peerleess Motor Car Corporation. Under the arrangement, the service parts for Peerless vehicles would be provided by the General Parts Corporation. While on the surface it might appear that this was a favorable financial deal, it was, in fact, an indication of how strapped Peerless was for operating capital.

The 1930 models were carried over into 1931, discontinuing the Standard Eight and concentrating on the Master Eight and Custom Eight. In early 1931 Alexis de Sakhnoffsky left Peerless and moved to Grand Rapids, Michigan, to join the Hayes Body Corporation as art director. It must have been a let-down for Sakhnoffsky, after designing the magnificent Peerless Master and Custom Eight models in 1930 and 1931, to channel his creativity into the designs of the DeVaux and subsequent Continental cars.

The Depression did not let up in 1931; it only became more severe. While the 1932 Peerless Master Eight and Custom Eight models were announced in June 1931, there are no records showing that they got into production. At the end of June the last production Peerless came off the line. While some publications may show Peerless ceasing operations at a later date, no production cars were built after June 30, 1931.

Certainly 1931 was a dismal year for Peerless, with sales amounting to only 1,249 units, providing a sales income of only $2,819,364 in 1931 against $6,478,047 in 1930. Peerless Motor Car Corporation suffered a net loss of $712,744 in 1931. Even though the Peerless pulse was weak, the breathing difficult and irregular, Peerless was not dead. Peerless would have one more surge of energy, a long shot, in an attempt to restore its prestige and desirability.

The Last Hope: The V-16

When Howard C. Marmon and a number of Marmon Motor Car Company officials formed the Midwest Aircraft Corporation of Indianapolis in 1926, it was generally assumed that Marmon Motor Car Company intended to enter the aircraft engine field. It was later disclosed that the aircraft company was used mostly as

1931 Peerless Master Eight cabriolet with rumble seat. (Source: Crawford Museum)

a cloak, under which Marmon developed the engine for the Marmon V-16. Design and development work started in 1927.

It may not be sheer coincidence that shortly after Owen Nacker left Marmon and joined the Cadillac engineering group, that Cadillac developed a V-16-cylinder engine for their top-of-the-line cars. Likewise, Bohannon (former vice president of Marmon), after he became president of Peerless, was infected with an extravagant passion for a V-16-cylinder engine to power an exclusive Peerless motor car.

Thus, in 1930 and 1931, despite financial difficulties, Peerless engineers, in collaboration with the Aluminum Corporation of America, designed and built one prototype aluminum V-16 engine, having a bore of 3-1/4 in. and a stroke of 3-1/2 in. The 45-degree V-type cylinder block and crankcase were cast integral, having steel cylinder liners. The connecting rods were forged ST-25 aluminum alloy with steel caps. The cylinder heads with overhead valves were cast of chrome-nickel-iron alloy. The engine power output was 173.3 hp at 3,300 rpm.

The chassis fabricated by Peerless on a wheelbase of 145 in. had all the components (except springs, brake drums, and rotating parts) of stamped or forged aluminum alloys. The 18-in. wheels of aluminum alloy were of the imported Bucciali design, mounted with 7.00 x 18 balloon cord tires. This beautiful chassis was driven to Pasadena, California, fitted with a temporary body for the trip. In Pasadena, the Walter M. Murphy Company designers created the most exquisite and flawlessly executed four-door sedan model, possessing the finest in materials and skilled workmanship. The finished product was a beauty to behold, giving pleasure to the observer in delicacy and excellence. The Peerless V-16 made some speed tests in excess of 100 mph in 1932 at Muroc Dry Lake, California, and returned to the factory in Cleveland. It was ironic that the Peerless V-16 should be completed at about the same time that the last production Peerless would come off the production line.

While the 1932 pilot models may have been on exhibit at the January 9, 1932, National Auto Show in New York, records do not indicate any sales in 1932. If there were any sales, they may have been included under miscellaneous sales. The 1932 National Auto Show Roster showed the Peerless Motor Car Corporation officers as follows: J.A. Bohannon president, George A. Ellis vice president, R.E. Wilcox treasurer, S.T. Creighton secretary, and E.C. Sudhoff director of purchasing.

1932 Peerless V-16.
(Source: Crawford Museum)

From Autos to Ale

Several months before the termination of production, the Peerless Motor Car Corporation executive committee headed by Bohannon realized that putting more money into automotive production would be counterproductive and only prolong the agony. About this time both major political parties in the United States were drawing up the battle lines for the presidential election in November 1932. It was obvious to almost everyone that the repeal of the Volstead Act and the legalization of alcoholic beverages would be just a matter of time. With the landslide victory of Franklin D. Roosevelt in November 1932, the return of alcoholic beverages was a foregone conclusion.

According to the information furnished by the New York Stock Exchange on July 1, 1933, the Peerless Motor Car Corporation had decided to enter the brewing business. It was proposed that Peerless would sell 178,150 shares of common stock held in its treasury at $3 per share, to furnish funds for plant alteration and operating capital. Additional funds were made available by the issuance of $300,000 in 6% notes to contractors and the suppliers of the needed equipment. A subsidiary company known as the Peerless Company would brew and distribute the ale.

The final move in the transformation of the Peerless Motor Car Corporation from the automotive scene was taken on June 15, 1934, when the company formally went into production of Carling's Canadian Ale, a beverage hitherto made only in Canada. The $1,500,000 equipment was on view for the first time at a reception, by invitation, from 4 to 10 p.m. on June 15, 1934. Production of the ale was under the supervision of Carling Breweries Ltd. of London, Ontario, Canada. The Peerless Motor Car Corporation became the Peerless Corporation, and its subsidiary, the Peerless Company, became the Brewing Corporation of America.

After the shutdown of automobile production, Bohannon purchased the Peerless V-16 and kept it as his personal property until 1946, when he donated it to the Frederick C. Crawford Auto-Aviation Museum of the Western Reserve Historical Society.

August Duesenberg. (Source: Indianapolis Motor Speedway Corp.) Fred Duesenberg. (Source: Indianapolis Motor Speedway Corp.)

Fred and August Duesenberg

Fred Duesenberg, a colorful, energetic product of the machine shop and the speedway, was democratic, world-renowned, and loved by everyone who was touched by his life. Fred had a noticeable limp as the result of a race car crash years before that ended his driving career, but it never diminished his cheerful disposition, nor did it ever take away his winning smile and sense of humor. He was truly the "cut diamond" in the age of iron men. A half-century after their passing, young automotive engineers are rediscovering innovations pioneered by the Duesenbergs many decades before.

* * *

Frederic Samuel Duesenberg was born in the county of Lippe, Germany, on December 6, 1876, one of a family of seven children, and was brought to the United States by his mother in 1885 when he was eight years old. The family settled on a farm near Rockford, Iowa. In 1894, at the age of 17, Fred went to work for a dealer in farm implements where he set up a successful bicycle business. This gave him a release for his boundless energy and passion for racing. Fred established two- and three-mile records, unpaced, and established the two-mile world record of 4 minutes, 24 seconds in 1898. Also in 1898, after having been paced by a running horse in a number of races, he designed and built a motorcycle pacing machine.

In 1902, enamored with the automobile, Fred sold his bicycle business and went to work as an engineer and test driver for the Thomas B. Jeffery Company in Kenosha, Wisconsin (manufacturer of the Rambler car). August S. Duesenberg, his younger brother, joined him there as a machinist and test driver a short time later. Fred and August Duesenberg were inseparable and were an unbeatable combination in years to come. Fred was the "engineer and designer" and August was the "builder" who could build anything Fred designed.

In 1903 Fred made his debut in auto racing in a rebuilt Marion at a midwestern fairground.

The Mason Motor Car Company

In May 1906, Fred and August convinced a young lawyer named Edward R. Mason to put up enough capital to establish the Mason Motor Car Company in Des Moines, Iowa. Incorporated in Iowa with capital stock of $25,000, the first production Mason cars were the 1906 models, car and engine designed by Fred Duesenberg.

In 1910 the Mason Motor Car Company was sold to Frederick I. Maytag, manufacturer of washing machines. The re-organized company, Maytag-Mason Motor Company, was moved to Waterloo, Iowa, with the focus of building and selling passenger cars. However, they did permit and encourage Fred and August Duesenberg to continue building racing cars.

Fred Duesenberg in the Marion race car, 1903. (Source: ACD Museum)

Part of the parking area at the 1912 Indianapolis 500. (Source: Indianapolis Motor Speedway Corp.)

Mason Racing Cars

The Mason racing cars successful at local race tracks and fairgrounds; however, the 1912 Mason racing car failed to qualify for the 500 due to a cracked cylinder block.

On April 27, 1913, Fred Duesenberg and Isle Denny of Runnells, Iowa, were married, and Isle was to share Fred's life with racing cars.

For 1913 Fred and August designed and built three new racing cars using Duesenberg's newly designed four-cylinder racing engine. The engine had horizontal valves actuated by long (12-in.) 90-degree rocker arms opening the valves into a rectangular combustion chamber. The three racing cars were entered in the Indianapolis 500 with Willie Haupt, Robert Evans, and Jack Tower as drivers. Tower qualified with the fastest time at 88.23 mph; Evans had the 4th and Haupt the 15th starting positions. Tower led the race for many laps, but on the 51st lap he flipped in the south chute. Haupt finished 9th and Evans 13th, not bad for first time entries.

For 1914 Fred and Augie built two new racing cars and entered them under the Duesenberg name in the 1914 Indianapolis 500, with Eddie Rickenbacker and Willie Haupt as drivers. While they qualified for 23rd and 28th positions, they finished 10th and 12th, respectively.

Pit area at the 1912 Indianapolis 500. (Source: Indianapolis Motor Speedway Corp.)

As an interesting sidebar, in 1960, in the presence of the author while walking in the Tower Terrace parking area, Gertie Duesenberg (wife of August), who was 80 years old at the time, asked Harry Hartz: "Harry, what

Gertie Duesenberg and the author, Michael Kollins, in 1960. (Source: Indianapolis Motor Speedway Corp.)

Left to right: Tom Alley, August Duesenberg, Eric Schroeder, Eddie O'Donnell, Vic Wells, Art Klein, Bob Burman, Eddie Pullen, Joe Thomas, and Glover Rockstell, in 1915. (Source: Indianapolis Motor Speedway Corp.)

year did Eddie Rickenbacker first drive for us at the Speedway? The way I remember it must have been 1914, because Fritz was just a little boy of 2, and Denny was just a baby still nursing at his mother's breast." What a wonderful way for a mother to remember a special event. Gertie lived to be 95 years old.

Duesenberg Motor Company

In early 1915 Duesenberg Motor Company was organized and incorporated in St. Paul, Minnesota, to build engines and racing cars. The address given was 2654 University Avenue. It was here that the famous Duesenberg "side-valve" engines were built. Three racing cars were built and entered in the 1915 Indianapolis 500. Of the three, Eddie O'Donnell finished 5th, and Tom Alley 8th. Ralph Mulford in the other Duesenberg car was forced out on the 124th lap with a connecting rod failure.

The year 1916 was somewhat better for Duesenberg racing activities. With only one car entered in the 500, Wilbur D'Alene finished 2nd, only 42 seconds behind the winning car. Racing at the Indianapolis Motor Speedway was suspended during 1917-18 because of the war.

Duesenberg marine engines powered many speed boats engaged in racing. The Disturber IV, powered by a pair of Duesenberg twelve-cylinder marine engines, was successful in racing during 1912-14, and was sent to England to compete in the Harmsworth Trophy race. The craft attained speeds of 62 mph in the trials, but the race was called off on account of the war. Duesenberg marine engines were used by the navies of Great Britain, Russia, and Italy. Large numbers of these engines were purchased by the United States government during World War I.

Duesenberg aircraft engine factory in Elizabeth, New Jersey. (Source: ACD Museum)

Aircraft Engines

During 1917-18, Duesenberg concentrated on aircraft engines as well as marine engines after moving into their new plant in Elizabeth, New Jersey, in early 1917. Their first war effort was a four-cylinder 125-hp aircraft engine for training planes. The aircraft engines used by Allied flyers in combat were the French-built Hispano-Suiza V-8 and the Bugatti and the Liberty V-12 built by Packard, Leland, and Nordyke and Marmon in the United States. While the Bugatti aircraft engines were designed and built in France by the enigmatic Ettore Bugatti, the U.S. War Department was impressed with Bugatti's newest U-16 engine. A sample of this engine was obtained by the United States government and sent to the Duesenberg plant in Elizabeth, accompanied by a Frenchman, Monsieur Ernest Friderick, to protect the French interests.

Charles Brady King A.M.E. of the U.S. Signal Corps was placed in charge of the study and investigation of the engine, and was to oversee the manufacture of the Bugatti engine in the United States. King thoroughly inspected and analyzed the French Bugatti engine which had undergone a 37-hour test run in Paris. King found that it had several mechanical faults that needed correcting and certain changes were required to adapt it to American manufacturing methods. King conducted his redesign and modification work at the Duesenberg plant. The King V-12 Aero engine was built at the plant of Brewster and Company in Long Island City, New York.

Tests of the Duesenberg-built Bugatti U-16 engines were conducted during early 1918. The new Bugatti engine had a number of important design changes brought about by King. The water-jacketing of the cylinders and the entire cooling system were simplified, eliminating the cooling problems. However, the most important change was the incorporation of a "full-pressure" lubrication system that greatly extended the longevity of the engine and the periods between overhauls. By mid-summer of 1918 the War Department had enough confidence in the Bugatti engine to place an order for 2000 to be built by the Duesenberg Motors

Dynamometer testing of the aircraft engine. (Source: ACD Museum)

Final assembly of the King-Bugatti U-16 aircraft engine. (Source: ACD Museum)

Duesenberg's Elizabeth, New Jersey, factory at the start of post-WWI production. (Source: Mrs. August Duesenberg Collection)

Corporation. By Armistice Day, November 11, 1918, 40 U-16 Bugatti aircraft engines were built. Shortly afterwards, the balance of the contract was cancelled, and no further work was done on this famous engine. Thus, the pooled engineering brilliance and manufacturing skills of Ettore Bugatti, Charles B. King, Fred and August Duesenberg were able to produce 40 of these masterpieces. Two of these engines are in existence today: one in the Smithsonian Institution and the other at the Wright-Patterson USAF Museum.

The Straight-Eight

While only 40 engines were the fruit of the effort of these brilliant men, Fred Duesenberg was fascinated by this work of art which would influence his and August's future in the automobile business and racing. Fred could envision an "eight-in-a-row" racing engine similar to one bank of this glorious U-16. Fred Duesenberg was an inventor and a man prolific in ideas, but they all converged on one objection to build the fastest, safest, and most nearly perfect automobile that could be put on the road. That was his passionate goal, and everything became secondary to the "eight-in-a-row" or "straight-eight" engine.

Tommy Milton (driving) and Jimmy Murphy in a Duesenberg at Altoona, 1919. (Source: Indianapolis Motor Speedway Corp.)

The first step toward Fred's goal was the selling of the Elizabeth, New Jersey, plant to John N. Willys (Willys Corporation) in August 1919. In late 1919 the rights to the old but popular four-cylinder racing engine were sold to the Rochester Motors Company, who manufactured and sold it under the Duesenberg name to power such cars as Revers, Roamer, Argonne, Kenworthy, and Meteor. Fred retained a room in his home as his engineering office and drafting room, and he set up a machine shop in a nearby building. The influence of the Bugatti engine was very conspicuous in Duesenberg's new engine design: overhead camshaft with three valves per cylinder with special attention to combustion chamber shape and valve porting. The designs included a 3-liter (183-cu.-in.) displacement for racing with three valves per cylinder actuated by a single camshaft and rocker arms and a 4.2-liter (260-cu.-in.) displacement for the Model A passenger car that had two valves per cylinder operated by a single camshaft and rocker arms.

The Duesenberg "straight-eight" racing engines were not ready for the 1919 Indianapolis 500 race. Fred had entered three racing cars using the engines of pre-WWI design with Wilbur D'Alene, Eddie O'Donnell, and Tommy Milton as drivers. They finished in disappointing 17th, 22nd, and 25th positions.

Duesenberg Automobile and Motors Company

In 1920 the fortunes of Duesenberg were more promising. On March 8, 1920, the Duesenberg Automobile and Motors Company was incorporated for the purpose of manufacturing automobiles and engines. In early 1920 Fred designed a special racing car for an attempt at the world speed record. He took two of his tremendously powerful and efficient eight-cylinder racing engines and placed them side by side, thus giving the car 16 cylinders. The car was equipped with two clutches, two propeller shafts, and two separate drive gears and pinions in the rear axle. There was a separate drive to each rear wheel so that each wheel received the power and torque of an eight-cylinder engine. Fred not only designed the engines for this car, but it was necessary for him to design and build the front and rear axles, clutches, and entire driving mechanisms. The engines were deliberately tuned and controlled by one set of levers. The success of this racing car confirmed the reputation he had for being America's master designer of engines and cars.

Tommy Milton in Duesenberg race car at Indianapolis, 1920. (Source: Indianapolis Motor Speedway Corp.)

World Speed Record

So fast that the eye could scarcely follow in a blur of indistinct motion, on April 27, 1920, Tommy Milton shot his "eight-in-a-row" Duesenberg along the smooth hard sands of Daytona Beach at a faster pace than machines had ever traveled on the earth before. Milton established a new world record of speeds of 156.04 mph for 1 mile and 149.95 mph average for 5 miles.

Success followed Duesenberg to Indianapolis in May. The four Duesenberg racing cars entered in the Indianapolis 500 driven by Tommy Milton, Jimmy Murphy, Eddie Hearne, and Eddie O'Donnell finished in the respectable 3rd, 4th, 6th, and 15th positions. Tragedy struck the Duesenberg team in November when colorful Eddie O'Donnell along with suave Gaston Chevrolet lost their lives in a crash at the Beverly Hills California Speedway on Thanksgiving Day, November 25, 1920.

Duesenberg Straight-Eight Tourer

The new Duesenberg Straight-Eight Tourer made its debut in a special Salon showing in the foyer of the Commodore Hotel in New York City in November 1920. The very fact that it was displayed alone in the foyer away from the competitors symbolized the exclusiveness of this famous motor car. The Duesenberg Straight-Eight featured the overhead-camshaft straight-eight engine and four-wheel hydraulic brakes. The advertising copy proclaimed boldly that "The Duesenberg Straight-Eight was 'an automobile for the connoisseur,' built to outclass, outrun, and outlast any car on the road." The requests for additional information such as prices, delivery dates, body types, and purchasing arrangements were greater than ever before for any luxury car that had been exhibited in any salon showing. The exhibit show cards emphasized that the Duesenberg Straight-Eight was being exhibited by Fred and August Duesenberg and would be built by the Duesenberg Automobile and Motors Company, Inc., Indianapolis, Indiana.

Early-1920s Duesenberg Model A straight-eight engine, overhead view, top cover removed. (Source: ACD Museum)

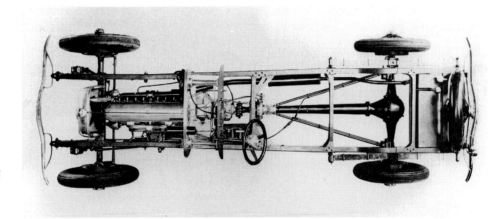

Early-1920s Duesenberg Model A chassis, overhead view. (Source: ACD Museum)

Fred and August Duesenberg had a new plant of modest size built on West Washington at Harding Street in Indianapolis which was officially opened on June 1, 1921. The reason it wasn't opened sooner was that Fred and August were at the speedway every day during May. Duesenberg racing cars successfully finished 2nd, 4th, 6th, and 8th in the 1921 Indianapolis 500. Two other Duesenberg racing cars were involved in separate crashes and did not finish the race.

It was in the plant on West Washington Street that the production of the Duesenberg Straight-Eight engines and chassis for the Model A, and later the J and SJ were built.

Victory at Last

The successes at the Indianapolis 500 and the opening of the plant were only the beginning of the good fortunes for Duesenberg in 1921. Duesenberg entered four 3-liter racing cars in the 1921 French Grand Prix at LeMans, France, to be driven by Jimmy Murphy, Joe Boyer, Albert Guyot, and Andre DuBonnet. The French Grand Prix course was a road circuit of 17 km (10.75 miles). This was the first Grand Prix of France since 1914 when the German team of Mercedes won 1st, 2nd, and 3rd places.

The race on July 25, 1921, started under partially cloudy skies. Joe Boyer immediately grabbed the lead with Ralph DePalma in a Ballot right alongside completing the lap in identical times. The French Grand Prix had

Duesenberg factory in Indianapolis. (Source: ACD Museum)

a starting grid different from other Grand Prix races in that they started the competitors in matched pairs of cars. Within 16 minutes after the start, Jimmy Murphy and Joe Boyer moved into 1st and 2nd positions, respectively. By the end of the 10th lap Duesenberg held 1st, 3rd, and 4th positions.

As the race ground on, Murphy tenaciously held on to the lead, relinquishing it only when he made a pit stop. Murphy handily won the race finishing a full 15 minutes ahead of Ralph DePalma in the Ballot racing car. There was joy and celebration in the Duesenberg pits; however, there was gloom and disappointment in the Ballot pits and the grandstands. The French lost the Grand Prix to a foreign car and driver again.

For 1922 Jimmy Murphy had his own racing car, the "Murphy Special," built using a Duesenberg chassis and body components but using an engine of another make. During the 1920s and 1930s, it was not uncommon to have a racing car identified as "_____ Special," eluding the true identity of the car or engine builder. In more recent years, however, racing cars are identified as to the engine builder as well as the chassis builder, for example: Lotus-Ford, McLaren-Offy, Penske-Cosworth, etc. Jimmy Murphy won the Indianapolis 500 in the Murphy Special using a Duesenberg chassis. Duesenberg racing cars finished 2nd, 4th, 5th, 6th, 7th, 8th, and 10th. Thus, Duesenberg finished with seven (eight) cars in the top ten.

Financial Troubles

The year 1923 turned out to be disappointing for Duesenberg. The lone Duesenberg entry finished 11th at Indianapolis. They also faced trouble with the manufacture and sale of Duesenberg passenger cars, and financial pressures were mounting.

Left to right: Louis Chevrolet, Harry Miller, Fred Duesenberg, and August Duesenberg, 1922. (Source: Indianapolis Motor Speedway Corp.)

On January 7, 1924, Duesenberg Automobile and Motors Company was placed in receivership with William T. Rasmussen appointed receiver in the state of Indiana. On January 14, 1924, Chester S. Ricker, General Manager of Duesenberg, was appointed receiver in Delaware. He was also co-receiver with Rasmussen in Indiana, and Rasmussen was named co-receiver in Delaware. Thus, the management of the company was placed in the hands of two co-receivers, who by court order were able to operate the plant jointly.

On April 21, 1924, at a meeting of the stockholders and creditors of Duesenberg, a committee of three prominent businessmen not financially connected with the company was appointed to draft a reorganization plan to bring Duesenberg out of receivership. The plans considered were to provide enough additional

1924 Duesenberg Model A phaeton. (Source: ACD Museum)

Joe Boyer, winner of 1924 Indianapolis 500. (Source: Indianapolis Motor Speedway Corp.)

working capital to meet the expenses for the cars to be built by the company. At that time the company was said to have assets of $1,358,000 and liabilities slightly in excess of $500,000.

While Duesenberg was experiencing financial problems in the manufacture and sale of passenger cars during 1924, they enjoyed much better success in racing. Four Duesenberg racing cars were entered in the Indianapolis 500. These cars had the new smaller straight-eight engine of 2-liter (121.8-cu.-in.) displacement with dual overhead camshafts. Three of these cars were the first racing cars at Indianapolis to be supercharged. Joe Boyer qualified for the outside of the second row at 104.480 mph. Ernie Ansterberg qualified in 10th position, Peter DePaolo in 13th, and L.L. Corum in 21st position.

On race day at the drop of the green flag Jimmy Murphy grabbed the lead on the first lap. From then on it was a duel between Murphy and Boyer. Boyer was experiencing handling difficulty, and was relieved by Ansterberg, who hit the southeast wall on his second lap. Joe Boyer replaced L.L. Corum on lap 109 when he made a pit stop. Within a few laps Boyer regained the lead and went on to win the race setting a new record of 98.23 mph average. Murphy finished 3rd, DePaolo 6th.

Reorganization

On February 20, 1925, Duesenberg was taken out of receivership, and the Duesenberg Motors Company was formally incorporated in Indiana, taking over the assets of the Duesenberg Automobile and Motors Company. During the year that Rasmussen was in charge of receivership, the company built and sold approximately 100 Duesenberg cars. The officers of the new company were: Fred Duesenberg president, James H. Dunn vice-president, Edward Widman secretary, and Harry B. Mahan treasurer. August Duesenberg was also on the board of directors.

The reorganization plan had worked successfully. Not only were the financial difficulties straightened out, the racing successes were even greater.

Pete DePaolo, winner of 1925 Indianapolis 500. (Source: Indianapolis Motor Speedway Corp.)

August Duesenberg, Pete DePaolo, and Fred Duesenberg. (Source: Indianapolis Motor Speedway Corp.)

For the 1925 Indianapolis 500, Duesenberg entered four racing cars driven by Peter DePaolo, Phil Shafer, Peter Kreis, and Wade Morton. In qualification, Leon Duray in a Miller was on the pole, Peter DePaolo was in the 2nd spot, the outside of the first row, Peter Kreis was in 9th, Wade Morton in 16th, and Phil Shafer in 22nd position.

At the drop of the green flag, DePaolo got the jump on Duray and widened his lead during the first lap. DePaolo hurled his Duesenberg through the 500 miles in 4 hours, 50 minutes, 39.47 seconds, for a winning average speed of 101.13 mph, a record that was not broken for seven years. DePaolo also established a one-lap record speed of 114.285 mph. Phil Shafer finished 3rd and Peter Kreis 8th. Jimmy Gleason, after relieving Mourre who had relieved Wade Morton, hit the wall on the backstretch on lap 165 but escaped injury.

Duesenberg entered two racing cars in the 1926 Indianapolis 500. Peter DePaolo finished 5th, and Ben Jones hit the wall after an axle broke on lap 54. In 1926 the displacement limits were lowered to 1.5 liters (91.5 cu. in.).

Cord Takes Over

On October 6, 1926, E.L. Cord (see his chapter in Volume 2) and associates completed negotiations through Manning and Company to take over the assets, control, and operations of the Duesenberg Motors Company of Indianapolis. The new Duesenberg organization was incorporated and became known as Duesenberg, Incorporated, a subsidiary of the Cord Corporation. Cord was elected president, and Fred Duesenberg vice president and chief engineer. Capitalization was at $1,000,000. The Duesenberg car was to be merchandised independently, which was a very positive move toward the production of a "superlative" car, giving Fred Duesenberg the latitude for the development of the magnificent J and SJ Duesenbergs that were to follow.

The year 1927 was better for racing. George Souders won the 1927 Indianapolis 500 in a Duesenberg, and Dave Evans finished 5th in their other entry. It was on lap 24 of this race that Norman Batten heroically drove his flaming racing car through the length of the pits before leaping to safety. (The irony of this incident was that Norman Batten survived the ordeal of the burning racing car only to lose his life in the sinking of the S.S. Vestris on November 12, 1928.)

The Magnificent Js

During 1927 most of Fred's time and effort was expended on the design and development of the Duesenberg J. Since the Model A was discontinued in August or September 1926, the Model X, of which only ten or

George Souders, winner of 1927 Indianapolis 500. (Source: Indianapolis Motor Speedway Corp.)

twelve were built, was to fill in until the magnificent J was ready. The Model X was a refined A, but still had rather stodgy closed-body style. Many engine designs were considered before finally deciding on the J with a bore and stroke of 3.75 x 4.75-in. The first J chassis was completed in the fall of 1928.

The development of the J was not without differences of opinion between Cord and Fred Duesenberg. Fred preferred a light, agile, high-performance sports car; however, Cord insisted that the new Duesenberg symbolize the Golden Age of America: "A car with the most of the best of everything," that it truly be "the world's finest motor car."

The new Duesenberg Model J made its debut in the Salon showing of the 29th National Auto Show in New York in December 1928, displayed in several body styles by as many "custom coach builders." Because of

Jimmy Gleason at the 1928 Indianapolis 500. (Source: Indianapolis Motor Speedway Corp.)

Ira Hall at the 1928 Indianapolis 500. (Source: Indianapolis Motor Speedway Corp.)

time limitations, the display models did not have completed operational chassis. The cars sold during the Salon showing were promised delivery in May or June of 1929.

Even though the Duesenberg J made its debut just before the coming of the Great Depression, every Duesenberg J built had a waiting buyer. So great was the desirability of the Duesenberg that in their advertising in prestigious publications such as *Vanity Fair* and *Harper's Bazaar*, the caption that accompanied the artwork had only four words: "He drives a Duesenberg," or "She drives a Duesenberg." That was all that needed to be said.

In 1928 Duesenberg was so involved with the development of the J that little effort was devoted to racing. Of the Duesenberg racing cars in the 1928 Indianapolis 500, Jimmy Gleason was forced out with magneto trouble on the 195th lap, and Benny Shoaf and Ira Hall were both eliminated in minor collisions. In the 1929 Indianapolis 500, with four Duesenberg cars running, Jimmy Gleason finished 3rd, Fred Winnai 5th, Ernie Triplett was forced out on lap 48 with connecting rod failure, and Bill Spence was fatally injured when he hit the southeast wall on lap 14.

The 1929 Duesenberg Js were sold as fast as they could be built. Generally, the deliveries were delayed because of the time the custom coach builders took to build the special bodies. Prominent and successful businessmen, movie stars and directors, political greats, and royalty were all interested in the prestige and greatness that Duesenberg denoted. Among the Duesenberg J and SJ owners were Philip E. Wrigley, Thomas Manville, Jr., Cornelius V. Whitney, William Randolph Hearst, Jr., Mayor Jimmy Walker of New York, Col. E.R. Bradley, Gary Cooper, Clark Gable, Tyrone Power, Lupe Velez, Dolores Del Rio, Marion Davies, James Cagney, Ben Blue, Paul Whiteman, movie directors Howard Hawks and Howard Hughes, King Alfonso of Spain, Victor Emmanuel of Italy, HRH Prince Nicolas of Romania, and many other great and noted personalities.

Since each chassis and special body was handcrafted, Duesenberg did not make yearly model changes. Instead the improvements and refinements were incorporated as they were developed. Unfortunately, the stock market crash on "Black Thursday" and the subsequent Depression that followed dampened all car sales. Even those who could and would be buying Duesenbergs who were not brought down to the poverty level had the character and consideration that they did not want to drive these magnificent Duesenbergs on city streets past the unfortunate hungry people standing in bread lines.

King Alfonso of Spain and his 1930 Duesenberg town car. (Source: ACD Museum)

1931 Duesenberg twin cowl phaeton. (Source: ACD Museum)

Unfortunately, 1930 was not a good year for Duesenberg racing activities either. Of the five cars entered in the Indianapolis 500, only the one driven by Bill Cummings finished (in 5th). The other four were destroyed in separate crashes. The displacement limits were increased to 6 liters (366 cu. in.) in 1930.

The following year was better for Duesenberg, not only in car sales and deliveries, but at the speedways as well. The 1931 Indianapolis 500 start was delayed two hours because of rain. At the halfway mark (100 laps), Fred Frame, driving a Duesenberg, was in 5th position with Billy Arnold in the lead, Tony Gulotta 2nd, and Louis Schneider 3rd. With less than 40 laps to go, and after crashes of Arnold and Gulotta over the wall at the northwest turn, Schneider was in the lead with Frame running second. A foul-up during Frame's last refueling pit stop cost him the race. Frame finished 43 seconds behind Schneider to take 2nd place. Jimmy Gleason finished 6th in a Duesenberg.

Duesenbergs fared better at the Altoona, Pennsylvania, board speedway. On July 4, 1931, Duesenbergs finished 2nd, 3rd, and 4th in the 100-mile race. On September 7, 1931, Jimmy Gleason won the first event of 25 miles of the 150-mile race at Altoona at a record speed of 117.75 mph. In the 1932 Indianapolis 500, Ira Hall driving a Duesenberg finished 7th, and Billy Winn in another Duesenberg finished 9th.

Egbert (Babe) Stapp at Indianapolis, 1932. (Source: Indianapolis Motor Speedway Corp.)

1932 Duesenberg SJ LeBaron tourster. (Source: ACD Museum)

James Cagney and his 1932 Duesenberg J Dietrich convertible sedan. (Source: ACD Museum)

1932 Duesenberg SJ torpedo phaeton. (Source: ACD Museum)

Supercharged Passenger Cars

During late 1931 Fred Duesenberg relinquished his racing activities to his son Denny and brother August to concentrate on the development of the supercharger for passenger cars. In May 1932, the supercharged Duesenberg SJ was announced as a companion to the J. The SJ in many ways eclipsed the famous J by providing 320 hp vs. the J's 265 hp, a speed of 109 mph in second gear and a top speed of over 130 mph, and bold continental styling of the racy exposed external exhaust system, with stainless-steel pipes blending in with the right front fender inner panel and the chrome mesh hood side panel. During the production years of 1929 through 1937, approximately 440 Js and 36 SJs were built. According to Lycoming Motors shipping records, 480 J and SJ engines were to have been built and shipped.

Fred Dies

The year 1932 was traumatic and tragic for the Duesenbergs. On July 2nd Fred Duesenberg was critically injured when the automobile he was driving overturned on Ligonier Mountain near Johnstown, Pennsylvania.

Marion Davies' 1933 Duesenberg J town car. (Source: ACD Museum)

1935 Duesenberg SJ speedster LWB. (Source: ACD Museum)

Clark Gable and his 1935 Duesenberg SJ Bowman & Schwartz convertible coupe. (Source: ACD Museum)

As he was recovering from an injured spine and dislocated shoulder, pneumonia set in. He was administered oxygen and thought to be out of danger; however, a relapse proved to be fatal. Fred Duesenberg died on July 26, 1932. His wife, Isle, and son, Denny, were at his bedside in the Johnstown hospital.

The loss of Fred was so profound that the Duesenberg family lost interest and enthusiasm for racing. The lone racing car entered by Denny Duesenberg with Ira Hall driving earned them a 7th-place finish in the 1932 Indianapolis 500. In 1934 Joe Russo finished fifth in the Indianapolis 500, and in 1935 Fred Winnai was out on lap 16.

Mormon Meteor

While 1935 was a dismal year at Indianapolis, the Duesenbergs were rewarded with the ultimate success of the "Mormon Meteor" later that year. Through the combined efforts of August and Denny Duesenberg and J. Herbert Newport, the Mormon Meteor was created. To obtain favorable publicity, the Mormon Meteor

Denny and Fred Duesenberg, 1932. (Source: Indianapolis Motor Speedway Corp.)

Mormon Meteor. (Source: Indianapolis Motor Speedway Corp.)

had to be a stock chassis, street-driveable vehicle. A 142.5-in. wheelbase stock chassis with an SJ engine was used. Newport created and fabricated an excellent aerodynamic design with pontoon-type fenders and all protruding parts fared into the coachwork or fenders. For the record run, the bumpers, headlamps, and special muffler fitted into the large exhaust were removed, not for aerodynamic reasons, but to avoid them coming off during the run.

Driven by Ab Jenkins, mayor of Salt Lake City, the Mormon Meteor attained speeds of 160 mph during practice on the 10-mile circular course of the famed Bonneville Salt Lake Flats in Utah. The record run was made on August 31, 1935, under the sanction and timing of the American Automobile Association Contest Board. The Mormon Meteor averaged 152.45 mph for the first hour and 135.5 mph for 24 hours elapsed time including stops for fuel and tires every 400 miles. While the record was subsequently broken by a special non-production vehicle powered by a massive aircraft engine more than double the size of the SJ, it did not belittle the accomplishment of the ultimate point of greatness in Duesenberg's career. It did once more confirm the status of Duesenberg as the most successful great car built by Americans.

The End of an Era

Considering the adverse business performance, E.L. Cord respected and accepted the suggestion of Lucius B. Manning to discontinue the manufacture of automobiles, and to sell his interest in the holdings of the Cord Corporation. On August 7, 1937, E.L. Cord agreed to sell his entire stock holdings in the Cord Corporation to the banking firms and Lucius B. Manning for $2,632,000. Automobile production stopped on that day.

Fritz Duesenberg, 1962. (Source: Indianapolis Motor Speedway Corp.)

The Duesenberg plant on Washington Street was sold to the Marmon-Harrington Company in October 1937. The availability of parts and service was made possible by the formation of the Auburn-Cord-Duesenberg Company (ACD) in Auburn, Indiana. The Auburn-Cord-Duesenberg Museum is still located in Auburn, Indiana, and is open for viewing daily.

After World War II, Denny Duesenberg became design engineer for the Ford Motor Company in Dearborn, Michigan. Fritz Duesenberg (son of August) became an executive in an electronics firm in Mooresville, Indiana. Fritz was also the Chairman of the Technical Committee of the United States Auto Club (who sanctions the Indianapolis 500) from 1956 through 1965.

1926 Kissel car.

Kissel Brothers

Some companies experience such short-lived success that they can easily be forgotten. Kissel existed for only 24 years, but produced some memorable automobiles.

* * *

The Kissel family migrated from Germany and settled in Washington County, Wisconsin. They engaged in agriculture and later expanded into hardware, builder's supplies, buggies, and carriages. By 1905, George A. and William L. Kissel became interested in horseless carriages and built their first automobile using a four-cylinder Beaver engine and a body fashioned by the Zimmer Brothers, carriage manufacturers.

Kissel Motor Car Company

In April 1906 the Kissel Motor Car Company was incorporated in Hartford, Wisconsin, under the statutes of the state of Wisconsin. Capitalized at $50,000 the incorporators were: George A. Kissel, William L. Kissel, Otto P. Kissel, Adolph P. Kissel, and H.K. Butterfield. The Kissel Motor Car Company got into production seriously in 1907, still using the Beaver four-cylinder engine and bodies by the Zimmer Brothers.

In 1908 Herman Palmer, a graduate engineer, and Freidrich Werner, a meticulous body designer and builder, joined the Kissel Motor Car Company. Thus for 1909, not only was Kissel building their own engines and bodies, they also added a new L-head, six-cylinder, 60-hp engine to their model availability. Kissel offered a 107-in. wheelbase chassis powered by a 30-hp four-cylinder engine, a 115-in. wheelbase chassis with a 40-hp four-cylinder engine, and a 128-in. wheelbase chassis powered by the 60-hp six-cylinder engine. Available in nine body styles, the 1909 Kissel cars were priced at $1,350 to $3,000.

The 1910 model designations were LD-10 (30 hp), D-10, F-10 (50 hp), G-6 (60 hp), and the engine for the 115-in. wheelbase chassis was increased to 50 hp. Eight different body types were available, priced at $1,500

1909 Kissel tonneau.

to $3,000. Kissel also introduced four truck chassis models, ranging from 1- to 3-ton capacity, powered by an L-head, four-cylinder engine, 3-7/8-in. bore and 5-1/2-in. stroke, rated at 35.7 hp and developing 50 brake hp. The 1-ton model was driven by a bevel gear rear axle, while the 2-ton and 3-ton were driven by dual chain drive. On November 27, 1910, Kissel added a 3/4-ton and 1-ton truck models to their truck line-up, featuring a worm drive rear axle, priced at $1,250. This provided Kissel dealers with seven truck models. Kissel continued building trucks through 1918.

The 1911 Kissel passenger car models were carried over from 1910, but were refined and designated the LD-11, D-11, and F-11. The wheelbases were increased to 116, 124, and 132 in., respectively. Thirteen body types were offered, priced at $1,500 to $3,200. In December 1911, a 1911 Kissel seven-passenger touring was used by United States President Howard Taft and his staff assistants on their western United States exploration trip, after leaving the train that brought them to Huron, South Dakota.

The 1911 Kissel model line-up was carried over for 1912, except the models were designated by the developed horsepower, and the body types were coded with capital letters: A for touring, B for semi-touring, C for semi-racer, D for coupe, and E for limousine. The touring had four doors, and the semi-touring was with rear doors only. Twenty-one body styles were offered, priced at $1,500 to $4,200. The technical specifications were the same as for the corresponding 1911 models.

Growth and Expansion

The sales of the 1912 Kissel cars exceeded production capacity at the Hartford plant. Thus, on October 3, 1912, Kissel Motor Car Company acquired a relatively new factory with approximately 200,000 sq. ft. of floor area at Center Avenue and 32nd Street in Milwaukee, Wisconsin. General offices were to be located there, and the additional factory area would supplement the assembly operations in Hartford.

The new 1913 Kissel models featured electric starting, charging, and lighting systems by Ward-Leonard. The ignition was still by Bosch magneto. A silent timing chain was used instead of timing gears. While the general styling of the 1912 Kissel models was carried over into 1913, the body and sheet metal were more streamlined. The model designations remained the same: 30, 40, 50, and 60; however, the body type availability was reduced to eight, priced at $1,700 to $3,150.

The 1914 Kissel cars announced on January 1, 1914, consisted of three models: the 4-40, 6-48, and 6-60. The Model 30 was discontinued and the Model 50 was superseded by the 6-48.

1914 Kissel 6-48

Wheelbase (in.)	132
Price	$1,850-4,000
No. of Cylinders / Engine	L-6
Bore x Stroke (in.)	4.00 × 5.50
Hp	48 adv, 38.4 SAE
Body Styles	11
Other Features	

The 1915 Kissel models introduced on January 2, 1915, consisted of four model lines: 4-36, 6-42, 6-48, and 6-60. The first digit represented the engine cylinder, and the following two digits the developed horsepower. A detachable sedan top was available on all touring models at $350 extra cost. Offered in 18 body styles, the Kissel 1915 price range was $1,450 to $3,150.

For 1916 the Kissel model availability was reduced to two lines: the 4-36 was superseded by the 4-32, built on a 115-in. wheelbase chassis powered by a smaller four-cylinder engine rated at 25.3 NACC hp. The 6-42 was built on a 126-in. wheelbase chassis, powered by a six-cylinder L-head engine rated at 31.45 NACC hp, available in ten body types priced at $1,095 to $1,750. During 1916 Kissel had over 350 direct dealers in the principal cities of the United States and Canada.

Kissel Double-Six

The 1917 Kissel Models 6-38 and 6-42 were refined carryovers of the 1916 models, but the Kissel 4-32 was discontinued. The 6-38 and 6-42 models were available in eight body styles, priced at $1,295 to $2,050. As the trend in the automobile industry moved toward multi-cylinder engine power, Kissel Motor Car Company released information on January 6, 1917, on the forthcoming Kissel V-12, although the actual production of the new Kissel V-12 (Double-Six) did not begin until April 1, 1917. At the start only five Kissel Double-Six cars were produced per day, while the 6-38 and 6-42 production was about 40 cars per day.

Pioneers of the U.S. Automobile Industry, Volume 2: The Small Independents

1917 Kissel Double-Six

Wheelbase (in.)	128
Price	$2,250-2,650
No. of Cylinders / Engine	V-12
Bore x Stroke (in.)	2.87 × 5.00
Hp	80 adv
Body Styles	seven-passenger touring and all-year sedan
Other Features	

The 1917 Kissel Double-Six was powered by the new Kissel V-12 valve-in-head engine manufactured by the Weidely Motors Company of Indianapolis, Indiana. The engine had an aluminum-alloy crankcase, onto which four blocks of three cylinders each were attached. The forged steel crankshaft was supported on four main bearings in the crankcase, and four valve-in-head cylinder heads were used. The engine was lubricated by a pressure oiling system. The styling, similar to the 6-38 and 6-42, was more streamlined, but still had the conventional radiator and shell (resembling a washboard).

As a result of material shortages caused by World War I, and Kissel being called upon to build military trucks, fewer civilian cars were built and sold during 1917 and 1918.

For 1918 the basic 1917 models were carried over with the addition of a "staggered" two-door sedan. The conventional front door on the left side provided entry for the driver and the door on the right side allowed access for the passengers in the rear seat by means of a passage between the front seats. The shape of the radiator and shell was changed to a classical, rounded, gothic-arch shape, which would continue to distinguish the Kissel hallmark though 1928. During 1918 the Kissel Model 6-42 was discontinued, the 6-38 was priced at $1,495 to $1,995, and the Kissel Double-Six at $2,250 to $2,650.

With World War I coming to an abrupt end on November 11, 1918, Kissel Motor Car Company had to readjust to peacetime production. The Double-Six and the 6-38 were discontinued, and the 6-42 was superseded by the Kissel Model 45, which was the only line produced in 1919 called Custom Built. However, due to the post-WWI depression and material shortages, fewer cars were built during 1919.

1919 Kissel Model 45

Wheelbase (in.)	124
Price	$2,450-2,875
No. of Cylinders / Engine	L-6
Bore x Stroke (in.)	3.18 × 5.50
Hp	26.3 NACC
Body Styles	8
Other Features	

Kissel Silver Specials

On January 23, 1919, the Kissel "Silver Specials" made their debut at the Chicago Auto Show. Two custom-built body styles were on exhibit: the Tourster in "Silver Blue" and the Speedster in "Chrome Yellow." The Speedster became affectionately nicknamed the "Gold Bug." Conover T. Silver, the New York distributor, and his stylists were responsible for the creation of the exciting Kissel Silver Specials. Both the Tourster and the Speeedster had a high cowl (emphasizing the high, gothic, cathedral arch-shaped radiator shell) and blending hood. The doors and rear of the body were lower, providing a pleasant streamline rearward. The

1919 Kissel Speedster.

most notable feature of the Kissel Silver Special Speedster was how the top folded, with the main bow being able to jack-knife, with the booted top nestled to the rounded body contour.

The rear portion of the Speedster model was ellipsoidal, in which the plane surfaces were elliptical or circular (resembling a jelly bean). This is the styling motif that contemporary stylists are attempting to accomplish 80 years later. The Kissel "Gold Bug" attracted a great number of purchasers who were dignitaries in aviation, automobile racing, and the theatre; among them were Amelia Earhart, Ralph DePalma, and Al Jolson. The Kissel Silver Specials were indeed exciting cars, in fact, they made up 60% of Kissel's sales in 1919.

The Kissel 45 was carried over through 1920-22. The prices for 1920 increased to $3,175 to $3,475 and spiraled up to $3,475 to $4,275, indicating the effect of the inflation that took place after World War I. The prices were slightly reduced in 1922 to a range of $2,675 to $3,475. The 1922 Kissel 45 featured full-crown individual front fenders and three-quarter running boards. The spare tire and rim were supported on a stanchion at the cowl side between the front fender and running board.

The 1923 Kissel Model 55 was introduced at the National Automobile Show in New York on January 6, 1923. While it was still considered a 1923 model, it was also carried over into 1924.

1923-24 Kissel Model 55

Wheelbase (in.)	124
Price	$1,485-2,585
No. of Cylinders / Engine	L-6
Bore x Stroke (in.)	3.18 × 5.12
Hp	26.3 NACC
Body Styles	6
Other Features	

On October 1, 1924, the Kissel Model 55 was re-introduced in two series as the 1925 Kissel models, offered in eleven body styles in the Standard and eleven body types in the DeLuxe series. The Standard was priced at $1,685 to $2,285 and the DeLuxe at $1,885 to $3,385 ($3,385 for the Berline sedan).

Pioneers of the U.S. Automobile Industry, Volume 2: The Small Independents

1923 Kissel brougham.

On December 17, 1924, Kissel announced the coming of the new Kissel Model 75, which was formally introduced at the National Automobile Show in New York on January 8, 1925.

1925 Kissel Model 75

Wheelbase (in.)	125/137
Price	$1,985-2,485 for Standard
	$2,185-3,585 for DeLuxe
No. of Cylinders / Engine	IL-8
Bore x Stroke (in.)	3.18 × 4.50
Hp	32.52 NACC
Body Styles	10 in Standard series
	13 in DeLuxe series
Other Features	

According to production serial number records, Kissel built and sold 4,433 Kissel 55 models, and 2,233 Kissel 75 models. Indeed Kissel must have enjoyed a prosperous year in 1925.

The 1926 Kissel models were announced on October 1, 1925. With 1925 and 1926 being "boom" business years, Kissel decided to carry over the 1925 Models 55 and 75 into 1926. Eleven body types were offered in

1926 Kissel "Gold Bug." (Source: Henry Ford Museum)

each of the Standard series, and twelve body styles in each of the DeLuxe series. Thus, for 1926 Kissel offered a grand total of 46 body-chassis combinations, priced at $1,585 to $3,585. The factory-installed front bumper had three face bars providing a wider impact area.

The Slump Begins

On August 19, 1926, the new 1927 Kissel models were announced, essentially a carryover of the 1926 models, with the addition of a new Model 65.

1927 Kissel Model 65

Wheelbase (in.)	125
Price	$1,995-2,495
No. of Cylinders / Engine	IL-8
Bore x Stroke (in.)	2.87 × 4.75
Hp	65 adv, 26.45 NACC
Body Styles	10
Other Features	

Stylewise, the fenders were wider and had a deeper crown. The spare tire option permitted the owner to choose between having a single spare mounted at the rear or having two spares mounted on the stanchions between the front fenders and the three-quarter-length running boards. The specifications and prices remained the same as for the corresponding 1926 models. Despite the proliferation of body-chassis combinations, Kissel car sales continued to drop, causing losses for the Kissel Motor Car Company in 1927.

On September 10, 1927, Kissel Motor Car Company announced their new 1928 models. A new and smaller Kissel Model 6-70 superseded the former 6-55. And, on December 17, 1927, the Kissel Model 80 was introduced.

	1928 Kissel Model 6-70	**1928 Kissel Model 80**
Wheelbase (in.)	117	125/132
Price	$1,495-1,595	$1,895
No. of Cylinders / Engine	L-6	IL-8
Bore x Stroke (in.)	2.87 × 4.75	2.87 × 4.75
Hp	19.8 NACC	26.5 NACC
Body Styles	four-door sedan and roadster coupe	four-door brougham-sedan
Other Features	adjustable clutch and brake pedals	

Other body types for the Model 80 were offered on January 1, 1928. The first-series Kissel 90 had been announced on August 15, 1927, essentially superseding the former 1927 Kissel Model 75, having the same specifications and prices. However, on January 1, 1928, the Kissel Model 90 engine bore was increased to 3-1/4 in., raising the piston displacement to 298.6 cu. in. and the rated horsepower to 33.8 NACC. The prices were increased to $2,185 to $2,295.

On January 1, 1928, the new Kissel 80S Silver Special was introduced; however, sales continued to slump during 1928.

1928 Kissel 80S Silver Special

Wheelbase (in.)	125
Price	$1,895
No. of Cylinders / Engine	IL-8
Bore x Stroke (in.)	2.87 × 4.75
Hp	26.5 NACC
Body Styles	brougham-sedan
Other Features	

Kissel White Eagle

On September 1, 1928, Kissel announced their new lines of cars with conventional radiator shell design. The radical changes in outward appearance were combined with numerous refinements in chassis design and increased engine power in the three new 1929 Kissel White Eagle series. Two eight-in-line and six-cylinder engines powered the new Kissel lines. The rounded, gothic-arch radiator shell was replaced by a distinctly foreign motif. The shell had a narrow rim in front but appeared wide and massive on the sides. The shell was peaked in the middle to blend with the contemporary hood design. The hood sides had two sets of vertical ventilating louvres. The Kissel badge or medallion, formerly mounted on the radiator core, was replaced by a large white eagle whose wings extended almost across the entire width of the radiator, mounted over a black panel covering the radiator upper tank. Another departure of Kissel was the use of a sweeping full fender, eliminating the individual self-supporting front fender which was featured on Kissel products for more than 15 years.

Twenty-one body styles were offered in the three new series, compared to 34 body styles in seven line classifications in 1928. The top of the line was the White Eagle Model 126, having two wheelbase lengths of 132 and 139 in. The Kissel White Eagle 126 was powered by an eight-in-line L-head engine rated at 33.9 NACC hp developing 126 brake hp. The Kissel White Eagle Model 95 superseded the Model 80; however, the developed hp was increased to 95. The White Eagle Model 73 superseded the former Model 70 with the same specifications. The price range was $1,595 to $3,885.

October 24, 1929, "Black Thursday," curtailed most automobile sales, and Kissel, despite its features and new designs, suffered along with the rest of the automotive industry.

The Final Blow

The 1929 Kissel White Eagle models were carried over into 1930, with the prices and specifications remaining the same as for 1929. In desperation in early 1930, the Kissel Motor Car Company became involved with Archie M. Andrews and the New Era Motors Incorporated to produce some of the Ruxton front-drive cars.

The dark clouds of doom and gloom began to accumulate on April 25, 1929, when Andrews announced the Ruxton Front-Drive car. A demonstration was given for reporters and trade magazine writers at the Columbia Yacht Club in New York City on May 28, 1929. The Ruxton Front-Drive car was built on a 130-in. wheelbase chassis, low-swung to provide about 63 in. overall height. It was powered by an eight-in-line Continental L-head engine of 3-in. bore and 4-3/4-in. stroke. The bodies were designed by Joseph Ledwinka and built by the E.G. Budd Manufacturing Company. The technical design by W.J. Muller left much to be desired. The design might have been better applied to the front end of a four-wheel-drive truck. The interior and color effects were credited to Joseph Urban, an internationally known theatre set designer.

1930 Ruxton.

The New Era Motors Inc. officers were: Archie M. Andrews president, W.J. Muller vice president and designer, E.H. Wills, and William V.C. Ruxton, financier. Temporarily, Andrews arranged for the building of the Ruxton cars in the Kissel, Gardner, and Moon automobile plants on a contract basis. Fortunately Russell E. Gardner was able to work out from under the contract with New Era Motors (see his story in this volume). Kissel Motor Car Company was saddled with building 25 Ruxton cars. However, on September 22, 1930, George W. Kissel was able to arrange a friendly receivership and to dissolve the agreement with New Era Motors Inc. Moon Motor Car Company ended up building approximately 475 Ruxton cars before its demise. Kissel discontinued building cars after 1930.

The final chapter in the history of the Kissel Motor Car Company was written by the Circuit Court in approving and initiating the lifting of the receivership on October 1, 1930, and allowing the sale of Kissel Motor Car Company assets.

The entire stock of replacement parts and materials were sold outright to the Stephens Corporation of Rockford, Illinois, who handled all parts and maintenance needs. The Kissel Hartford plant remained intact. It embraced a large group of buildings, so arranged that it was an easy plan to convert them to a large number of lines of diversified industries. The Kissel plants and assets were purchased by Hartford Industries Incorporated on August 24, 1932, a group organized to represent the interest of the bond holders.

The Kissel plants were later sold to the Outboard Marine Corporation, builders of the Elto and Evinrude outboard marine engines for the United States Government during World War II. The plants were later sold and used by the Chrysler Corporation to build marine engines. They have since been sold to an independent marine engine manufacturer.

Hupp/Drake/Hastings/Young and the Hupp Motor Car Corporation

1939 Hupp Skylark

In trying to determine the cause of Hupp's demise, no single factor can be pinpointed as responsible. Hupp Motor Car Corporation had all the ingredients of a successful corporation: great talent, good reputation, excellent engineering, and a reliable product. Then, what was wrong? It may have been the accumulation of many reasons: national economy, wars, depressions, fierce competition, labor strife, and political turmoil in the management.

* * *

The Players

R.C. Hupp

Robert Craig Hupp was born in 1876 in Grand Rapids, Michigan, and went to Detroit at an early age. In 1897 he connected with the Olds Motor Works in Detroit and later in Lansing, Michigan. He joined Ford Motor Company in 1905 to learn all he could about automobile manufacture, including what not to build. After leaving Ford in 1908, he built a prototype runabout with an oval gas tank mounted transversely behind the two bucket seats, powered by a four-cylinder engine rated at 16.9 hp ALAM. It was completed in October 1908, at which time he organized the Hupp Motor Car Company.

R.C. Hupp. (Source: NAHC)

J.W. Drake. (Source: NAHC)

J. Walter Drake

J. Walter Drake was born in Sturgis, Michigan, on September 27, 1875. He earned his law degree in Detroit, Michigan, where he then practiced law. He joined the Hupp Motor Car Company on October 8, 1908, as vice president and general manager.

Charles Hastings

Charles Douglas Hastings was born on August 25, 1858, in Hillsdale, Michigan. After graduating from high school in Detroit in 1876, he began as a traveling salesman, and then was a railroad accountant from 1890 to 1894, and a wholesale dealer from 1894 to 1902. In 1902 he joined the Olds Motor Works in Detroit and Lansing, Michigan. Hastings joined the Hupp Motor Car Company on October 8, 1908, as one of its founders, and was appointed vice president.

DuBois Young

DuBois Young was born in Greenville, Ohio, on July 7, 1879. In 1893 he went to Indianapolis and found employment at the Atlas Engine Works, makers of steam engines and later gasoline-fueled engines. By 1895 he had qualified as a journeyman machinist, earning $1.75 per day. With the lure of other cities paying higher wages, Young moved to Chicago in 1901. But he was fascinated with automobiles, so he went to Detroit in 1902. At that time Olds Motor Works was the only active plant, so Young was unable to get employment. Humbly, he returned to Indianapolis and joined the Premier Motor Car Company. He worked earnestly from 1902 to 1909 and he rose rapidly through promotions to become general foreman.

C.D. Hastings. (Source: NAHC)

D. Young. (Source: NAHC)

In 1909 he founded the American Gear and Manufacturing Company in Jackson, Michigan. In 1915 the automobile industry had a bad year, and Hupp Motor Car Company was in need of a good production administrator. They made him an offer that he could not refuse; they even bought the American Gear and Manufacturing Company. For two years DuBois Young had no title at Hupp, but in 1917 he was made works manager, and later, vice president of manufacturing.

The Birth of the Hupmobile

On October 8, 1908, Robert C. Hupp took the responsibility to organize the Hupp Motor Car Company as president of the company. Joining him as officers of the company, and providing financial assistance, were: J. Walter Drake vice president and general manager, Charles D. Hastings vice president, and Emil Nelson and Otto von Bachelle, engineers.

The first Hupmobile car was introduced in February 1909 and was in production in March 1909. The production Hupmobile was a handsome runabout, similar to Hupp's prototype. It was built on an 86-in. wheelbase chassis powered by a four-cylinder L-head engine rated at 16.9 hp ALAM, available as a runabout, priced at $750. The sales were phenomenal; 1618 cars were built and sold in the remainder of 1909. The same model was continued for 1910 and 1911. In 1911 three additional body types were offered: a torpedo roadster, a touring, and a coupe priced at $750 to $1,100. The sales were 5,340 cars in 1910 and 6,079 cars in 1911. The 1911 Hupmobile models were carried over into 1912, with the specifications and prices unchanged. Approximately 6,000 cars were built and sold domestically in 1912.

First Hupmobile prototype in 1908.

1910 Hupmobile. (Source: NAHC)

In the fall of 1912, a new and more powerful Hupmobile Model 32 was introduced. Domestic sales were 11,649 units.

1913 Hupmobile Model 32

Wheelbase (in.)	106
Price	$975–1,350
No. of Cylinders / Engine	L-4
Bore x Stroke (in.)	3.25 × 5.50
Horsepower	32 adv, 16.9 NACC
Body Styles	3
Other Features	

In 1912 Emil Nelson and Otto von Bachelle resigned from the Hupp Motor Car Company, and Frank E. Watts was appointed chief engineer.

The new Hupmobile factory erected in 1911, and occupied in 1912, was a model to be copied, with its three, large, two-story main buildings, convenient administration building, and about a half-mile of railroad tracks within the property. The additional structures included a transformer house to furnish electrical power, and a boiler house to furnish steam power and heat. It housed a well-equipped kitchen and dining hall, with a seating capacity of 400. Through this plant, Hupp was able to triple its productive capacity.

R.C. Hupp Breaks Away

In the meantime, even before the Hupp Motor Car Company had a chance to establish itself firmly, Robert Hupp, being the visionary, was restless and was thinking of his further ventures. In August 1911 he announced the organization of the new Hupp Corporation to build the Hupp-Yeats electric car. The incorporators were R.C. Hupp and R.T. Yeats. Drake and Hastings were not pleased with the announcement, to say the least. They had no intention of allowing the Hupp name to be exploited by another corporation.

In September 1911, R.C. Hupp resigned, and sold his interest in the Hupp Motor Car Company to Drake and Hastings. After Hupp's departure from the Hupp Motor Car Company, under not the best circumstances, he was enjoined by Drake and Hastings to not use the name Hupp. The matter was resolved in a lawsuit in October 1911. The court ruled in favor of the Hupp Motor Car Company in February 1912, and R.C. Hupp was ordered to not use the name Hupp. Therefore, Hupp's new corporation was called the R.C.H. Corporation.

1910 Hupp-Yeats electric car.

Because the popularity of electric-powered cars was disappearing rapidly, the emphasis was to build a gasoline-engine-powered car as a companion to the Hupp-Yeats Electric. The new R.C.H. Model F was introduced in late 1911, and deliveries were made in early 1912. The car was in fact a copy of the successful Hupmobile Model 20, with a few added improvements. The R.C.H. Model F was built on an 86-in. wheelbase chassis, powered by a four-cylinder L-head engine of 3.25-in. bore by 5-in. stroke giving a displacement of 165.9 cu. in., rated at 16.9 hp NACC, but developing 22 brake hp. It was offered originally in a runabout only, priced at $775. While the sales reports from the R.C.H. factory were glowing, the author believes the sales figures were exaggerated. For 1913 the runabout was continued, and touring and coupe models were added, priced at $900 and $1,300, respectively. The wheelbase was increased to 110 in. For 1914 only the touring model was offered, still priced at $900.

The exaggerated sales reports could not hide the truth, though. In 1913 R.C. Hupp stepped down, and was replaced by J.F. Hartz as president. R.T. Yeats also resigned and was replaced by Charles Seider as vice president. A friendly reorganization left R.C. Hupp without a job. In July 1913 R.C.H. Corporation went into receivership and ended production in 1914.

Meanwhile, Back at Hupp Motor Car Company

The departure of R.C. Hupp, and the organization and operation of the R.C.H. Corporation, had no effect on the Hupp Motor Car Company except for the nuisance. The 1914 Hupmobile introduced on September 25, 1913, had only the Model 32 which was now designated the Model H. It had numerous minor styling changes, and most of the specifications were the same as for 1913. Available in three body styles, it was priced at $1,050 to $1,200. The Model H was continued into 1915.

In the fall of 1914, Hupp Motor Car Company introduced a new, larger, and more powerful car, the Model K. Unfortunately, 1915 was not a productive year, as less than 7,000 Model Ks were built and sold, and less than 700 Model Hs were sold.

	1915 Hupmobile Model H	1915 Hupmobile Model K
Wheelbase (in.)	106	119
Price	$1,050	$1,200-1,365
No. of Cylinders / Engine	L-4	L-4
Bore x Stroke (in.)	3.25 × 5.50	3.37 × 5.50
Horsepower	32 adv, 16.9 NACC	32 adv, 16.9 NACC
Body Styles	touring, roadster	4
Other Features	electric starting, charging, and lighting systems by Westinghouse ignition by Bosch magneto, mounted at rear of cylinder block	

Young Joins Hupp

However, 1915 was a more profitable year for the Hupmobile in a different way. When Hupp Motor Car Company bought the American Gear and Manufacturing Company of Jackson, Michigan, they obtained the services of DuBois Young as a bonus. While DuBois Young had no title at Hupp Motor Car Company for two years, he was made works manager in 1917, and shortly afterward became vice president of manufacturing.

Growing Into a Corporation

On November 24, 1915, the Hupp Motor Car Company was reincorporated under the statutes of the state of Virginia, capitalized at $8,000,000 and the name became Hupp Motor Car Corporation. In 1914 Hastings retired, ostensibly to get a little rest and recreation, but not for long. Hastings would soon be recalled by the board of directors.

For 1916, only one Hupmobile car line was offered, in two wheelbase lengths: the Model N was introduced in June 1915. The early introduction was prompted by the dismal sales performance of the Model K, which was scheduled to be discontinued in 1916.

	1916 Hupmobile Model N	1916 Hupmobile Model N
Wheelbase (in.)	119	134
Price	$1,185-1,340	$1,225-2,365
No. of Cylinders / Engine	L-4	L-4
Bore x Stroke (in.)	3.75 × 5.50	3.75 × 5.50
Horsepower	22.5 NACC	22.5 NACC
Body Styles	5 (two-and five-passenger)	seven-passenger sedan and limousine
Other Features		

Hastings Returns

In late 1916 the board of directors convinced Charles Hastings to return to the Hupp Motor Car Company. Hastings rejoined Hupmobile in January 1917, supposedly on a temporary basis, but as it turned out he was president and general manager again, and J. Walter Drake continued as chairman of the board.

The 1916 Hupmobile model line-up continued through 1917, except that the limousine model was discontinued, and the prices were increased to $1,285 to $1,735. The domestic sales of the Hupmobile Model N were over 27,500 vehicles for 1916 and 1917.

The completely new 1918 Hupmobile Model R was introduced on November 15, 1917.

1918 Hupmobile. (Source: NAHC)

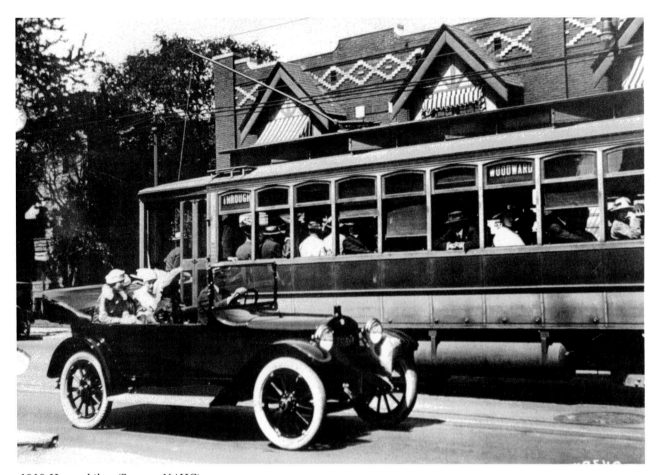

1918 Hupmobile. (Source: NAHC)

1918 Hupmobile Model R

Wheelbase (in.)	
Price	$1,250
No. of Cylinders / Engine	L-4
Bore x Stroke (in.)	3.25 × 5.50
Horsepower	35 adv, 16.9 NACC
Body Styles	touring and roadster
Other Features	

The main feature of the new engine was the en-bloc cylinder casting with upper crankcase, full-length water jackets, thermo-syphon cooling, and a removable cylinder head. The chassis was new, having four, parallel, semi-elliptical springs, and a Hotchkiss shaft drive with two Spicer universal joints. The engine had a vertical Atwater-Kent battery-powered distributor, and Westinghouse starting and charging systems. The styling of the Model R was much improved, following the trend of the industry but still retaining its own characteristics, such as the tall radiator filler neck.

The domestic sales of the Hupmobile Model R were very encouraging, as approximately 10,800 units were built and sold through 1918. For 1919 the Model R was unchanged except the prices were increased to $1,500 for the touring and roadster, and a five-passenger sedan and a four-passenger coupe were added to the line-up, priced at $2,185 and $2,135, respectively. The success of the Model R was self-evident, as 17,442 cars were built and sold in 1919, increasing to 19,225 cars in 1920.

Into the Body Business

On September 14, 1919, the Hupp Motor Car Corporation and the Mitchell Motors Company of Racine, Wisconsin, entered into an agreement to organize the H and M Body Corporation, to build bodies for Hupmobile and for Mitchell. Subsequently, in 1921 Hupmobile purchased the Mitchell interests and the factory became a subsidiary of the Hupp Motor Car Corporation, to build bodies for Hupmobile exclusively. Hupp Motor Car Corporation continued to purchase fenders, steel stampings, and sheet metal parts in Detroit.

1921 Hupmobile. (Source: NAHC)

The 1921 Hupmobile Models R-4-5-6 were introduced on July 15, 1920, available in a five-passenger touring, a three-passenger roadster, a four-passenger coupe, and a five-passenger sedan, priced at $1,250 to $2,150. The domestic sales in 1921 were 13,326 cars. For 1922 a sport phaeton was added to the line-up. In 1922 Hupmobile Models R-7-8-9-10 offered five body types priced at $1,250 to $2,150. Domestic sales amounted to 34,168 units. The Hupmobile Model R was continued with minor body changes and added body styles through 1925. The prices fluctuated as follows: 1923 - $1,150 to $1,785; 1924 - $1,175 to $1,750; 1925 - $1,225 to $1,800.

1924 close-coupled Hupmobile. (Source: NAHC)

Federal Appointment for Drake

In 1923 J. Walter Drake, continuing as a board member and chairman of the board of Hupp, was prevailed upon by president Calvin Coolidge to accept the appointment as assistant Secretary of Commerce under Herbert Hoover. Drake served our country faithfully in that capacity until June 30, 1927. One of his outstanding accomplishments as assistant Secretary of Commerce was the establishment of the aeronautical division of the Department of Commerce, with an assistant secretary as the foremost officer of that department. While he had planned to leave the department earlier, he loyally remained until the air activities of the department were firmly established.

On January 1, 1925, Hupmobile introduced a new, eight-cylinder-engine-powered car identified as Model E-1.

1925 Hupmobile Model E-1

Wheelbase (in.)	125
Price	$1,975-2,375
No. of Cylinders / Engine	IL-8
Bore x Stroke (in.)	2.87 × 4.75
Horsepower	60 adv, 26.45 NACC
Body Styles	4
Other Features	hydraulic four-wheel brakes
	balloon tires mounted on wood-spoke or steel-disc wheels

The eight-cylinder engine apparently boosted Hupmobile sales, as 29,101 cars were sold domestically. The styling was attractive, according to the trend of the times, with a taller nickel-plated radiator, drum-shaped headlamps, nickel-plated, two-piece bumper face bars in front, and two-face bar bumperettes in the rear.

Out of the Body Business

On May 25, 1925, the Hupp Motor Car Corporation sold its body plant to the Murray Body Corporation, who took over the Racine, Wisconsin, plant immediately after the sale. The terms of the sale also included a contract for Murray to build Hupmobile bodies, open and closed, for the next five years. The Hupp Motor Car Company discontinued the finishing and trimming department in Detroit, devoting that area to chassis manufacturing. The plan at the time was to have Murray continue the building of Hupmobile bodies

*1925-26 Hupmobile Eight.
(Source: NAHC)*

temporarily in Racine, gradually moving to Murray's large plants in Detroit. The Racine plant, with a capacity of 60,000 bodies annually, devoted most of the area to bus body manufacturing.

The New Models

On September 22, 1925, sensing the demand for a popularly priced, six-cylinder-engine-powered car, Hupp announced the new Hupmobile Model A for 1926.

1926 Hupmobile Model A

Wheelbase (in.)	114
Price	$1,225-1,285
No. of Cylinders / Engine	L-6
Bore x Stroke (in.)	3.12 × 4.25
Horsepower	49 adv, 23.4 NACC
Body Styles	touring and sedan
Other Features	mechanical four-wheel brakes

The engine featured four crankshaft main bearings, silent chain camshaft drive, and a Lanchester vibration damper. The chassis had mechanical four-wheel brakes, 30 x 5.25-in. balloon tires, four tubular frame cross-members, and threaded "U" spring shackles. Apparently the Hupmobile Six Model A was a success, because domestic sales of Hupmobile in 1926 were 37,778 cars, including those of the Hupmobile Eight Model E-2 of which 14, 822 cars were sold. The Hupmobile Eight Model E-2 engine bore size was increased to 3 in., developing 63 brake hp.

In 1926, DuBois Young took over as president of Hupp Motor Car Corporation.

The 1927 Hupmobile models, the Six Series A, were introduced on August 1, 1926, and the Eight Series E-3 were introduced at the National Automobile Show in New York on January 8, 1927. The Six Series A added a five-passenger brougham to the model line-up. The Series E-3 offered three custom body styles by Dietrich: a five-passenger Victoria, a seven-passenger limousine sedan, and a seven-passenger Berline. The 1927 improvements focused on the engine and chassis, including the pushbutton-operated manifold heat control for faster warm-up, and two-filament headlamp bulbs for tilting the beam. The Six Series A was available in five body types priced at $1,325 to $1,385. The Eight Series E-3 was priced at $1,945 to $2,595. The total domestic sales amounted to 33,809 cars.

1926 Hupmobile Six.
(Source: NAHC)

During 1927 it was due to the untiring effort of new president DuBois Young that the productivity and sales of Hupmobile cars produced record profits in 1926 and 1927. The next important event of 1927 was the return of J. Walter Drake from his tenure as assistant Secretary of Commerce on July 1, 1927. After a short rest, he resumed activity in the interests connected with the automobile industry.

On October 29, 1927, Hupp introduced the first of the new 1928 Hupmobile creations. The 1928 Hupmobile "Six of the Century" had new styling by Amos Northup of the Murray Body Corporation, which inherited many designs from the famous Hispano-Suiza. The taller peaked radiator shell and the hood followed the shell distinctive lines. The long, sweeping, one-piece, front fenders were heavily crowned and graceful, and offered a wheel well for spare wheel and tire in the left front fender. The truncated, bullet-shaped headlamps were nickel-plated, as were the cowl lamps, mounted on a bracket of the surcingle.

Leading the mechanical chassis changes was the adoption of the Midland "Steeldraulic," four-wheel, "self-energizing" brakes, actuated by shielded steel cables. They were pioneered and perfected in cooperation with Hupp Motor Car Corporation, using a one-piece circular shoe with molded composition lining. The engine of the Hupmobile Century Six Series A was a development of the previous Hupmobile six-cylinder engine, but embodied a number of important changes. The increase in power was due partially to the displacement increase, but also to the new cylinder head design. The connecting rods were rifle-drilled to provide pressure lubrication to the piston pins.

	1928 Hupmobile Century Six Series A	1928 Hupmobile Century Eight Series M	1928 Hupmobile Century Eight Series 125
Wheelbase (in.)		120	125
Price	$1,325-1,455	$1,825-2,085	$1,795-2,520
No. of Cylinders / Engine	L-6	IL-8	IL-8
Bore x Stroke (in.)	3.25 × 4.25	3.00 × 4.75	3.00 × 4.75
Horsepower	57 adv, 25.3 NACC	80 adv, 28.8 NACC	80 adv, 28.8 NACC
Body Styles	5	7	6
Other Features			

The new 1928 Hupmobile Century Eight Series M and Series 125 were introduced at the National Automobile Show in New York on January 7, 1928. The specifications and appearance of the Hupmobile Century Eight were impressive and attractive. In many respects it carried the popularity and distinction of the Century Six to greater heights.

The in-line, eight-cylinder, L-head engine achieved greater power by dual carburetion and the use of two manifolds. With any grade of fuel the engine provided rapid acceleration, great speed, remarkable hill-climbing ability, and unusual smoothness achieved by the Lanchester vibration damper on the front end of the engine, and a damper in the clutch.

All Hupmobile Century Eight models were strikingly low, made possible by utilizing the double-drop frame. In accordance with modern features of design, the radiator was deep and shapely, peaked in the center, carrying the peaked line rearward on the hood. The Hupmobile Century Eight had the same unique paneling at the body rear, concealing the gasoline tank, frame, and spring ends. No mechanical parts obtruded at any point, enhancing the clean appearance at the lower rear. Subsequently, competition called this unique design the "beaver tail" rear end.

The clamshell-like sweeping front fenders were one-piece, fully crowned, and seamless. Chromium-plating was utilized for all exterior bright metal parts because of its superior resistance to film and corrosion. The

1928 Hupmobile. (Source: NAHC)

closed bodies were fitted with a one-piece safety glass windshield, designed to be opened forward and upward by its crank mechanism, allowing unrestricted ventilation.

The domestic sales record of 55,550 cars in 1928 may have been misleading as to the economic condition of the United States, and perhaps prompted the Hupp management to over-extend its resources. During 1928, the Hupp Motor Car Corporation earned a profit of $8,790,510.

Hupp Purchases Chandler-Cleveland

On November 30, 1928, the officers of the Hupp Motor Car Corporation negotiated a deal to purchase the assets of Chandler-Cleveland Motors Corporation (see their story in Volume 3). At a meeting on December 10, 1928, all officers of the Chandler-Cleveland Motors Corporation resigned and were succeeded by the officers and directors of the Hupp Motor Car Corporation. DuBois Young was elected president of the Chandler-Cleveland Motors, and Charles Hastings and J. Walter Drake and other directors of Hupp were elected to the Chandler-Cleveland board of directors. The new management at that time continued the line of Chandler cars which were sold by the Chandler-Cleveland distributors.

For 1929 the Hupmobile consisted of two model lines: the Century Six Series A and the Hupmobile Century Eight Series M.

	1929 Hupmobile Century Six Series A	1929 Hupmobile Century Eight Series M
Wheelbase (in.)	114	120/130
Price	$1,425-1,550	$1,825-2,475
No. of Cylinders / Engine	L-6	IL-8
Bore x Stroke (in.)	3.25 × 4.25	3.00 × 4.75
Horsepower	57 adv, 25.3 NACC	80 adv, 28.8 NACC
Body Styles	8	10 (seven-passenger sedan and limousine used 130-in. wheelbase)
Other Features		chromium-plated folding trunk rack

Mechanical improvements in all cars included the adoption of Monroe shock absorbers (hydraulic). Externally, the appearance was improved by chromium-plating on all bright metal parts with the higher belt molding extending around the rear of the body. The interior featured new instrument panels of oxidized bronze finish on the Series M and oxidized silver on the Series A, both with individual openings in the face panel to observe the gauges and speedometer. The instruments were horizontally revolving drums, pivoted vertically (similar) to the speedometer.

The Depression Hits

While the desirable features and equipment would indicate an increase in sales, it was not to be because of "Black Thursday," October 24, 1929. Still, Hupp enjoyed domestic sales of 44,337 units in 1929.

The 1930 Hupmobile Model S, introduced on August 10, 1929, was a new, lighter, lower-priced model built in the Cleveland plant. It was hoped that the price structure would increase sales. The 1930 Hupmobile Model C, which superseded Series M, was introduced on September 18, 1929. In addition to the two models announced in the fall, the Hupmobile line-up added two new eight-cylinder models: Model H and Model U.

	1930 Hupmobile Model S	1930 Hupmobile Model C	1930 Hupmobile Model H	1930 Hupmobile Model U
Wheelbase (in.)	114	121	125	137
Price	$995-1,160	$1,605-1,805	$2,080-2,265	$2,495-2,645
No. of Cylinders / Engine	L-6	IL-8	IL-8	IL-8
Bore x Stroke (in.)	3.25 × 4.25	3.00 × 4.75	3.50 × 4.75	3.50 × 4.75
Horsepower	70 adv, 25.35 SAE	100 adv, 28.9 SAE	133 adv, 39.2 SAE	133 adv, 39.2 SAE
Body Styles	6	6	6 (standard and custom)	seven-passenger sedan and limousine
Other Features				

Optimistic expansion of the model availability did not improve sales, as the cruel Depression was taking its toll, and the domestic Hupmobile sales dropped to 24,307 units in 1930.

Still the over-optimism persisted, as the press releases and radio talk from the politicians predicted "Two cars in every garage, two chickens in every pot." Who could expect the Depression to get even worse? Even cutting the prices of the Model S did not help the 1931 sales. Feeling there was too great a price gap between the Model S and the Model C, Hupmobile added a lower-priced Century Eight Model L to its 1931 model line-up on June 14, 1930.

1931 Hupmobile Century Eight Model L

Wheelbase (in.)	118
Price	$1,295-1,350
No. of Cylinders / Engine	IL-8
Bore x Stroke (in.)	2.87 × 4.37
Horsepower	87 adv, 26.45 SAE
Body Styles	6
Other Features	"Ray Day" aluminum-alloy pistons

Perhaps, the most outstanding feature of the Model L engine, as well as those of the other Hupmobile models, was the incorporation of the "Ray Day" aluminum-alloy pistons, with longer skirts and increased length above the piston pin. The next most important feature was the transmission with "constant-mesh" helical silent-second gears, and a Warner Gear Corporation free-wheeling unit at the rear of the transmission. The free-wheeling unit could be locked out by depressing a button on the gearshift lever. The body lines, a development of the previous models, had been refined. A larger peaked hood centerline, starting with the broader, deeper, chromium-plated radiator shell, gave a more gracefully tapered front end.

During 1931 Hupmobile proliferated into five model lines, built on five different wheelbase lengths, offering 27 body-chassis combinations, priced at $995 to $2,445. Hupp management must have figured that more was better. Nearly everyone underestimated the profound effect of the Great Depression. Only 17,427 cars were sold domestically in 1931, and Hupp suffered a great financial loss. Certainly, the styling could not be blamed for the poor sales results, since Amos Northup styled the 1928-31 Century models as attractively as any on the market at that time.

On December 7, 1931, Robert Craig Hupp, the founder of the Hupp Motor Car Company and the R.C.H. Corporation, died from a stroke.

A New Design

DuBois Young engaged Raymond Loewy, a former French fashion designer, to design an entirely new car for 1932. After showing Young and Hastings his prototype, they were convinced that this design was what Hupmobile needed, and they were right. With assistance from Amos Northup to put on the finishing touches, the car finally got into production on January 1, 1932. The 1932 Hupmobile consisted of three new model lines: the 216, 222, and 226. The model numbers represented the wheelbase length plus 100. The Models 214, 218, 221, and 225-237 were carryover models from 1931.

The new 1932 Hupmobile exterior design represented a rectilinear variant to the existing V-front radiator motif. The accepted common rounding and sloping radiator shell had been replaced by a new perpendicular shell with built-in vertical louvres, giving a massive effect that was entirely distinctive. A new broad double bar of a characteristically Hupmobile design protected the front and rear of the vehicle. On Models 222 and 226, a new radiator ornament of original design adorned the top of the radiator shell. The ornament offered a new emblem which consisted of a circled H in which a barb was mounted to form the cross-bar of the H.

1932 Hupmobile. (Source: NAHC)

1932 Hupmobile. (Source: NAHC)

	1932 Hupmobile Model 216	1932 Hupmobile Model 222	1932 Hupmobile Model 226
Wheelbase (in.)	116	122	126
Price	$895	$1,295	$1,595
No. of Cylinders / Engine	L-6	IL-8	IL-8
Bore x Stroke (in.)	3.37 × 4.62	2.93 × 4.62	3.06 × 4.75
Horsepower	75 adv, 27.3 SAE	93 adv, 27.6 SAE	103 adv, 30.0 SAE
Body Styles	6	4	4
Other Features			

The sloping windshield corner posts (A-pillars), as well as the hood-side rear edge and the leading edge of the front doors, enhanced the appearance. The unusual low silhouette was achieved by the double-drop frame and construction methods that gave it a total overall height of 66 in. Despite the low silhouette, head room was not sacrificed. Contributing to the beauty of the 1932 Hupmobile were the wheel-cowl fenders that the envious competitors dubbed "cycle fenders." The fenders accomplished the purpose for which they were intended, which was to prevent road dirt from soiling the sides of the car without exposing the unattractive but necessary undercarriage. The front and rear sheet-metal tailoring covered all unsightly projecting parts: only the bumper and fuel tank filler cap projected. The hood was long, with the radiator cap placed under the left hood upper panel; its usual place was taken by the smart-looking chromium-plated ornament. On the 226, the hood louvres were replaced by rectangular doors. The 1932 Hupmobile was offered in 14 body styles, priced at $995 to $1,695, represented by 854 dealers who sold 10,794 units in 1932.

The engines of the 1932 models had full-flow oil filtration and an engine oil cooler. The engine coolant water temperature was controlled by a thermostat. The Hupmobile Models 214, 218, 221, 225, and 237 were discontinued after January 1, 1932.

In 1932 Hupmobile pioneered the "X" member frame, which was further complemented by the chassis torsional stabilizer. The stabilizer was formed by a series of connected V members made of tubular steel and

NEW YEAR... NEW CAR... NEW CONTRACT, TOO

Black Ink... not Red... for 1932!

It takes more than a new leaf on the calendar to make that new leaf on your ledger sheet show black ink — not red.

Hupmobile meets the new year with two things the automobile merchant needs most if he is to make money in these times.

A new car. And a new sales agreement.

The car is a whole new sales idea on four wheels. Beauty, design, distinction that suggest the highest priced custom cars. Increased power, speed, performance. Lowest prices in Hupmobile history.

And the agreement — we've been told — is the fairest, squarest factory-dealer document ever offered.

It talks hard facts — such vital issues as forced shipments, car repurchase, service warranties, fleet agreements — everything that has a bearing on the important question of "will you or won't you make money this year?"

Maybe you've seen the new car already. But have you seen the sales agreement? It takes both to complete the Hupmobile picture for 1932.

Hupp Motor Car Corporation... Detroit, Michigan.

angle-iron bracing. Structural engineers recognized the merits of the triangular brace as the strongest structure known. Further driving stability was obtained by the use of a tubular front axle with underslung front springs, the new rubber "silent-bloc" spring shackles. Automatically controlled, thermostatically compensated Houdaille shock absorbers, that required no attention for road or temperature changes, were standard on all models. "Steeldraulic" four-wheel brakes, synchro-silent transmission, and free-wheeling were standard.

On January 9, 1932, the Hupp Motor Car Corporation management was: Charles D. Hastings chairman of the board, DuBois Young president, R.S. Cole vice president of sales, W.S. Graham vice president of manufacturing, and R.P. Lyons treasurer. During 1932 Hupmobile suffered a net loss of $4,515,482 against a loss of $4,249,128 in 1931.

In May 1932, a championship racing car, called the "Hupp Comet," was entered in the Indianapolis 500, driven by Russell Snowberger, powered by a Hupmobile eight-in-line L-head engine of 3-5/16-in. bore and 4-3/4- in. stroke. Snowberger qualified at 114.326 mph for 4th position (inside of the second row), and finished in 5th place at an average speed of 100.791 mph.

The 1933 Hupmobile models were announced on November 30, 1932, consisting of three model lines: the 321, 322, and 326.

	1933 Hupmobile Model 321	**1933 Hupmobile Model 322**	**1933 Hupmobile Model 326**
Wheelbase (in.)	121	122	126
Price	$995-1,095	$1,195-1,245	$1,445-1,545
No. of Cylinders / Engine	L-6	IL-8	IL-8
Bore x Stroke (in.)	3.37 × 4.25	2.93 × 4.62	3.06 × 4.75
Horsepower	90 adv, 27.3 SAE	96 adv, 27.6 SAE	109 adv, 30.0 SAE
Body Styles	4	4	4
Other Features			

The Models 322 and 326 were refined versions of the 1932 Models 222 and 226. However, the Hupmobile Model 321 was a complete modernization and redesign of the Hupmobile Six, on a 5-in. longer wheelbase

Hupp Comet at the 1932 Indianapolis 500. (Source: Indianapolis Motor Speedway Corp.)

chassis. The appearance of the Model 321 follows that of the Models 322 and 326, with the same massive, sloping V radiator shell, sloping windshield A-pillar lines, a short cowl, and long hood with sloping louvres on the hood sides. The fenders are of the same wheel-cowl type as the Models 322 and 326. Larger valves and ports, larger manifold for increased volumetric efficiency, and change in valve timing were responsible for the increased performance on the 321.

The Aerodynes

The new 1934 Hupmobile "Aerodyne" models were introduced at the National Automobile Show in New York on January 6, 1934, consisting of the new Hupmobile Models 421J and 427.

	1934 Hupmobile Model 421J	**1934 Hupmobile Model 427**
Wheelbase (in.)	121	127
Price	$1,095	$1,395
No. of Cylinders / Engine	L-6	IL-8
Bore x Stroke (in.)	3.50 × 4.25	3.31 × 4.75
Horsepower	93 adv, 29.4 SAE	115 adv, 32.5 SAE
Body Styles	3	3
Other Features	Aerodyne styling	Aerodyne styling

1934 Hupmobile Aerodyne. (Source: NAHC)

1935 Hupmobile Aerodyne. (Source: NAHC)

The chassis features and specifications were the same as for the corresponding 1933 models except that a new radiator dual-valve cap, the first in the industry, was used to form a closed cooling system that prevented the loss of coolant due to surging. The new universal joints by Universal Products Corporation incorporated needle bearings at the cross and yokes. While the chassis features were commendable, the greatest 1934 Hupmobile attraction was the new "Aerodyne" styling. While the styling was controversial and appealed to many, it was not the author's favorite. The 1934 Hupmobile bodies were also designed by Raymond Loewy, director of design of the Murray Corporation of America, working in cooperation with the Hupmobile engineers and chief stylist Amos Northup.

The Aerodyne body design provided an individual appearance exclusive to Hupmobile that distinguished their cars from any others offered the following years. By providing a steep slope of the grille, adjacent panels and headlamps in front of the hood, the front was widened, causing it to taper off into the body lines at the rear. This allowed the front seat to accommodate three adult passengers, and offered less wind resistance. The three safety-glass windshields were so located that they gave a panoramic or "control tower" appearance, with curved corner glass panels. (The use of three windshields was formerly a feature of the 1914 Peerless Six-48.)

Fortunately for Hupp the management exercised prudent judgment in deciding to keep the Model 417 as a price leader. Designed in its entirety by Amos Northup and his staff designers, Northup retained the desirable features of the former Hupmobile Model 321, adding attractive features such as the gradual slope of the radiator shell and grille, free-standing chromium-plated, bullet-shaped headlamps, horizontal louvres in the hood side panels, and an aero-propeller-shaped front bumper. Offered in sedan and coupe body styles, priced at $795, the sales of the Model 321 bolstered Hupmobile enough to avoid a total sales disaster. The 1934 Hupmobile domestic sales amounted to 6,566 cars.

The "Predator"

It would seem that neither the Depression nor the lack of appeal of the Aerodyne models alone could destroy the Hupp Motor Car Corporation. It took the help of a "predator" in the form of a human being. Archie Moulton Andrews egotistically portrayed himself as an entrepreneur; however, his definition of entrepreneur was one of expediency above morality, and he countenanced the use of craft and deceit to maintain control. Normally, entrepreneur meant an honorable employer of productive labor to achieve prosperity. Instead, Andrews haughtily used people, their money, someone else's reputation, someone else's factory and employees, by devious methods to attain his ends. In his march through the automobile industry, he left many casualties: Ruxton, Gardner, Moon, Kissel, and New Era Motors Corporation. Now his target was the Hupp Motor Car Corporation.

Andrews had convinced himself that he could run the Hupp Motor Car Corporation better than the present management. His first move was to have a close friend, Frederick W. Burnside, send a letter to all Hupp Motor Car Corporation stockholders on September 3, 1933, asking for proxies for the annual meeting to be held on September 13, 1933, at which three directors were to be elected. The letter also criticized current management policies, and alleged that over $1,000,000 had been illegally transferred to certain individuals. On September 8, 1933, Andrews brought a legal suit against the Hupp directors and Landenberg, Thalman and Company of New York, and the A.G. Becker and Company of Chicago banking houses; depositaries for Hupp Motor Car Company were also named defendants in the suit.

Archie M. Andrews.

The litigation continued until September 15, 1934, when a settlement was reached in the New York State Superior Court with Judge Edgar J. Lauer presiding. The settlement ordered the proceedings to stop, and the terms of the settlement disclosed that $250,000 was to be paid to the Hupp Motor Car Corporation by the Landenberg, Thalmann and Company, and $262,500 by the A.G. Becker and Company. As the result of negotiations, prior to the settlement of the suit, Hupp Motor Car Corporation received $800,000 from the New York banking firm on total deposits of $1,882,037 held for the company. With funds obtained by the settlement of the suit, Hupp Motor Car Corporation now had in excess of $2,394,000 in operating capital.

The next board of directors meeting was held on October 18, 1934, at which Archie Andrews was elected chairman of the board, J. Walter Drake was elected vice president, Arthur J. Brandt was named assistant general manager, and Rufus S. Cole executive vice president and general manager. Charles D. Hastings, who had been serving in a three-fold capacity of chairman, president, and general manager since DuBois Young retired in 1933, resigned as chairman of the board and president of the company. Other officers elected included A.W. Bangham assistant treasurer, and G.E. Roehm assistant secretary. The offices of president and treasurer were not filled.

William J. McAneeny. (Source: NAHC)

William J. McAneeny, former president of Hudson Motor Car Company, was elected president and general manager of Hupp Motor Car Corporation on December 1, 1934. Wallace Zweiner became treasurer. It was understood that Rufus Cole, who had been executive vice president, remained as vice president of sales. On January 1, 1935, Francis H. Fenn was appointed Hupmobile sales manager, succeeding Cole, who resigned. While plant operations were going smoothly under his leadership, sadly, on March 24, 1935, William McAneeny passed away.

On March 29, 1935, J. Walter Drake filed suit against Archie Andrews, seeking his removal from office, and restraining him from presenting himself as board chairman of Hupp Motor Car Corporation. The suit also alleged that a contract was pending between Andrews and the Hupp Motor Car Corporation, under which the board chairman would be given an option to purchase 100,000 shares of Hupp stock, and would receive a monthly salary of $1,500.

Wallace Zweiner testified at the hearings before Federal Judge Arthur J. Tuttle in the suit of J. Walter Drake against Archie M. Andrews, that the Hupp Motor Car Corporation had bought $220,000 worth of new tools and dies for the 1936 cars, of which $145,000 had been paid on account. Zweiner also testified that the corporation had paid $332,000 on the $400,000 of notes and other debts that existed in April 1935. He pointed out that the Hupp Motor Car Corporation needed $500,000 additional working capital, and could obtain it if the litigation were out of the way.

A temporary restraining order enjoining Andrews and other members of the board from effecting certain stock options and commission contracts was granted on March 30, 1935, by Judge Edward J. Moinet sitting in the District Court at Detroit.

The position of president was not filled until May 7, 1935, when Vern R. Drum was elected president of Hupp Motor Car Corporation. Drum held executive positions at the Houdaille-Hershey Company and the Chrysler Corporation. The position of chairman of the board was abolished. The Hupp board of directors was composed of Archie Andrews, J. Walter Drake, Frank S. Lewis, Vern Drum, William B. Mayo, Harvey Campbell, Alex J. Groesbeck (former governor of Michigan), Hal Smith, and Seward J. Merriam.

On June 30, 1935, Hupp Motor Car Corporation reported a deficit of $2,090,456 compared to a net loss of $1,479,367 in the corresponding period the year before. Net working capital was down from $2,462,925 in 1934 to $1,399,747 on June 30, 1935. On November 2, 1935, William R. Hurlbut, vice president and director of sales of Hupp Motor Car Corporation, died suddenly in Brookside, South Wilton, Connecticut.

The game of "musical chairs" by Hupmobile management, and the strain of court litigation, resulted in very little attention being paid to product development. Despite distressing conditions, the engineering and design personnel loyally proceeded to design the new 1936 Hupmobile cars, revisions of the 1935 Hupmobile Models 518 and 521 Aerodyne.

	1936 Hupmobile Model G-618	1936 Hupmobile Model N-621
Wheelbase (in.)	118	121
Price	$795-1,175	$995-1,175
No. of Cylinders / Engine	L-6	IL-8
Bore x Stroke (in.)	3.50 × 4.25	3.31 × 4.75
Horsepower	101 adv, 29.4 SAE	120 adv, 32.5 SAE
Body Styles	two-door sedan and touring, four-door sedan and touring, business coupe, coupe with rumble seat, in Special and Custom lines	6 (no business coupe)
Other Features	touring sedans had built-in trunks	

Particular attention was paid to the styling. The exteriors were the work of Raymond Loewy and Amos Northup, while the interiors were by Adolph Lichten. Most notable was the single, wider, one-piece safety-glass windshield. The wide doors, front and rear, were hinged at the center body (B-pillar), and the usual quarter window was eliminated. The body was all-steel construction, except the floorboards, roof rails, ribs, and a few trim items. The entire body was moved forward, allowing the rear seat to be entirely ahead of the rear axle. Avoiding the mounting of the spare tire and wheel on the rear deck lid provided a generous luggage compartment, and more on those with the built-in trunk. Despite the features offered, the 1936 Hupmobile domestic sales were a dismal 1,556 units. Thus, the remainder of the 1936 models built were carried over into 1937.

In November 1935 W.A. MacDonald was appointed director of sales, succeeding the late William R. Hurlbut. The attrition at Hupmobile continued when Wallace Zwiener died on July 7, 1936. Zwiener joined Hupp Motor Car Company as treasurer in 1934, and had been looked upon as a principal executive in the company's activities since that time. Thomas M. Bradley was appointed president (without title) on July 8, 1936, and his presidency was confirmed on March 8, 1937, by the board of directors. In October 1936, Hupp sold three of its plants in Cleveland to the Weatherhead Company.

Despite the handicaps, the designers, with the encouragement of Amos Northup, toiled feverishly to design a new car for 1938. The mission was accomplished: in September 1937, the 1938 models of the Hupmobile were reviewed by the news media.

	1938 Hupmobile Model E-822	1938 Hupmobile Model H-822
Wheelbase (in.)	122	122
Price	$1,180-1,340	$1,325-1,485
No. of Cylinders / Engine	L-6	IL-8
Bore x Stroke (in.)	3.50 × 4.25	3.31 × 4.75
Horsepower	101 adv, 29.4 SAE	120 adv, 32.5 SAE
Body Styles	3	3
Other Features		

While the chassis and running gear specifications were the same as for the previous models, the styling was entirely new and in refreshingly good taste, differing from the Aerodyne models. The headlamps were vertically fared into the front of the hood side panels, as was the horizontally louvred front grille. The hood was of the alligator-type, hinged at the rear. The front doors were hinged at the windshield A-pillar. The two

1938 Hupmobile. (Source: NAHC)

safety-glass windshields were fitted into a V form, with a division bar in the center. The fenders were a modified pontoon type, concealing all chassis components. The rear body design was offered in either a sloping fast-back or built-in trunk notch-back.

The effects of the 1938 depression were showing, as Hupmobile was sharing their branch facilities with other independent car makers. The total industry domestic sales plunged from 3,483,752 units in 1937 to a dismal 1,891,021 units in 1938. The 1938 Hupmobile domestic sales fell to approximately 1,020 cars in 1938, not enough to keep the plants in operation. Thus, the Hupmobile plants were shut down during the largest part of 1938. Dismayed with the sales performance, Thomas Bradley resigned as president in August 1938, and was succeeded by Samuel L. Davis, a director.

A Final Effort

On July 27, 1938, the Hupp Motor Car Corporation was offered a deal that neither Samuel Davis nor Norman DeVaux could refuse. Norman DeVaux, general manager of Hupp Motor Car Company (former builder of the DeVaux car in 1932), managed to acquire the patent rights, drawings, dies, jigs, and tooling for the 1936-37 Cord 810 and 812 for a reported $45,000. DeVaux engaged John T'jarda (famed stylist) to restyle and redesign the Cord front end to conform to the proposed Hupp Skylark 115-in. wheelbase and rear-wheel drive. The redesign was excellent and a restored Hupp Skylark (if one could be found) is still handsome even today. That part was fine, and the first pilot model was built in 1939, but Hupp Motor Car Company had no operating plant, nor the money to finance the building of the car.

The demise of Hupmobile began in 1939, when the appeal of Hupp Motor Car Corporation for an extension to pay their delinquent tax bill, offering a payment of $125,000, and the balance when Hupp would get the loan they expected. The appeal was turned down by the City Council in June 1939. The time extension was requested by Davis, Hupp president, who said it would enable the company to resume full production. The

1940 Hupp Skylark.

request was opposed on the grounds that it would establish a precedent that other delinquent taxpayers might follow. Hupp Motor Car Corporation paid the taxes in full, but then had no operating capital left.

DeVaux was able to interest Robert Graham in the project (see his story in this volume). An agreement was reached between DeVaux and Graham, by which Graham-Paige Motors would build the cars for Hupp Motor Car Company in exchange for the licensed right to build the Graham Hollywood models on the same assembly line. Graham used its six-cylinder, naturally aspirated as well as the supercharged engines. The Hupp Skylark was powered by Hupp's own 101-hp, six-cylinder engine.

The Hupp Skylark was available in a four-door sedan model priced at $1,145. The first production Skylark came off the assembly line in early May 1940. But, the production consisted of each car being practically hand-built. The last Hupp Skylark was built during the second week of July 1940, and no more Hupmobiles would ever be built again. During 1940, 211 Skylarks were sold, and 103 were sold in 1941.

On November 1, 1940, the Hupp Motor Car Corporation filed for reorganization under the Chapter 11 Bankruptcy Act. Upon reorganization, it became the Hupp Corporation and was engaged in sub-contracting defense work from 1940 through 1945. In 1946 the Hupp Corporation became a components supplier to automobile manufacturers. They supplied window regulators and other body hardware, steel stampings, and machined forgings and castings to the automobile industry and appliance manufacturers. The surplus buildings and warehouses were leased to other suppliers.

Walter Flanders

From the very beginning, Walter Flanders stood out as a skilled and efficient organizer of automotive enterprises, when, as a young salesman, he shored up the operations of the Ford Motor Co. From there he went on to found or fix numerous companies with one common principle: The future belonged to the concern that could produce on the largest scale, most accurately, most consistently, and economically. Flanders was the very keystone to the arch of his organizations.

* * *

Walter E. Flanders was born in Rutland, Vermont, on March 4, 1871. He was a machinist by trade, and later became a machinery and forgings agent for the J&L Machine Company of Cleveland, Ohio. In 1905, when calling on Henry Ford to sell forgings and machine tools, as they walked through the plant on Piquette Avenue in Detroit, Flanders pointed out to Ford where he was wasting money. He also suggested other ways in which Ford could increase production and reduce costs. Ford became very interested and made a deal with Flanders to hire him based on cost savings and increased production. Ford also agreed to pay Flanders a bonus of $20,000 if Flanders could make good his promise to produce 10,000 cars during the first year. Flanders immediately got busy, rearranging the machinery, getting new equipment, and setting up the procedures for time-saving methods. Needless to say, Flanders accomplished the goal of building the 10,000 cars on time, and got the bonus. He then left the Ford Motor Company in 1908 to join Byron Everitt and William Metzger to form the E-M-F Company.

Byron P. Everitt (1872–1940)

Byron (Barney) Everitt was born in Ridgetown, Ontario, Canada, in 1872, and was the automobile industry's pioneer body builder. He started work under Hugh Johnson, a carriage maker, but in 1899 set up his own carriage shop and built the first automobile bodies for R.E. Olds. He subcontracted some of the work to a friend who founded the Wilson Body Company. He then called in his friends, Frederick and Charles Fisher of Norwalk, Ohio, and thus, set the stage for "Body by Fisher." He also built the first 10,000 car bodies for Ford Motor Company. Everitt organized the Wayne Manufacturing Company in 1904, and engaged in the building of the Wayne car from 1905 to 1908.

William E. Metzger (1868–1933)

William E. Metzger was born in Peru, Illinois, on September 30, 1868. He was a salesman of remarkable ability. In 1898 he established the first automobile dealership in Detroit, Michigan. He was distributor of the Pope-Waverly cars in 1899, and Oldsmobile in 1901. In 1902 Metzger became one of the organizers of the Northern Manufacturing Company, and by 1908 he owned 25% of the Northern stock. Metzger also became general sales manager of the Detroit Automobile Company (Cadillac) in late 1902. At the National Auto Show in New York in 1903, he obtained orders for 2,700 Cadillac cars, even before the Cadillac had actually been introduced and only three cars had been built. Metzger resigned from Cadillac in 1908 to join Everitt and Flanders in the organization of the E-M-F Company on June 2, 1908.

Byron Everitt. (Source: NAHC)

William Metzger. (Source: NAHC)

Everitt-Metzger-Flanders (E-M-F) Company

The organization of the "million-dollar" Everitt-Metzger-Flanders Company was announced on June 2, 1908, at a formal dinner at the Cafe des Beaux Arts in New York City. The company was incorporated under the laws of the state of Michigan on June 4, 1908, with a capital stock of $500,000, increased to $1,000,000 on October 7, 1908. The officers elected were: Byron P. Everitt president, W.T. Barbour vice president, William E. Metzger secretary and sales manager, and Walter E. Flanders general manager. In addition to the officers, the board of directors included Charles E. Palms, J.B. Book, and Albert D. Bennett.

Thomas Walburn was appointed general superintendent, E. LeRoy Pelletier (former Ford advertising manager) was appointed advertising manager, and William E. Kelly was named chief engineer. Barbour and Metzger were principal stockholders of the Northern Manufacturing Company. Everitt was president of the Wayne Automobile Company.

The "All-Star" Cast

The organization—a merger of capital, brains, and experience—is essential to success on a large scale. This fact was fully appreciated by E-M-F organizer and president Everitt. As one of the largest manufacturers in the parts and accessories field, Everitt's prior acquaintance with men who did the things others got credit for eminently qualified him to select the right manpower. In framing up his company he decided on an "all-star" cast. The result was a group of men, every one of whom was a specialist and acknowledged leader in his own field. All had made good to a noticeable degree; the history of the industry was their history, and the credit for its achievements was theirs, too.

Everitt-Metzger-Flanders (E-M-F) Company plant in 1910. (Source: Studebaker Corp.)

Metzger, as secretary and sales manager, had no peer. No man in the automotive world was better qualified, nor more favorably known. As sales manager of the Cadillac (Detroit Automobile Company) from its inception, he had run the whole gamut of experience, held a prominent role in every event of importance, held many offices, and been the recipient of many honors. The name Metzger always stood for a broad policy, a solid sales organization, and, to the buyer of E-M-F cars, ensured liberal and fair treatment at all times.

Flanders, as general manager, was recognized as the greatest factory organizer, the keenest buyer, and the most-capable producer the industry had known. As manufacturing manager of Ford Motor Company his record was evident to the world; it stood unequalled and unchallenged. On Flanders' team were: Thomas Walburn, factory manager, Max Wallering, general superintendent, and several department heads, forming an "officers corps." Perpetual harmony prevailed because each was a specialist, an expert in his field, and none conflicted with another. It was only with such an organization that great accomplishments were possible.

Chief engineer Kelly had been in the automotive business since its infancy. He antedated some of the pioneers who received greater recognition. Like many others, he had designed some cars that did not come up to his expectations; however, in later years with greater knowledge and experience, he had produced some of the very best. In the E-M-F 30, Kelly and his corps of designers had realized a long-cherished ambition to build the very best.

Pelletier, advertising manager, was known by his work and its results as former advertising manager of the Ford Motor Company. In fact, there wasn't a weak link in the management chain: No man of mediocre ability filled an important position anywhere in the E-M-F Company organization.

Getting Things Moving

Immediately after incorporation, in order to obtain the necessary facilities, E-M-F purchased the plants, properties, and assets of the Wayne Automobile Company of Detroit and the Northern Manufacturing Company of Detroit and Port Huron, Michigan. E-M-F made contractual arrangements with the Studebaker Brothers Manufacturing Company under which Studebaker obtained exclusive rights to sell the entire output of the E-M-F Company. This arrangement at the time proved to be mutually satisfactory to both Studebaker

1910 E-M-F 30 touring car. (Source: NAHC)

E-M-F 30. (Source: NAHC)

and E-M-F because Studebaker had approximately 4,000 sales outlets. Relieved of the retail sales burden, E-M-F could concentrate on engineering, purchasing, and production.

Flanders placed Thomas Walburn in charge as manufacturing manager, and operations began smoothly and on time with the first E-M-F 30 being completed on August 4, 1908. In Kelly, chief engineer, Flanders had some of the best engineering talent available. As an adjunct to the engineering department, a chemical and metallurgical department was established, and put under the charge of the best metallurgist that could be hired.

The production of the E-M-F 30 was referred to as a "jig job." The part on which an operation was being performed was held tightly clamped in place in a frame or support called a jig during the machining or assembly operation. The result was accuracy and uniformity of the product, with no dependence on the human element. To get the maximum productivity at minimum cost, Flanders had arranged many unusual and versatile types of special machinery. Among these were milling machines, multiple spindle drills, gear cutters, shapers, automatic screw machines, and die presses.

The E-M-F 30 proved to be a good-selling product and a tremendous money-maker. The directors and financial backers were convinced of the soundness of Flanders' ideas. They showed their readiness to back him in his plant expansion plans, and the request to purchase a large bathtub factory on the Detroit riverfront. Flanders equipped this plant for the production of pressed-steel parts to replace costly machined castings. The growth of business was so rapid, that the E-M-F Company was compelled to increase its facilities.

By October 1, 1909, the E-M-F Company owned and operated the following plants: Wayne Automobile Company of Detroit, Michigan; Northern Manufacturing Company of Detroit and Port Huron, Michigan; DeLuxe Motor Car Company of Detroit, Michigan; Monroe Manufacturing Company of Pontiac, Michigan; Western Malleable Steel Company of Detroit, Michigan; Pressed Steel Sanitary Manufacturing Company of Detroit, Michigan; E-M-F Company of Windsor, Ontario, Canada.

With adequate production facilities, a second car, the Flanders 20, was added to the line. The Flanders 20 produced in 1910 was powered by a four-cylinder engine, had a three-speed sliding gear transmission, seating for five passengers, and was priced at $750.

Flanders 20. (Source: NAHC)

Flanders Takes Over

In 1909, Everitt and Metzger left E-M-F and organized Metzger Motor Car Co. This left the total management of the E-M-F Company to Walter Flanders.

Late in 1909 the Studebakers shocked the E-M-F organization by serving notice that they would no longer accept deliveries of E-M-F cars. Although not a salesman, Flanders' skill guided him through his response. He immediately took out large advertisements in newspapers, magazines, and automotive trade journals, informing automobile dealers that E-M-F products would no longer be handled by a jobber, but through direct dealers. Dealers responded instantly and flocked to the E-M-F Company sales office, and literally wore down the office's doorstep! Often there were two or more dealers bidding for the same territory. The program was a tremendous success.

The Studebakers fought the matter in legal courts, trying to stop the organization of an independent E-M-F sales staff. Flanders' tenacity prevailed again in the legal battles. Frustrated by their failure to stop the E-M-F sales organization, the Studebakers, with the financial backing of J.P. Morgan and Company, in March 1910 made Everitt, Metzger, and Flanders a cash offer for their stock holdings in the E-M-F Company.

It was understood that the offer was at the rate of $7.00 for each dollar of par value stock held by the trio. As a result of the purchase of the remainder of E-M-F Company stock, Studebaker became the sole owner of the E-M-F Company, and on December 31, 1910, the E-M-F Company merged with the Studebaker Brothers Manufacturing Company. With the proceeds of the sale of E-M-F Company stock, Everitt and Metzger were able to finance the Metzger Motor Car Company, which had been organized in September 1909 to build the Everitt car. Technically the Everitt car was similar to the E-M-F 30.

Flanders Motor Company

After the purchase of the balance of the E-M-F stock by Studebaker Corporation, Flanders stayed on with Studebaker as vice president and general manager with a three-year contract. However, in January 1911,

Reputation Earned

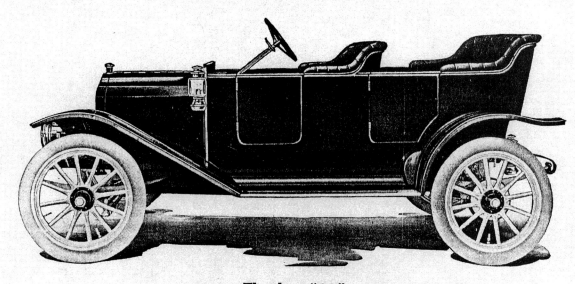

Flanders "20"
Three Speed, Fore Door Touring Car $800

The reputation of the E-M-F Factories is due to the high standard of quality maintained in the E-M-F product. Fame for unexcelled Durability and Satisfactory Service is not an accident. Nor can you buy it. You must earn it. That is the reason why thousands today are buying E-M-F "30" and Flanders "20" cars.

They know automobilists everywhere will tell you that these cars are steadily doing their work and giving a vast amount of pleasure. No wonder the factories are unable to cope with all their orders. There is no argument so strong as the test of time grown into the approval of Public Opinion.

The E-M-F "30" $1000 and Flanders "20" $800 are the world's standard of value.

 The Studebaker Corporation
E-M-F FACTORIES
Detroit, Mich.

E-M-F chassis assembly. (Source: Studebaker Corp.)

Flanders organized the Flanders Manufacturing Company of Chelsea and Pontiac, Michigan, to build the Flanders bi-mobile (motorcycle) and the Flanders electric car. When this became known to the Studebaker officers, they felt that it was a conflict of interest, so in August 1911 the Studebakers reinstated Flanders' right to the Flanders name and released him from his contract. Upon leaving the Studebaker Corporation, Flanders was asked by Everitt and Metzger to join them in their automotive venture. By this time the Metzger Motor Car Company had become the Everitt Motor Car Company. With the entrance of Flanders and his additional capital, the company name was changed to Flanders Motor Company.

The Flanders Motor Company was properly organized and the factory was adequately equipped, similar to that of the former E-M-F Company. LeRoy Pelletier left Studebaker and joined the Flanders Motor Company as advertising manager, and Paul Smith, sales manager of Studebaker, also joined them as sales manager. For a while it was believed that Flanders Motor Company might be successful, with the combined talents of Everitt, Metzger, Flanders, Pelletier, and Smith. A new Flanders Six was being prepared to be introduced at the National Auto Show in New York in January 1913. However, by mid-1912, conditions had changed drastically, which caught Flanders Motor Company floundering as the result of the debts incurred during the unsuccessful operation of Metzger Motor Car Company. Flanders Motor Company was about to go into receivership.

Walter Flanders' skill in meeting situations and his resourcefulness triumphed again. Flanders immediately abandoned all relationship with his Flanders Manufacturing Company of Pontiac, Michigan, which had gone into receivership with the Detroit Trust Company named receiver. He then negotiated a merger with the receivers of the United States Motor Company to trade the (failing) Flanders Motor Company (which was about to go into receivership) for $1,000,000 in cash and $2,750,000 in stock of the Standard Motor Company, a Delaware corporation. (The story of the United States Motor Company is detailed in the chapters on Benjamin Briscoe in Volume 3 and Jonathon Maxwell in Volume 4.)

In the negotiated arrangement, Flanders became president of the Standard Motor Company. Everitt and Metzger, realizing a sizeable return on their investment in the Flanders Motor Company, left and engaged in their individual enterprises. Everitt returned to the body-building business, establishing the Trippenzee Body Company, and Metzger formed new companies, among them the Columbia Motors Company of Detroit, organized in 1916.

Maxwell Motor Company

While the Standard Motor Company was incorporated in the state of Delaware on January 11, 1913, it was found that the name "Standard" conflicted with titles of existing corporations in other states, and therefore could not be used. Thus, on January 16, 1913, the name of the corporation was changed to Maxwell Motor Company (Maxwell had been part of United States Motor Company). On February 6, 1913, a formal announcement of the incorporation and the name change was issued by Maxwell Motor Company. The officers of the Maxwell Motor Company, whose names appeared on the statement, were: Walter E. Flanders president, W.F. McGuire vice president, W.B. Anthony comptroller, and Carl Tucker treasurer.

Upon assuming the presidency, Flanders quickly recognized the problems that brought about the demise of United States Motor Company, agreeing with the findings of a report by Roberts Walker and W.S. Strong on October 28, 1912. One of the problems in particular that he perceived was the over-extension of facilities; it had more than a dozen manufacturing plants, reaching over six diverse and faraway states. He discovered there were no less than 28 body styles spread over many product lines, most of which were certainly not profitable. Flanders knew that the number of plants would have to be reduced, and the product line be confined to one chassis, for a car of popular price.

Flanders Reorganizes Maxwell

After the formal merger of Maxwell Motor Company and Flanders Motor Company, Flanders, using blank sheets of accounting forms, listed all the units, subsidiaries, affiliates, plants, and facilities, listing their costs, assets, liabilities, and performance. After carefully reviewing and studying the alternatives, he crossed off all the company affiliate names that he felt were definite losers.

While Flanders felt that Maxwell was the only product and name worth saving, at the same time he did not want to alienate the former buyers of the products of United States Motor Company. So Maxwell Motor Company Incorporated ran full-page ads in automotive journals and other publications, with a Declaration of Policy. The advertisement outlined the relationship of Maxwell Motor Company Incorporated with its Maxwell and Flanders dealers, assuring the owners of Brush, Columbia, Courier, Stoddard-Dayton, Flanders, Maxwell, and Sampson cars and trucks of the continued availability of service parts, for which the Maxwell Motor Company felt itself morally bound.

The 1912 Maxwell four-cylinder models were carried over in 1913. The Flanders Six became the new Maxwell Six for 1913. On April 15, 1913, the new Maxwell 25 was announced and deliveries began on May 1, 1913. During the years 1913-15, Flanders spent a great amount of time and effort in restoring Maxwell's image and its proper place in the sales market. He did so by disposing of unneeded facilities and rearranging others for useful purposes. He also decided on sale or abandonment of a number of former United States Motor Company properties, as recommended by Walker and Strong. Flanders concentrated the manufacturing operations in the Midwest, with company headquarters in Detroit. The manufacturing plants in Dayton, Ohio, and Newcastle, Indiana, were both within a short commuting distance from Detroit, and in cities where Maxwell could benefit from a favorable labor market.

The Newcastle, Indiana, plant was rearranged as a "sixty-hammer" forge plant in which all forgings for the Maxwell car were produced. A large section of the plant was converted into a service parts supply center. A complete machine shop was outfitted to manufacture parts for obsolete cars as needed. At Dayton, Ohio, Flanders arranged for a body-building plant, as well as facilities for the manufacture of transmissions and rear axle assemblies. In Detroit, Michigan, three manufacturing plants were maintained: the former Briscoe Manufacturing Company plant at Woodward Avenue and the Michigan Railroad for the production of radiators and sheet metal parts; the former Alden Sampson plant on Rhode Island Avenue as the engine factory; and the former Brush Runabout plant for the assembly center. These plants were located on outskirts of Detroit (where later the city of Highland Park was incorporated), which is where the Chrysler Corporation Headquarters and General Offices were later located.

Growth and Change

On April 15, 1913, the much-heralded Maxwell 25 small touring car was introduced, at a starting price of $725. It was exhibited to Maxwell dealers, although deliveries did not begin until May 1, 1913. The features of this new car were in keeping with those found only in more expensive cars. The hood had a gentle slope from the dash to the radiator shell. The car had a wheelbase of 102 in. and a standard tread of 56 in., providing a mounting for a body that was roomy and seated five adults comfortably. It was powered by a four-cylinder, L-head engine of 3-5/8-in. bore and a stroke of 4-1/2 in. The cylinders were cast-en-bloc with a removable cylinder head.

The three-speed sliding-gear transmission was mounted directly at the rear of the engine, making a compact power plant unit. The rear axle was a three-quarter floating type, and the propeller shaft was enclosed in a torque tube. The front axle suspension was by semi-elliptical leaf springs, and the rear axle by three-quarter elliptical. The frame was constructed of pressed-steel channel section rails and cross-members. The steering

Ted Tetzlaff in a Maxwell at the 1914 Indianapolis 500. (Source: Indianapolis Motor Speedway Corp.)

gear mechanism was located on the left side with the gearshift and hand brake levers at the center of the floor boards. A Stewart-Warner speedometer with odometer was standard equipment.

In September 1913 the Courier plant in Dayton, Ohio, was sold for $50,000. On November 25, 1913, J.A. Vail was elected chairman of the board, succeeding H.C. Bronner. Ray Harroun (winner of the 1911 Indianapolis 500) was named chief engineer of the racing activity in early 1913. Three new Maxwell racing cars were designed, built, and entered in the 1914 Indianapolis 500. Two of them qualified: Ted Tetzlaff for the middle of the front row at 96.36 mph and Willie Carlson in fifth position (middle of the second row) at 93.36 mph. Tetzlaff's car was forced out of the race on the 39th lap by a broken valve rocker arm. Carlson finished in ninth position at an average speed of 70.97 mph.

In 1914 the Columbia Motor Car Company plant in Hartford, Connecticut, was sold, and the Maxwell-Briscoe plant located at Kingsland Point, Tarrytown, New York, was sold to the Chevrolet Motor Company for $267,000. The other plant in Tarrytown, New York, known as the Beekman Avenue plant, was sold in March 1915 to Chevrolet Motor Company for approximately $250,000. The Pawtucket, Rhode Island, plant had been previously sold to a machine tool manufacturer.

During 1914, with Charles F. Redden as general sales manager, Flanders concerned himself with the improvement of the Maxwell 25. The six-cylinder Maxwell 50 was continued, and a new larger Maxwell 35 was introduced, replacing the former four-cylinder Maxwell 30 and 40. By late 1914 the effects of World War I were being felt by the automobile industry in terms of export sales, and the availability of and increased cost of raw materials. Tire prices increased by 20-30% on August 8, 1914, and other material costs increased by an average of 5-20%.

The Maxwell 50 and the four-cylinder 35 were discontinued in 1915, and all effort of manufacture and sales was concentrated on the Maxwell 25. The new 1915 model offered an electric starter, lighting, and full electrical equipment at $35 extra. However, the base price of the Maxwell 25 was reduced to $695. Thus, the Maxwell 25 touring equipped with electrical starting and lighting could be bought for $750. The two-passenger roadster, the cabriolet, and the town car completed the full line of four Maxwell models priced from $695 to $920.

For the 1915 Indianapolis 500 race, three new Maxwell racing cars were designed and built. While they had the same appearance as the 1914 racing cars, the engine piston displacement was reduced to 298 cu. in. to

Willie Carlson in a Maxwell at the 1915 Indianapolis 500. (Source: Indianapolis Motor Speedway Corp.)

conform to the new rule limits of 300 cu. in. The engines featured a new, two-piece, ball-bearing crankshaft, permitting the use of ball bearings at all three main bearing positions. The rigid joint at the center of the crankshaft was accomplished by telescoping construction whereby one part slipped within the other. Four Morse tapers were used to key the two parts together, and a 3/4-in.-diameter steel-alloy cap screw held the two halves of the crankshaft tightly together. The engine lubrication was by pressure and splash.

Willie Carlson qualified car #19 for the 16th starting position, Tom Orr qualified car #21 for the 17th spot, and Eddie Rickenbacker qualified car #23 for the 19th position. Carlson finished 9th in the race, Orr's car experienced a bearing failure on the 169th lap, and Rickenbacker's car was forced out of the race with a connecting-rod bearing failure on the 104th lap.

Orland F. Weber was elected vice president of Maxwell Motor Company on May 1, 1916. During 1916 and 1917 the Maxwell 25 was continued with practically no changes except for minor price variations. Due to the shortages of raw material for domestic car production, Flanders sought United States Government contracts for production of defense equipment to keep the plants operating fully. After war was declared on April 6, 1917, Maxwell Motor Company received many U.S. Government contracts for the production of cars, trucks, munitions, etc., so that by the end of 1917, Maxwell plant production was 80% for war materiel and 20% for civilian usage.

Maxwell Takes Over Chalmers

In the meantime, Chalmers Motor Company was experiencing cash flow and other financial difficulties. Hugh Chalmers, Walter Flanders, and other officers of the Chalmers and Maxwell organizations held many conferences with the result that, on September 10, 1917, the stockholders of the Chalmers Motor Company ratified a refinancing plan. Maxwell Motor Company would advance Chalmers $3,000,000 and Chalmers would lease the plant and assets of Chalmers Motor Company to Maxwell Motor Company Incorporated for five years. Maxwell Motor Company Incorporated assumed complete control of Chalmers Motor Corporation on September 17, 1917, under the terms of the lease arranged the week before.

Following the signing of the lease and contract, Hugh Chalmers resigned as president, but remained chairman of the board of directors of the Chalmers Motor Corporation. (See the chapter on Chalmers in Volume 3.) Flanders was elected president, Carl H. Pelton assistant, Walter M. Anthony treasurer, John F. Flint comptroller, and Thomas J. Toner, Maxwell sales manager, became sales manager and advertising manager of both companies. Maxwell Motor Company Incorporated, after taking over the Chalmers plant, continued to build Chalmers cars, and used the surplus plant area to build Chalmers trucks for commercial usage and four-wheel-drive trucks for the United States Army. In late 1917, Flanders was elected board chairman, and W. Ledyard Mitchell was elected president of Maxwell.

For 1918 the Maxwell 25 was changed only slightly: the wheelbase was increased from 103 to 109 in. to provide more interior room. Demountable, artillery-type, wood spoke wheels were standard, with the wheel and tire mounted in a carrier at the rear. All previous body types were continued, and a new five-passenger touring car with an all-weather top was added to the line, priced at $855; the touring car and roadster were $745; the sedan, coupe, and Berline limousine were priced at $1,095. The electrical units of the car remained the same, but the Simms-Huff two-wire system was replaced by a 12V single-wire system. The 1918 Maxwell automobile production continued along with war materiel. Nicholas R. Feltes was elected treasurer of Maxwell during 1918.

1917-18 Maxwell truck assembly. (Source: Chrysler Corp.)

After the production of war materiel was curtailed, the production of cars and trucks for civilian use gradually proceeded. However, it was not until May 1, 1919, that the company's plants were completely cleared of war work and available for car and truck production. The result was that the number of cars and trucks sold during 1919 was substantially less than that of 1916 and 1917. The net income for 1919 dropped to $1,529,499 against a net of $2,292,202 for the preceding year. However, the assets of Maxwell Motor Company Incorporated increased from $50,804,146 in 1917 to $62,419,064 in 1918, and to $70,251,021 for 1919.

Flanders Retires

On May 12, 1919, Walter E. Flanders advised the board of directors of his intention to retire as board chairman, effective July 31, 1919. Thus, upon retiring, Flanders left the Maxwell Motor Company Incorporated in good financial condition, and in the capable hands of W. Ledyard Mitchell. Flanders planned a long cruise down along the Atlantic coast and the Caribbean Sea in his new yacht, crossing through the Panama

Canal and up the Pacific coast to California. He was out of the automobile business, and would not again be involved in the automotive industry until late 1920, when he was enticed by Everitt and Metzger into the venture of the Rickenbacker Motor Company. (See the Rickenbacker chapter in Volume 3.)

Meanwhile, the departure of Flanders and the post-war depression of 1920 placed Maxwell and Chalmers organizations in grave financial difficulties. Several reorganization attempts were made. W.R. Wilson was elected president and Mitchell became board chairman. In August 1920 both Maxwell and Chalmers organizations went into receivership. The financial firms to whom Maxwell and Chalmers were indebted formed a management committee, headed by vice chairman J.R. Harbeck. The banking firms sought out Walter P. Chrysler to join them as chairman of the management committee, with the prime responsibility of straightening out the tangled affairs of Maxwell and Chalmers. (See the Chrysler chapter in Volume 1 for the continuation of this story.)

Postscript

Byron P. Everitt retired from business after the demise of the Rickenbacker Motor Company in 1927. After declining health for about one year, he died in Detroit on October 5, 1940, at the age of 67.

William E. Metzger retired in 1929 from his dealerships after the discontinuance of the Wills Sainte Claire car, the stock market crash, and the Depression. He died in Detroit on April 11, 1933, at the age of 64.

The passing of Walter E. Flanders on June 16, 1923, as the result of an automobile accident, had a profound and immediate effect on the Rickenbacker Motor Company, and perhaps may have precipitated its subsequent demise.

Joseph Hudson driving. (Source: NAHC)

Chapin/ Coffin/ Bezner/ Jackson/ Hudson/ McAneeny and The Hudson Motor Car Company

From a modest start, five brave automotive pioneers and a greatly respected merchant embarked on a business venture that destined them to manufacture and merchandise some of the most popular cars of the time. The famous marque may be gone, but those who have driven and enjoyed these fine vehicles will never forget them.

* * *

The background of Roy Chapin, Howard Coffin, Frederick Bezner, as well as Hugh Chalmers and the creation of the Chalmers Motor Company, is detailed in the chapter on Chalmers in Volume 3.

Three new members joined the Chapin-Coffin-Bezner team of entrepreneurs in 1909; they were Roscoe B. Jackson, Joseph L. Hudson, and William J. McAneeny.

Roscoe B. Jackson

Roscoe B. Jackson was born in Ionia, Michigan, in 1879. It was while attending the University of Michigan that he met Chapin and Coffin, who would later become his business associates and closest friends. After graduating in 1903, Jackson moved to Lansing, Michigan, and joined the Olds Motor Works engineering department. In 1907 he married Miss Louise Webber, niece of Joseph L. Hudson, Detroit merchant. They moved to Buffalo, New York, where Jackson became factory manager of the E.R. Thomas Motor Company.

Joseph L. Hudson (1846–1912)

Joseph Lothian Hudson was born on October 17, 1846, in Newcastle on Tyne, England, the son of Richard Hudson, who conducted a tea and coffee business. In 1853 Richard Hudson emigrated to Hamilton, Ontario, Canada, where he was employed by the Grand Trunk Railway. His wife and young Joseph joined him in 1855. In 1860 the Hudson family moved to Grand Rapids, Michigan, where young Joseph worked on a fruit farm, and then was employed by Christopher R. Mabley. Around 1863 Richard Hudson and Christopher Mabley jointly opened a clothing store in Ionia, Michigan. Two years later Hudson bought out Mabley's interest and Joseph became his father's partner. They operated as partners for eight years until the death of Richard Hudson in 1873. Joseph made a go of it alone; however, the economic panic of 1873 so destroyed him financially that, in 1876, he was forced into bankruptcy, closing the store, able to pay his creditors only 60 cents on the dollar.

Christopher Mabley, now operating a clothing store in Detroit, asked Joseph Hudson to run his store for two months while he was away. During this time the business took a strong upturn, and Hudson was rehired at the end of the period, at $50 per week plus 10% of the profits. He saved his earnings and, on April 2, 1881, opened a men's and boys' clothing store of his own. Twelve years after his store failure in Ionia, Hudson personally went to the creditors of that business, asked them for their statements, and reimbursed them fully for their losses.

William J. McAneeny (Source: J.A. Conde)

Hudson's business expanded so that by 1891 he had one of the finest stores in the Midwest. In 1907 he added new buildings, and in 1911 he had erected the famous Hudson eleven-story building on Woodward Avenue.

It was in 1909 that Joseph Hudson backed Chapin, Coffin, Bezner, and Jackson in the founding of the Hudson Motor Car Company. The fact that Jackson was married to his niece, Louise Webber, might have had some influence on Hudson's providing the financial backing for this new company.

William J. McAneeny

William J. McAneeny was born in Newport, New York, on November 21, 1872. After his formal education, McAneeny entered the automobile industry in 1899 as purchasing agent for the Riker Motor Vehicle Company of Elizabeth, New Jersey. He moved to Hartford, Connecticut, in 1903 and became the purchasing agent for the Electric Vehicle Company. However, he was convinced that Detroit was and would remain the center of the automobile industry, so he went there in 1908 and joined the Chalmers-Detroit Motor Company. There McAneeny met a group of executives who were organizing a new company, the Hudson Motor Car Company, which he joined in October 1909 as a purchasing agent.

The Hudson Motor Car Company

On March 6, 1909, the first board of directors meeting was held, electing the following officers of Hudson Motor Car Company: Joseph L. Hudson, president (prominent merchant of Detroit); Hugh Chalmers, vice president (remained president of Chalmers Motor Co.); Roscoe B. Jackson, treasurer and general manager; Roy D. Chapin, secretary (remained treasurer and secretary of Chalmers); Howard E. Coffin, vice president of engineering (remained vice president and chief engineer of Chalmers Motor Company); George W. Dunham, chief engineer and designer (formerly of Olds Motor Works).

Immediately after the board meeting a meeting of the stockholders was held, at which time the stockholders holdings were disclosed: Joseph L. Hudson - 1,584 shares, Hugh Chalmers - 1,334 shares, Roscoe B. Jackson - 1,083 shares, Howard E. Coffin - 1,083 shares, Frederick Bezner - 1,083 shares, Roy D. Chapin - 1,083 shares, James J. Brady - 1,083 shares, Lee Counselman - 666 shares, George W. Dunham - 1 share.

When the articles of incorporation were filed in the Wayne County, Michigan, Registrar's Office on February 24, 1909, the incorporators were listed as follows: Joseph L. Hudson, Roscoe B. Jackson, Roscoe B. Jackson, Trustee, and George W. Dunham. The Jackson trusteeship was filed on behalf of Chapin, Coffin, Bezner, Brady, and Counselman. Hugh Chalmers, while providing financial support, remained a silent

The early Hudson plant. (Source: J.A. Conde)

349

partner. The Hudson Motor Car Company, initially capitalized at $100,000, by November 1910 brought the capitalization to $1,000,000 due to a 900% stock dividend.

While Hudson Motor Car Company was originally associated as an offshoot of Chalmers Motor Company, with Hugh Chalmers as a silent investor, on January 1, 1910, it became an entirely separate organization, complete in itself, and not allied with any other concern. The controlling interest was held by the organizing officers.

The actual manufacturing operations of the Hudson Motor Car Company began on July 3, 1909, in the plant formerly occupied by the Northern Manufacturing Company at Champlain Street and Canton Avenue in Detroit, Michigan. Approximately 1,000 open vehicles were built during the remainder of 1909, which were marketed and registered as 1910 models. Most of the vehicles were powered by L-head, Buda, four-cylinder engines of 3.75-in. bore and 4.5-in. stroke, while a smaller number of cars were equipped with an Atlas engine during the early stages of production.

Early Changes

After building and selling approximately 7,000 cars in 1910, Hudson Motor Car Company officials realized that the plant facilities were inadequate to produce enough vehicles to meet the anticipated demand. By January 1911 plans were well underway to build a new modern plant of approximately 225,000 sq. ft. of floor area. This new plant was built on a tract of land from the D.J. Campau farm of approximately 68 acres, plus additional acreage acquired for a total of 106 acres, at the corner of Jefferson and Conner Avenues in

Roy Chapin in a 1910 Hudson. (Source: J.A. Conde)

Detroit, Michigan. The new factory was completed in mid-1912. Production for the calendar year of 1912 was approximately 5,708 cars.

On July 15, 1912, Joseph Hudson died after serving as president for one year and about one year as chairman of the board of Hudson Motor Car Company.

Early Models

On July 6, 1912, the Hudson Motor Car Company announced their new 1913 models, adding a new six-cylinder-engine-powered model to the line. The engine power was transmitted through a multiple-disc clutch of 15 plates, of which eight driven plates had cork inserts. The entire clutch driven in oil was contained within an oil-tight case as part of the flywheel. The rotating force was then transmitted by means of a three-speed transmission and propeller shaft with two exposed universal joints to the full-floating rear axle.

The fuel tank was at the rear with pressure fuel feed and the engine cranking pedal was in the floor of the driver's compartment. Approximately 6,401 Hudson cars were built and sold during 1913.

	1913 Hudson 37	1913 Hudson 54	1914 Hudson 6-40	1914 Hudson 6-54
Wheelbase (in.)	118	127	123.5	135
Price	$1,875	$2,450-3,750	$1,375-1,950	$2,350
No. of Cylinders / Engine	L-4	L-6	L-6	L-6
Bore x Stroke (in.)	4.12 × 5.25	4.12 × 5.25	3.50 × 5.00	4.12 × 5.25
Horsepower	37 adv, 26.6 NACC	54 adv, 40.84 NACC	40 adv, 29.4 NACC	54 adv, 40.84 NACC
Body Styles		11	3 (roadster, phaeton, carbriolet)	
Other Features	right-hand steering, Delco starting, charging and ignition system		Delco starting, charging and ignition system	four-forward-speed transmission

The new 1914 Hudson models were introduced on August 7, 1913, concentrating on building six-cylinder-engine-powered models only. The steering mechanism moved to the left side of the tapered frame, with the driver controls moved to the center of the driver's compartment floor. The 1914 Hudson models featured a four-forward-speed transmission, in which the third speed was a direct drive of 3.66:1 final drive ratio, and the fourth speed an "indirect overdrive," providing a 2.99:1 final drive ratio. The fuel tank was mounted under the cowl, with gravity feed to the carburetor.

In early November 1913, Hudson Motor Car Company announced the addition of the new Model 6-40, a companion model to the Hudson Model 6-54 introduced in August 1913. The six-cylinder engine aluminum upper crankcase had an extended partial flywheel housing. With the Hudson 6-40 prices starting at $1,750, Hudson production and sales climbed to approximately 10,261 units during 1914.

Introduced in the fall of 1914, the 1915 Hudson models 6-40 and 6-54 were basically 1914 models with improvements and refinements. Available in nine body styles, the 1915 Hudson production and domestic sales amounted to approximately 12,864 cars.

The 1916 Hudson models announced in the fall of 1915 featured the new "yacht" body line styling, with the door upper edges upholstered on a level with the rest of the design lines. The most important feature of the

6-54 was the new L-head Super Six engine built by Hudson, which developed 76 hp. Among the other mechanical improvements was that the cork inserts of the oil-lubricated clutch discs were placed in the driving discs.

During 1916 Hudson cars were engaged in many automobile races and other competitive events. On April 10, 1916, Ralph Mulford, driving a Hudson Super Six, covered a mile over the Daytona Beach sands in 39.11 seconds, for an average speed of 102.5 mph, a new stock car record. On May 13, 1916, a car driven by Ira Vail (a chassis of which was just two weeks before attached to a touring car body, and was performing as a demonstrator) crossed the finish line of a 150-mile race at an average speed of 91 mph, finishing in third place 5 minutes and 13 seconds behind the winner. The splendid performance of the 1916 Hudson cars may have been responsible for the brisk sales of approximately 25,772 cars domestically.

Among the important features offered by Hudson in their 1917 models, introduced in late 1916, were: the horizontal radiator shutters and the Boyce Moto-Meter. The adjustable horizontal radiator shutters used to help warm the engine during cold weather were controlled by a cable and knob on the left end of the instrument panel. The Boyce Moto-Meter mounted on the radiator filler cap indicated the engine temperature visibly. It was priced at $10, while the shutter assembly was a $15 option.

World War I

On April 4, 1917, Howard Coffin, vice president of engineering, requested a leave of absence from the Hudson Motor Car Company to assist the United States Government in the production of military equipment. He was named chairman of the United States Board of Aeronautics. After war was declared on Germany on April 17, 1917, Hudson Motor Car Company made all their plant facilities and manpower available to the United States Government for the production of cars, trucks, ambulances, special vehicles, and other war needs.

Hudson Motor Car Company concentrated on the production of war needs, including the manufacture of the Curtiss O-X-5 aircraft engine, during the balance of 1917 and 1918. However, the United States Government allowed Hudson Motor Car Company enough raw materials to build and sell approximately 20,926 civilian cars in 1917.

The basic 1917 Hudson models were carried over into 1918, with the addition of two new body types: a five-passenger full folding to Landau, and a two-passenger runabout Landau. The restricted availability of materials reduced Hudson production and sales for 1918 to approximately 12,526 civilian vehicles.

Essex Motor Company

In October 1917 the officers and directors of the Hudson Motor Car Company incorporated the Essex Motor Company in order to manufacture a light car. They tentatively leased the Studebaker #5 plant on Franklin Street in Detroit, but it was engaged in war materiel production, so in early February 1918, they decided to build the Essex car at Hudson's relatively new plant on Jefferson Avenue in Detroit, Michigan.

The Essex car was introduced in early 1918 but, because of material shortages, it did not get into production until late 1918 as a 1919 model. The F-head engine design permitted the exhaust valves to be located in the

Hudson F-head six-cylinder engine.
(Source: J.A. Conde)

cylinder block while the large intake valves were incorporated in the cylinder head, actuated by pushrods and rocker levers. This arrangement allowed the use of larger valves, and with the machined combustion chambers the breathing of the engine was so favorable that the Essex F-head engine power output far exceeded the output of competitive engines of the same displacement capacity. The F-head design remained important to Hudson, as Hudson would reintroduce the design in the 1927 Super Six.

The Essex car was reasonably priced and had many desirable features, making it an immediate hit with the car-buying public. Approximately 21,879 Essex cars and 18,175 Hudson cars were built and sold domestically in 1919. The acceptance of the 1919 Essex and Hudson cars was so favorable that the 1919 designs were carried over into 1920.

	1919 and 1920 Essex	1921 Essex	1921 Hudson
Wheelbase (in.)	108.5	108.5	125.5
Price	$1,395-1,795	$1,195-1,995	$1,875-3,495
No. of Cylinders / Engine	F-4	F-4	L-6
Bore x Stroke (in.)	3.37 × 5.00	3.37 × 5.00	3.50 × 5.00
Horsepower	55 adv, 18.2 NACC	55 adv, 18.2 NACC	76 adv, 29.4 NACC
Body Styles	3	5	7 (including seven-passenger limousine)
Other Features	five passengers, light weight (2400 lb)		

To publicize the performance of the 1920 Essex cars, the Hudson Motor Car Company engaged a team of professional drivers for four Essex production cars to establish transcontinental records from San Francisco to New York, the first arriving in New York in 4 days, 14 hours, and 43 minutes.

The post-WWI depression and accompanying inflation did not immediately affect Hudson and Essex sales; they built and sold a combined total of 45,937 Hudson and Essex cars in 1920. However, the spiraling inflationary costs cut deeply into company profits.

1920 Hudson Super Six limousine, sedan, and coupe. (Source: J.A. Conde)

Production of 1921 Hudsons. (Source: J.A. Conde)

For 1921 it was a different story, however, as all businesses suffered. Hudson car sales dropped to about 13,721 units, and the Essex car domestic sales amounted to 13,422 cars. The increased prices reflected the high manufacturing costs of building closed body types, and high inflationary costs of material. At that time closed car bodies were still constructed to a great extent of wood and metal composites, requiring a large amount of hand labor.

For 1922 the Hudson and Essex models were continued as a carryover of the 1921 models. Through aggressive selling, improved manufacturing methods and stringent cost control, Hudson Motor Car Company was able to show a profit of approximately $7,242,677 for 1922, on domestic sales of 28,242 Hudson cars and 35,222 Essex cars.

	1922 Essex	1922 Hudson	1923 Essex	1923 Hudson
Wheelbase (in.)	108.5	125.5	108.5	125.5
Price	$1,145 for touring	$1,275 for speedster	$1,045-1,895	$1,275-3,495
No. of Cylinders / Engine	F-4	L-6	F-4	L-6
Bore x Stroke (in.)	3.37 × 5.00	3.50 × 5.00	3.37 × 5.00	3.50 × 5.00
Horsepower	55 adv, 18.2 NACC	76 adv, 29.4 NACC	55 adv, 18.2 NACC	76 adv, 29.4 NACC
Body Styles	4 (roadster discontinued)	8	4	8
Other Features				

New Management

On January 22, 1923, at the annual Hudson Motor Car Company board of director's meeting, Roy Chapin, president, was elected chairman of the board, a newly created position. Roscoe Jackson, vice president and general manager, was promoted to the presidency of Hudson Motor Car Company. William McAneeny, secretary, was named vice president, Howard Coffin and O.H. McCormack were re-elected vice presidents, and Abraham Edward Barit was chosen as secretary. Hudson Motor Car Company and Essex Motor Company enjoyed a profitable year in 1923, building and selling a combined total of 88,914 cars.

The Record Years

For 1924 the Hudson Super Six was basically unchanged except for improvements and refinements. The new 1924 Essex models were introduced in December 1923. The new peaked radiator shell blending into the straight body lines stood out as a prominent design feature. A new sales record was reached during 1924, when a combined total of 133,950 Hudson and Essex cars were built and shipped, of which 100,472 cars were registered domestically in the United States.

Roy D. Chapin (Source: J.A. Conde)

	1924 Essex	**1924 Hudson**
Wheelbase (in.)	110.5	127.5
Price	$850-975	$1,295-2,145
No. of Cylinders / Engine	L-6	L-6
Bore x Stroke (in.)	2.62 × 4.00	3.50 × 5.00
Horsepower	50 adv, 17.32 NACC	76 adv, 29.4 NACC
Body Styles	2 (five-passenger phaeton and two-door coach)	6 (limousine discontinued)
Other Features		

The introduction of the 1925 Hudson and Essex models in December 1924 was met with much excitement. Not so much because of the design changes, which were minor, among which was making balloon tires standard, but because of the pricing structure, whereby the two-door coach model was priced below that of the touring models. The price changes took effect on October 27, 1924, for the Hudson Coach, and on November 26, 1924, for the Essex Coach.

Because these high-quality cars were reasonably priced, Hudson and Essex enjoyed record sales, combining for a total of 269,474 cars in 1925. This placed Hudson Motor Car Company third in sales behind Chevrolet and Ford. As a result Hudson Motor Car Company profits for the year of 1925 hit an all-time high of approximately $21,300,000.

The 1926 Hudson and Essex models were announced in December 1925 with only minor changes, among which were: the nickel-plated radiator shell, and balloon tires of a smaller cross-section diameter. The Hudson priced at $1,150 to $1,650, and the Essex priced at $695 to $795 kept up the 1925 sales momentum, permitting the domestic sales to climb to 70,247 Hudson cars and 157,247 Essex cars during 1926. Needless to say, Hudson Motor Car Company enjoyed a profitable year in 1926. Perhaps the most important event during 1926 was the completion of the $10,000,000 exclusive Hudson and Essex body plant by July 1, 1926. This plant enabled the Hudson Motor Car Company to build all of the all-steel closed bodies for Hudson and Essex cars.

The Super Sixes

The new Hudson and Essex models introduced at the National Automobile Show in New York in January 1927 were of an entirely new appearance and embodied important mechanical changes in both lines. The cars were named the Hudson Super Six and the Essex Super Six.

Fully rounded body panels, curved roof lines, and smooth streamlining characterized the new 1927 bodies. The rounded hood and cowl lines of both car lines flowed gracefully into the curved base of the new one-piece windshield. Fully crowned fenders were employed on both lines. The bullet-shaped headlamps and cowl lamps were nickel-plated on the Hudson models, and lacquer-finished on the Essex. The rounded radiator shell of both the Hudson and Essex models were nickel-plated.

A full-length belt line molding, finished in contrasting color, extended from the radiator shell around the rear of the body, and back to the radiator shell on the opposite side. The smaller 18-in.-diameter wheels lowered the car considerably, improving the appearance. The 1927 Hudson and Essex models were equipped with Bendix mechanical four-wheel brakes.

1927 Hudson in front of Hollman building. (Source: NAHC)

The most outstanding feature of the 1927 Hudson Super Six was the performance characteristics of the F-head six-cylinder engine, developing 92 hp. To demonstrate the engine's potential, Hudson Motor Car Company engaged Barney (Eli) Oldfield, famed racing driver, to pilot a 1927 Hudson Super Six coach around the Culver City, California, track for 1,000 miles at an average speed in excess of 76 mph.

Apparently the Hudson and Essex sales promotion paid off well, as sales continued to climb, building and shipping 66,034 Hudson cars and 210,380 Essex cars during 1927, earning profits of $14,042,536 during the first nine months.

The Hudson models introduced in 1928 were indeed handsome vehicles having a higher rounded radiator shell surrounding the vertical radiator shutter louvres. Mechanically, the Hudson was a refined version of the 1927 models. Sales emphasis centered on the performance characteristics of the Hudson Super Six engine. The 1928 Essex models were much refined, closely resembling the 1928 Hudson. Production and car shipments to dealers amounted to 52,316 Hudson cars and 229,889 Essex cars in 1928.

	1928 Essex Super Six	1928 Hudson Super Six	1929 Essex Challenger	1929 Greater Hudson
Wheelbase (in.)	110.5	118.5/127.5	110.5	122.5/139
Price	$735-795	$1,250-1,950	$695-965	$1,095-$1,500 for 122.5 $1,600-2,200 for 139
No. of Cylinders / Engine	L-6	F-6	L-6	F-6
Bore x Stroke (in.)	2.75 × 4.50	3.50 × 5.00	2.75 × 4.50	3.50 × 5.00
Horsepower	55 adv, 19.4 NACC	92 adv, 29.4 NACC	55 adv, 19.4 NACC	92 adv, 29.4 NACC
Body Styles	6	8 plus custom-built seven-passenger sedan	9	9 on 122.5 wheelbase 5 on 139 wheelbase
Other Features			Bendix four-wheel brakes	non-shatterable windshield, hydraulic two-way shock absorbers, electric fuel and oil level gauge, adjustable seats

The Hudson and Essex cars introduced in January 1929 abandoned the time-honored "Super Six" designation, and instead were called the "Greater Hudson" and the "Essex Challenger." The Greater Hudson's bright metal parts were chromium including the newly shaped radiator shell and double face bar bumpers, which had a dip in the lower face bar to spread the contact area. The styling was contemporary, following the trend of the time.

The Essex Challenger possessed many of the Hudson features and the resemblance won the buying public's acceptance. While the 1929 models carried the features of the 1928 cars, the former boat-tailed roadster was

1929 five-passenger phaeton. (Source: J.A. Conde)

now called "Speedabout," and featured new styling including a rumble seat in the boat-tail rear. All 1929 Essex Challenger models had a three-speed transmission and a single-disc clutch plate with cork inserts, operating in oil in a leak-proof housing. The Speedabout had a three-speed transmission which incorporated an "overdrive" instead of the conventional intermediate set of gears. This provided a 0.7796:1 clutch shaft to output shaft ratio, resulting in greater propeller rotation speed in relation to engine crankshaft rotation speed. In operation of the overdrive, a clutch-operated secondary interlock was provided. This was perhaps the first application of the overdrive technology to a three-speed transmission.

Apparently the buyers liked what they saw, as Hudson Motor Car Company was able to build and ship 71,179 Greater Hudson cars and 227,653 Essex Challenger cars in 1929.

McAneeny Takes Over

While the Hudson Motor Car Company was rejoicing in the public acceptance and popularity of the Greater Hudson and Essex Challenger, and the increased sales, the entire Hudson Motor Car Company organization was greatly saddened by the sudden and untimely death of Roscoe B. Jackson on March 19, 1929. Jackson was stricken with influenza only two days earlier while he was vacationing in Mentone, France. Jackson, who had been president since 1923, was credited with being the motivating force behind the Hudson Motor Car Company as it became one of the largest independent automobile manufacturers.

William J. McAneeny, who had been identified with the Hudson Motor Car Company since its formation in 1909, and served as first vice president and treasurer since 1923, was elected president of Hudson Motor Car Company on April 10, 1929. McAneeny's remarkable ability and energy earned him rapid advancement and ever-widening management influence.

The Great Eight

To meet the buyers' desire and the trend of the automobile industry, Hudson entered the eight-cylinder-engine field with the announcement of the Hudson "Great Eight" on January 1, 1930. The styling of the 1930 Hudson Great Eight was pleasing, with deeply flanged long sweeping front fenders, cadet-type sun visor, slightly sloping windshield, hinged rectangular ventilating doors on the hood side panels, and a single belt molding.

The 1930 Essex Challenger was also introduced on January 1, 1930, with many styling improvements similar to those of the Hudson Great Eight. The Essex models had a cadet-type sun visor, single belt molding, but retained the two rows of vertical louvres on the hood sides. Prices of the "second series" Essex Challenger, as well as the Hudson Great Eight, were reduced on August 14, 1930.

	1930 Essex Challenger	1930 Hudson Great Eight	1931 Essex Challenger	1931 Hudson Great Eight
Wheelbase (in.)	113	119/126	113	119/126
Price	$695-995 reduced to $650-795	$1,050-1,650 reduced to $885-1,295	$595-895	$875-1,095
No. of Cylinders / Engine	L-6	L-8	L-6	L-8
Bore x Stroke (in.)	2.75 × 4.50	2.75 × 4.50	2.87 × 4.50	2.87 × 4.50
Horsepower	55 adv, 18.1 SAE	80 adv, 24.2 SAE	60 adv, 19.8 SAE	87 adv, 26.4 SAE
Body Styles	9	11	8	17
Other Features				

Max Wollering (Source: NAHC)

The Depression Takes Its Toll

Despite the great values Hudson and Essex offered, the economic downturn took its toll in plunging automobile sales. Hudson Motor Car Company built and shipped 36,674 Hudson cars and 76,158 Essex cars during the 1930 calendar year.

During November 1930 Max F. Wollering, former Studebaker vice president of manufacturing, joined Hudson Motor Car Company as director of manufacturing.

On November 17, 1930 Hudson Motor Car Company broke the long tradition of announcing car prices at the time of introduction, or shortly after, by announcing the 1931 car prices in advance of the new model showing. The Hudson Great Eight and the Essex Challenger were offered at the lowest prices in Hudson Motor Car Company history up to that time. Production of 1931 Hudson and Essex models began on November 26 and formal announcement on November 29, 1930.

In August 1931, Hudson Motor Car Company was first to introduce "Startix" automatic starting on 1931 Hudson and Essex cars. Stylewise, the contours were

1931 Hudson touring sedan. (Source: NAHC)

smoother versions of the 1930 models. The vertical radiator shutter blades were replaced by a chromium-plated, grid-type grille. Two chromium-plated, diagonally placed tension members meeting near the grille center served as fender braces and provided a mounting location for the vibrator horn. In order to try to stimulate car sales, Essex offered three commercial vehicle types: a panel delivery, sedan delivery, and a pickup model, economically priced. At the same time Hudson discontinued the Biddle and Smart custom-body offerings as an economic move.

Due to the deepening of the Depression, Hudson and Essex sales dropped to 17,487 Hudson cars and 40,338 Essex cars built and shipped in 1931. On a sales dollar volume of about $38,235,636, Hudson Motor Car Company sustained a loss of $1,991,198.

With the disappointing sales performance in 1931, car manufacturers placed all their chips on the 1932 models, and gave their best. Hudson Motor Car Company was no exception, and with the new model announcement on January 1, 1932, provided America with the greatest values in Hudson history. Engine power was increased and the transmission was a three-speed with silent second speed.

The styling was new under the direction of Frank S. Spring, who gave the 1932 Hudson and Essex a new flair. The new V-shaped radiator shell with vertical louvres extended downward over the front slash shield,

1932 Hudson Eight touring sedan. (Source: J.A. Conde)

concealing the frame and suspension members. The front fenders were of the long clamshell type with a deeper crown. The bodies had graceful lines with a sloping shatterproof windshield. The exterior sun visors were omitted, replaced by adjustable interior visors. Chromium-plated, single-face bumpers were used. The Hudson retained the rectangular hood side ventilating doors, and a chromium-plated Griffin mascot adorned the radiator filler cap. In May 1932 the Greater Essex Super Six was renamed the Essex Pacemaker.

	1932 Greater Essex Super Six (Pacemaker)	1932 Greater Hudson Eight	1932 Essex Terraplane
Wheelbase (in.)	113	119/126/132	106
Price	$795	$1,095-1,445	$425-610
No. of Cylinders / Engine	L-6	L-8	L-6
Bore x Stroke (in.)	2.93 x 4.75	3.00 x 4.50	2.93 x 4.75
Horsepower	70 adv, 20.7 SAE	101 adv, 28.8 SAE	70 adv, 20.7 SAE
Body Styles	8	14	12
Other Features			four-wheel brakes, wire spoke wheels

The new Essex Terraplane went into production on June 15, 1932, and was formally introduced in July 1932. Because of the Terraplane's light weight of only 2,010 lb. for the roadster, it gave a very lively performance. During 1932 the Essex Terraplane established no less than fifty AAA hill climb records, in addition to a class record of 85.6 mph for a flying mile. Beautifully styled, the Terraplane accounted for 16,581 cars built and shipped during the balance of 1932.

In August 1932 Roy Chapin obtained a leave of absence from the Hudson Motor Car Company to carry out the duties of United States Commerce Secretary, by appointment by President Herbert Hoover, succeeding Robert P. Lamont.

Just as every American automobile manufacturer, the Hudson Motor Car Company suffered loss of sales in the Hudson and Essex lines during 1932. Only 7,777 Hudson cars and 17,425 Essex cars were built and shipped in addition to the 16,581 Terraplane models. As a result, Hudson suffered a loss of $5,429,350.

	1933 Hudson Super Six	1933 Hudson Standard Eight	1933 Hudson Major Eight	1933 Essex Terraplane Six	1933 Essex Terraplane Eight
Wheelbase (in.)	113	119	132	106	113
Price		$1,045	$1,250	$555	$675
No. of Cylinders / Engine	L-6	L-8	L-8	L-6	L-8
Bore x Stroke (in.)	2.93 x 4.75	3.00 x 4.50	3.00 x 4.50	2.93 x 4.75	2.93 x 4.50
Horsepower	70 adv, 20.7 SAE	101 adv, 28.8 SAE	101 adv, 28.8 SAE	70 adv, 20.7 SAE	94 adv, 27.6 SAE
Body Styles	6	4	4	18	14
Other Features					

The new 1933 Hudson and Essex Terraplane models were announced in January 1933, consisting of five model lines: the Hudson Super Six (superseding the Essex Pacemaker), the Hudson Standard Eight, and the Hudson Major Eight, the Essex Terraplane Six, and the Essex Terraplane Eight. The Essex Terraplane Eight developed 94 hp, compared to the 101 hp of the Hudson Eights, due to a bore 1/16-in. smaller.

The styling of the 1933 Hudson models reached the apex of Classic Hudson styling. The broad sweeping front fenders gracefully merged at the front grille area. The wheel openings of both the front and rear

1933 Hudson Model L.
(Source: J.A. Conde)

following the tire curvature concealed all chassis and undercarriage components. The rear fenders curved gently following the body rear panel contours. The front fenders extended farther down in front, concealing all front suspension parts.

The chromium-plated, V-shaped, sloping radiator grille, the triangular headlamp rims, and the dual horns certainly presented an impressive front end. Essex Terraplane styling was similar to that of the Hudson, except where the Hudson Eight and Terraplane Eight had adjustable hood side ventilating doors, the Hudson Super Six had a single row of vertical hood side louvres, and the Essex Terraplane Six had two rows.

The most impressive feature of the Essex Terraplane Eight was its phenomenal performance. During 1933 the Essex Terraplane Eight captured over fifty AAA-sanctioned and supervised hill climb records, including the prestigious Penrose Trophy by winning the Pike's Peak Hill Climb in 19 minutes, 52.2 seconds, driven by Al Miller, famous race driver. The Essex Terraplane Eight also attained a top speed record of 85 mph on the sands of Daytona Beach, Florida.

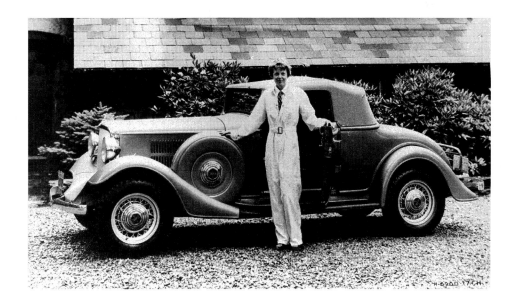

Amelia Earhart and the 1933 Terraplane.
(Source: J.A. Conde)

Chapin Returns

After concluding his service as Commerce Secretary of the United States, Roy Chapin returned to the active management of the Hudson Motor Car Company on May 27, 1933. Chapin was elected president of Hudson Motor Car Company, succeeding McAneeny who was elevated to chairman of the board of directors, the position formerly held by Chapin prior to his tenure as Commerce Secretary.

In spite of the superior features, luxurious interiors, fine engineering and quality manufacture, Hudson Motor Car Company was able to build and ship only a dismal 2,946 Hudson cars during 1933. But, because of its brilliant performance and low price, Essex Terraplane did somewhat better, accounting for 38,150 cars built and shipped in 1933. However, Hudson Motor Car Company experienced another loss of $4,409,903.

The 1934 Hudson and Terraplane models were introduced in January 1934, the Terraplane omitting the Essex portion of its former name. All 1934 Hudson and Terraplane models were extensively restyled, with a broader slanting, V-shaped radiator grille and wide flowing front fenders. The Terraplane had slender chromium-plated grille bars in the form of a fan extending upward from the bottom. A built-in trunk was provided as an option on the coach and sedan models, or a curved lid that concealed the spare tire within the body rear.

	1934 Hudsons (5 lines)	**1934 Terraplanes**
Wheelbase (in.)	116/123	112/116
Price	$695-1,145	$565-780
No. of Cylinders / Engine	L-8	L-6
Bore x Stroke (in.)	3.00 x 4.50	3.00 x 5.00
Horsepower	108/113 adv, 28.8 SAE	80/85/89.5 adv, 21.6 SAE
Body Styles	31	26 plus 4 commercial
Other Features		

With competitive cars featuring independent front suspension, Hudson countered by offering Hudson's "Axle-Flex," with the front axle constructed in two pieces. Connected by two parallel rubber-bushed links, it permitted individual front wheel vertical movements. The 1934 Terraplane models closely resembled the Hudson line.

Hudson Motor Car Company was able to increase the 1934 Hudson and Terraplane sales; they built and shipped 27,130 Hudson cars, and 56,804 Terraplane passenger cars and 1,902 commercial vehicles. Despite the apparent improvement in car sales during 1934, Hudson suffered a loss of $3,239,202.

McAneeny Leaves

During October 1934, Max Wollering resigned as factory manager of Hudson Motor Car Company, and was succeeded by I.B. Swegles, former superintendent of inspection. Then in December, William McAneeny resigned as chairman of the board to become president of Hupp Motor Car Company. Unfortunately he was forced to resign this position at Hupp in February 1935 for health reasons.

On March 24, 1935, the entire Hudson Motor Car Company organization was saddened to hear of the sudden and untimely death of William J. McAneeny. McAneeny had been largely responsible for the growth and success of Hudson Motor Car Company.

1935 Hudson steel roof panel press. (Source: J.A. Conde)

New Life for Hudson

During 1935 Roy Chapin was able to raise $6,000,000 of new capital through the sale of notes to the Federal Reserve Banks in New York and Chicago. These notes were secured by the company's manufacturing, payable in various amounts from August 1, 1936, through March 20, 1940. This financing showed the faith of the banking interests in the integrity of Chapin and the confidence in the management of the Hudson Motor Car Company. H.M. Northrup was appointed chief engineer of the Hudson Motor Car Company in October 1935.

For 1935 Hudson Motor Car Company announced their new models about a month earlier on December 5, 1934. Starting the production at least a month before, they introduced the industry's first one-piece

*1935 Hudson Eight
(Source: J.A. Conde)*

pressed-steel roof panel. Perhaps of equal importance was the elimination of the intruding gearshift lever from the driver's compartment. Hudson offered the "Electric Hand," an electro-vacuum-powered gearshift mechanism which performed the gear changes, controlled by a finger-operated selector on the steering column.

Another model lineup change was the reintroduction of a new Hudson Six.

	1935 Hudsons	**1935 Hudson Six**
Wheelbase (in.)	117/124	116
Price	$710-1,125	$770
No. of Cylinders / Engine	L-8	L-6
Bore x Stroke (in.)	3.00 x 4.50	3.00 x 5.00
Horsepower	121 adv, 28.8 SAE	93 adv, 21.6 SAE
Body Styles	25	13
Other Features		

Along with the refined styling and mechanical innovations, Hudson management leaned heavily on the no less than 43 performance records established by the Hudson Eight sedans. These record-breaking sedans were driven by professional racing drivers Sir Malcom Campbell, Wilbur Shaw, Egbert (Babe) Stapp, and Al Gordon.

While records for domestic car sales indicate that 21,587 Hudson cars and 53,838 Terraplane vehicles registered domestically, the export sales boosted the total cars built and shipped by Hudson during 1935 to over 100,000 units. Thus, the company was able to show a profit ($584,749) for the first time in many years.

The 1936 Hudson and Terraplane models introduced in October 1935 followed the styling trend of the times. This included the two fixed, V-type, laminated, safety glass windshields, rounded front grille (waterfall motif), and front fenders of teardrop (pontoon) shape that extended downward to meet the narrow portion of the front grille. The radiator shell was omitted as such, having an extension of the hood side panel to support the headlamps and chromium-plated horns.

Perhaps the most important feature of the 1936 Hudson and Terraplane models was the adoption of hydraulic four-wheel brakes; however, Hudson Motor Car Company accepted hydraulic brakes with "tongue in cheek" by providing an auxiliary mechanical brake system on the rear wheels, should the hydraulic brake system fail. The Axle-Flex front suspension was replaced by the conventional front axle, with two radius arms attached to front axle and pivoted at the frame to control lateral axle movement.

	1936 Hudson Six	1936 Hudson Eights	1936 Terraplanes
Wheelbase (in.)	120	120/127	115
Price	$785	$710-975	$595-760
No. of Cylinders / Engine	L-6	L-8	L-6
Bore x Stroke (in.)	3.00 x 5.00	3.00 x 4.50	3.00 x 5.00
Horsepower	93/100 adv, 21.6 SAE	113/124 adv, 28.8 SAE	88/100 adv, 21.6 SAE
Body Styles		25	15 including station wagon
Other Features	hydraulic four-wheel brakes		hydraulic four-wheel brakes

The Hudson, available in five series, offered 25 body types; this proliferation into so many styles apparently helped sales, as the company was able to build and ship a total of 118,718 Hudson and Terraplane cars in 1936, resulting in a profit of $3,305,616.

Chapin Dies

Sadly, on February 16, 1936, Roy D. Chapin died of pneumonia at the Henry Ford Hospital in Detroit, Michigan. Chapin was ill for just one week and would have been 56 years old on February 23rd. Following Chapin's passing, in the interim, A.E. Barit was placed in control of Hudson Motor Car Company as vice president and general manager. Then on February 25, 1936, Barit was officially elected president and general manager of Hudson Motor Car Company by the board of directors.

At the same meeting the board of directors promoted and elected the following officers: Stuart G. Baits (who joined Hudson as a draftsman in 1915) to first vice president and assistant general

Abraham Edward Barit (Source: J.A. Conde)

Stuart Baits (Source: J.A. Conde)

manager, W.R. Tracy vice president of sales, I.B. Swegles vice president of manufacturing, C.A. Oostdyk vice president in charge of purchasing, and H.M. Northrup vice president of engineering.

The 1937 Hudson and Terraplane models were announced on September 28, 1936, with the Terraplane now being called Hudson Terraplane. In addition to the front grille styling, the most significant change was the hinging of the front doors at the windshield pillar (A pillar). A new body style was added to both the Hudson and the Hudson Terraplane: a four-passenger convertible brougham.

The chassis specifications were almost the same as for the 1936 models (except for increased wheelbase length), drop center type disc wheels were standard, and the "Hydraulic Hill Hold" was an option. Approximately 19,848 Hudson cars, 83,436 Hudson Terraplane passenger cars, and 8,058 commercial vehicles were built and shipped during 1937.

	1937 Hudson Eights	**1937 Hudson Terraplanes**	**1938 Hudson Terraplanes**
Wheelbase (in.)	122/129	117	117
Price	$695-1,060	$595-850	$881
No. of Cylinders / Engine	L-8	L-6	L-6
Bore x Stroke (in.)	3.00 x 4.50	3.00 x 5.00	3.00 x 5.00
Horsepower	122 adv, 28.8 SAE	96/107 adv, 21.6 SAE	96 adv, 21.6 SAE
Body Styles	31	17	16
Other Features			

	1938 Hudson Eight	**1938 Hudson Super Six**	**1938 Hudson 112**
Wheelbase (in.)	122/129	122	112
Price	$1,060	$984	$625-800
No. of Cylinders / Engine	L-8	L-6	L-6
Bore x Stroke (in.)	3.00 x 4.50	3.00 x 5.00	3.00 x 4.50
Horsepower	112 adv, 28.8 SAE	96 adv, 21.6 SAE	83 adv, 21.6 SAE
Body Styles	14	8	19
Other Features			

The 1938 Hudson and Hudson Terraplane models made their debut on October 6, 1937, consisting of the Hudson Eight, Hudson Super Six, and the Hudson Terraplane. The new radiator grille had more massive horizontal louvres, giving a more rounded appearance of the front end.

In January 1938 at the National Automobile Show in New York, the long-awaited Hudson "112" was unveiled, a model fitting into the lowest price market. A.E. Barit emphasized the significance of the move, asserting that the expanded activity and employment it created were far more important than the preview of the car. Hudson Motor Car Company expended approximately $11,000,000 in tooling costs for this new car model. Powered by an L-head, six-cylinder engine of 175 cu. in. displacement, it was certainly an economical car to drive. The Hudson 112 was selected by the Indianapolis Motor Speedway to pace the start of the 1938 Indianapolis 500, to be driven by vice president Stuart Baits during pace lap.

Despite the sales effort and the proliferation of models to give the buyer a wide selection, Hudson was able to build and ship only 51,078 cars, resulting in a loss of $4,670,004 for 1938. The entire automobile industry experienced a great plunge in automobile sales during 1938 to about one-half of the sales for 1937. This was no doubt caused by a depressed economy and crippling work stoppages.

For 1939 Hudson Motor Car Company made earlier new model announcements. The Hudson 112 was introduced on September 29, 1938, while the Hudson Six and Eight made their debut on October 21, 1938. The Hudson Terraplane model was discontinued. Essentially, the 1939 technical specifications were the same as those for the 1937 models, including the "Auto Poise" torsion bar sway eliminator.

In styling, the front grille was smaller with narrower chromium-plated horizontal grille bars. The Hudson Six and Eight had an additional grille of chromium-plated vertical bars on each side at the front of the cat walk. The Hudson 112 did not have the side grilles, and still had the headlamps mounted on the radiator support, while the Hudson Eight and Six had fared-in, almond-shaped headlamps at the leading area of the front fenders. The hood side panels had a single bright metal horizontal molding. Nine body types were available in the Hudson 112 and 22 body styles in the Hudson Six and Hudson Eight models. Hudson pioneered the use of foam rubber padding in the seat cushions and backs on the 1939 models.

On March 13, 1939, Hudson Motor Car Company introduced a new line called the Hudson Pacemaker Six Series, superseding the Hudson Terraplane. Coincidentally, on May 4, 1939, Hudson celebrated the 30th anniversary of its founding in 1909, by building the 2,614,165th Hudson car. The year 1939 was better for the automobile industry in general, as the economy improved somewhat. The Hudson Motor Car Company built and shipped 82,161 cars, selling 62,855 units domestically.

	1939 Hudson Pacemaker Six	1940 Hudsons (7 models)
Wheelbase (in.)	118	118/125
Price	$790-855	$670-1,330
No. of Cylinders / Engine	L-6	L-8
Bore x Stroke (in.)	3.00 x 5.00	3.00 x 4.12 / 3.00 x 4.50
Horsepower	96 adv, 21.6 SAE	102/128 adv, 28.8 SAE
Body Styles	4	30
Other Features		sealed beam headlamps independent front-wheel suspension

The 1940 Hudson models were announced on August 25, 1939. The Hudson 112 was superseded by the Hudson Traveler and DeLuxe Six. Other Hudson models were the Hudson Super Six, the Country Club Six, the Hudson Eight, Eight DeLuxe, and Country Club Eight. The most salient features were sealed beam headlamps, and a new independent front-wheel suspension system using coil springs in front, and semi-elliptical leaf-type springs in the rear. The automatic clutch control and automatic overdrive were continued as options.

The 1940 Hudson styling was changed considerably, with fared-in sealed beam headlamps and no radiator center grilles. Instead, two sections of wider grille, having seven horizontal bars, extended across to meet the front fenders. Mechanically, the cars were practically the same as the 1939 models.

1941 Hudson Commodore Eight. (Source: J.A. Conde)

During 1940 Hudson Motor Car Company engaged in numerous contests for fuel economy, performance, and reliability, establishing over 120 AAA-sanctioned records. In spite of improved production to 86,865 units, and domestic sales of 79,979 cars, Hudson still sustained an operating loss of $1,507,780 for 1940.

The 1941 Hudson models announced on August 1, 1940, consisted of six model lines and were substantially redesigned, retaining only the hood, front end sheet metal, and the power train components of the 1940 models.

	1941 Hudson Traveler and DeLuxe Six	**1941 Hudson Super Six and Commodore Six**	**1941 Hudson Commodore Eight and Commodore Custom Eight**
Wheelbase (in.)	116	121	128
Price	$695-1,438	$880-1,385	$975-1,485
No. of Cylinders / Engine	L-6	L-6	L-8
Bore x Stroke (in.)	3.00 x 4.12	3.00 x 5.00	3.00 x 4.50
Horsepower	92 adv, 21.6 SAE	102 adv, 21.6 SAE	128 adv, 28.8 SAE
Body Styles	9	11	10
Other Features			

Consistent with the longer wheelbase lengths, the body interior lengths were increased, providing more leg room, and the overall lengths were increased by 5 in. The overall height was reduced by flattening the curvature of the roof panel. The horizontal front grille bars were lengthened and the pressed steel grille section divider width was decreased. Those changes gave the Hudson a long, low, and graceful look. Hudson Motor Car Company built and shipped 79,529 cars during 1941. With domestic sales of 73,261 units, they were able to show a profit of $3,756,418.

Hudson's 1941 sales promotion publicized economy of operation rather than performance. They won the Gilmore Grand Canyon Economy Run with both the Hudson DeLuxe Six and the Hudson Commodore Eight in their classes, averaging 24.6 and 20.1 mpg, respectively.

The WWII Years

In 1941, Hudson Motor Car Company negotiated with the United States Government to manufacture aircraft fuselage and wing sections, Naval ordnance, and other defense materials.

The 1942 Hudson Six (Traveler name was dropped), and the Hudson Six DeLuxe production started on August 1, 1941, and the Hudson Super Six, Commodore Six, and Commodore Eight began on October 1, 1941. The introductions were made without fanfare, as the war became more fierce in Europe, and the United States involvement was more imminent. Only minor styling changes were made, most significantly the flaring out of the lower portion of the door outer panels to conceal and protect the running boards against rain and snow. During the balance of 1941, Hudson built and shipped 5,463 cars and commercial vehicles, and 35,998 cars were built and shipped by February 5, 1942, at which point all automobile production was stopped, and all manufacturing effort was concentrated on war materiel. The cars built between January 1 and February 5, 1942, were devoid of bright metal ornamentation and replaced by enamel-finished parts.

During 1942, 1943, 1944, and 1945, Hudson Motor Car Company was the prime contractor and operator of the United States Naval Ordnance plant, erected during 1941 in Warren (Centerline), Michigan. During World War II, Hudson Motor Car Company produced the folding wings of the U.S. Navy "Hell Diver" aircraft, and pilot compartment cabins and ailerons for the Bell "Airacobra" P-39 fighter planes. They built fuselage sections for the U.S. B-29 bombers and other military aircraft. Hudson built a great quantity of the big guns for the U.S. Navy fighting ships. They also built the "Invader" engines for landing boats, anti-aircraft guns, and numerous other war armament and ordnance items for the United States Navy.

Because Hudson Motor Car Company was ahead of its schedule on its production of war materiel, the United States War Production Board allowed them to convert to civilian car production by mid-1945. Thus, Hudson was able to build its first 1946 model by August 30, 1945, with formal announcement on October 1, 1945. The 1946 Hudson models were confined to two series: the Hudson Super Six and the Hudson Commodore Six, offered in eight body styles, priced at $1,480 to $2,050. The most noticeable change was a concave area in the center of the horizontal grille bars. Because of the limitations of production material, Hudson was able to build and ship only 5,005 cars during the balance of 1945.

The Hudson Commodore Eight production was added to the lineup starting in January 1, 1946. The Hudson Commodore Eight was offered in three body styles, priced at $1,630 to $2,150. Because of the steel shortage for the automobile industry, Hudson was able to build and ship about 93,870 cars during 1946. The steel shortage was so acute that Hudson shipped a great number of cars without bumper face bars, substituting wooden impact beams for protection. The chromium-plated steel face bars were later installed at the dealerships, at no cost to the owners. The same basic 1946 models were carried over through 1947, and 142,454 cars were built and shipped through December 1947. In the meantime, Frank Spring and his designers were not idle: In January 1948 the "all new" Hudson models would be introduced.

The New Models

The new 1948 Hudson was introduced in December 1947, featuring the recessed floor panel and the famous "step-down" design, a creation of Frank Spring and his designers. The bodies were much lower with smooth flat sides and truly aerodynamic shapes. The new front end sheet metal was wider with a two-piece grille of full-width horizontal louvres. The bodies were all steel unit construction. The 1948 Hudson was available in four series: Super Six, Super Eight, Commodore Six, and Commodore Eight, all built on a 124-in. wheelbase chassis, with both the six- and eight-cylinder engines developing 121 hp. Sales were very good but the limitations of material allowed only 109,497 cars to be sold domestically, on a calendar year production of 142,454 units for 1948.

The 1949 Hudson models were introduced in November 1948, with practically no noticeable changes. Hudson discontinued the operation of the steel mill in New Castle, Pennsylvania, which they operated as an emergency measure to overcome the steel shortages after WWII. The year 1949 saw record production and sales of 137,987 cars, which also reflected a record profit.

The 1950 Hudson models made their debut on November 18, 1949, introducing the new Hudson Pacemaker series which sold at a price $200 less than the other models. The year 1950 was indeed profitable for Hudson as they sold 134,219 cars domestically, on a production of 143,586 units for 1950 calendar year. President Barit reported a profit of about $12,000,000 for 1950.

	1950 Hudson Pacemaker	1951 Hudson Hornet
Wheelbase (in.)	119.8	123.8
Price	$1,933-2,444	$2,543-3,099
No. of Cylinders / Engine	L-6	L-6
Bore x Stroke (in.)	3.56 x 3.87	3.81 x 4.50
Horsepower	112 adv, 30.4 SAE	145 adv, 34.9 SAE
Body Styles	5	4
Other Features		

In April 1950 Hudson resumed the operation of their Tillbury, Ontario, Canada plant, which was suspended during World War II.

The 1951 Hudson was introduced in September 1950, featuring the new Hudson Hornet series, powered by a new high compression H145 engine with a displacement of 308 cu. in., and equipped with dual carburetors. The performance of the Hudson Hornet was so phenomenal that it captured twelve victories out of the 41 Nascar stock car races in 1951. While the car performance was good in racing it was not in sales and production. The Korean War had an adverse effect on the economy, and the material shortages resulted in reduced domestic sales of 96,847 cars on a production of 131,915 units for the 1951 calendar year. Hudson Motor Car Company sustained a loss of $1,125,210.

AMC and the Decline of Hudson

The 1952 Hudson was introduced in January 1952 amid the material shortages caused by the Korean War. Hudson relied on its racing victories in Nascar stock car racing, winning 27 of the 34 races held. That alone was not enough to overcome the drop in domestic sales to 78,509 cars on a calendar year production of 79,117 units. In May 1952 Hudson announced the coming of a low-priced light car to be introduced in 1953.

The 1953 Hudson lines were introduced in November 1952, with the Hudson Jet, the long-awaited, low-priced car making its appearance in early 1953. The Jet was in fact a scaled-down Hudson Wasp, with a simple grille of one horizontal bar, and the body built by the Murray Corporation. Despite the 35 Hudson Hornet victories of the 53 races, and the debut of the Hudson Jet, sales continued to fall, amounting to 66,797 cars domestically on a calendar year production of 67,089 units for 1953. The reported loss was $10,411,060, amid rumors of a merger with another independent automobile manufacturer.

The dealer introductions of the 1954 Hudson model lines were on October 2, 1953, with the Hudson Jetliner announced on October 12, 1953. Hudson struggled with shrinking sales the latter part of 1953 and 1954.

On May 1, 1954, Hudson officially became a partner of the American Motors Corporation, with George W. Mason as chairman of the board. George Romney was elected executive vice president, and A.E. Barit retired as Hudson president. With the sudden death of George Mason on October 8, 1954, George Romney became president of American Motors. Production at the Hudson plant on Jefferson Avenue in Detroit was suspended on October 29, 1954, and it was announced that future Hudsons would be built in the Nash plant in Kenosha, Wisconsin. The domestic sales of the 1954 Hudson was 34,806 cars, on a calendar year production of 51,314 units.

	1954 Hudson Jet	**1955 Hudson Super V-8 and Custom V-8 Series**
Wheelbase (in.)	109	121.25
Price	$1,856	$2,825
No. of Cylinders / Engine	L-6	V-8
Bore x Stroke (in.)	3.00 x 4.75	3.81 x 3.50
Horsepower	104 adv, 21.6 SAE	208 adv, 46.5 SAE
Body Styles		
Other Features		Packard-built V-8 engines

The 1955 Hudson car production, transferred to Kenosha, Wisconsin, was built on the basic Nash design of Pinin Farina styling. The 1955 Hudson featured a distinctive front end and grille, styled in good taste, and offered a V-8 engine and automatic transmission, built by Packard. Hudson also inherited the Hudson Rambler, which it shared with Nash Rambler. The 1955 Hudson lines consisted of Hudson Rambler, Hudson Wasp, Hudson Hornet, Hudson Super V-8, and the Hudson Custom V-8 series.

Roy Abernathy joined American Motors Corporation in late 1955, appointed vice president of sales. Despite Hudson's great sales effort, only 20,522 cars were sold domestically, on a calendar year production of about 30,100 units. American Motors Corporation sustained a great loss in 1955.

The 1956 Hudson Wasp and Hudson Hornet V-8 were announced on November 30, 1955, and the Hudson Rambler announcement followed on December 15, 1955. The most noticeable styling feature of the 1956 Hudson was the pronounced triangular front grille frame and side molding treatment. The Hornet V-8 still

1956 Hudson Hornet.
(Source: J.A. Conde)

offered the Packard-built engine and automatic transmission. While industry car sales were climbing, Hudson sales plunged further to 11,822 cars, on a calendar year production of 22,588 units. With Nash sales also falling, American Motors Corporation sustained a loss of $19,746,234 for 1956.

The 1957 Hudson models, the last to be built, were announced on October 25, 1956, and consisted of the Hudson Rambler, Hudson Hornet, Hudson Super V-8, and the Hudson Custom V-8. The new American Motors designed and built V-8 engine of 190 hp was used. The 1956 Hudson also had 12V electrical system. The 1957 Hudson styling followed that of the 1956 Hudson.

1957 Hudson Hornet

Wheelbase (in.)	121.25
Price	$2,821-3,101
No. of Cylinders / Engine	V-8
Bore x Stroke (in.)	4.00 x 3.25
Horsepower	255 adv, 51.2 SAE
Body Styles	4
Other Features	Packard-built V-8 engine

Because sales dipped further to a dismal 4,596 units domestically in 1957, with Nash accounting for only 9,474 sales (with a reported loss of $11,833,200), American Motors Corporation decided to drop the Hudson and Nash lines in 1958, and concentrate on the Rambler models only. Thus, the last Hudson car was built on June 25, 1957, and the famous triangle Hudson badge descended into oblivion.

Harry Stutz

In the silence of that warm afternoon in 1934, somber-faced executives and workmen gathered to solemnize the occasion, "that of building the last of the great Stutz cars." This was really the end of a magnificent automobile, the car that made good in a day, a car with guts, a way of life, the end of the great Stutz enterprise, and the end of Indianapolis' long-established and enviable position as a great automotive production center.

* * *

Automobiles from the Start

Harry Clayton Stutz was born on a farm in Darke County, Ohio, near Ansonia, approximately 45 miles northwest of Dayton, on September 12, 1876. He worked on his father's farm, doing the usual farm chores along with repairing all sorts of implements and farm machinery. In 1894 Stutz went to Dayton and worked for the Davis Sewing Machine Company and the National Cash Register Company. In 1897 he established his own machine shop, and built his first gasoline-engine-powered vehicle using an engine of his own manufacture and other machine parts then available.

Around 1900 Harry Stutz married Clara Deitz, and in 1901 they moved to Indianapolis where their daughter was born in 1902. Harry's first employment in Indianapolis was for the Lindsay Automobile Parts Company, builders of the Lindsay car and Russell axles. The Lindsay car was the first to feature the Russell rear axle, with a differential of the modern type, similar to those used on cars today. His next employment was for the Central Motor Car Company, which lasted for only about one year. Stutz then joined the G. & J. Tire

Company as a test driver of new tires. He was assigned the responsibility for the monstrously huge and heavy car built by the Premier Motor Manufacturing Company. This vehicle, nicknamed Betsy, was so heavy and powerful that it could wear out and shred almost any set of tires in only a few days. Through testing the G. & J. Tire Company found exactly what kind of torture and abuse their tire could endure. While this was exciting work, it did not provide the adventure and challenges that Harry Stutz longed for.

In 1905, when V.A. Longaker and D.S. Menasco organized the American Motor Car Company, they sought out Harry Stutz and appointed him chief engineer and factory manager. It was here that Stutz was able to realize his great ambition, to design and build one of the best and fastest cars in America, the American Tourist. Stutz designed and built several exciting and fast cars at American, but the company had a weak financial structure and was continually on the brink of financial disaster. (Stutz did not claim credit for the design of the American Underslung, which was designed by Frederick I. Tone, who joined the American Motor Car Company as chief engineer on January 17, 1906, after Stutz's departure.)

In January 1906 Stutz left the American Motor Car Company and joined the Marion Motor Car Company of Indianapolis as chief engineer. At that time Marion was under a wholesale marketing contract with the American Motor Sales Company of Elmira, New York, and Indianapolis, Indiana, owned by John N. Willys (see the Willys chapter in Volume 3). Willys then bought the Marion Motor Car Company and plant in Indianapolis in early 1909. Two of the most significant changes Stutz made while at Marion were the change from an air-cooled to a water-cooled engine, and the extensive use of aluminum.

During the period of 1906 to 1910, Stutz had become renowned as the designer of the Marion Flyer and the Marion Bobcat, both of which created sensations at the New York, Chicago, and Atlanta Auto Shows. It was with those two cars that the excellent reputation of Stutz and Marion cars was established. What was more important was that Harry Stutz became acquainted with Henry F. Campbell, whose father, Eben B. Campbell, was at that time associated with John Willys and the Willys-Overland Company.

Stutz also came to know Gilbert Anderson, a splendid test driver and mechanic, who, with other skillful race drivers, later piloted Stutz to fame and glory on the American automobile race tracks. This was the beginning of a fruitful and long-lasting association and friendship. Harry Stutz, himself driving a Marion racing car, finished in sixth place in the Inaugural race at the Indianapolis Motor Speedway in 1909.

In early July 1910, Stutz resigned from his position as chief engineer and factory manager of the Marion Motor Car Company. Even though he was able to save a considerable amount of money over the years, he was also able to convince Henry Campbell to help him raise the additional money to finance his newly planned automotive venture. Apparently Campbell must have thought it was a good idea too, because his father had severed his connections with Willys-Overland and the American Motor Sales Company on June 2, 1910.

Stutz Auto Parts Company

On July 13, 1910, *Horseless Age* announced the organization of the Stutz Auto Parts Company, with Harry C. Stutz president and general manager, and Henry F. Campbell his partner. The new enterprise was located in the Industrial Building in Indianapolis, for the purpose of manufacturing transmission-axle assemblies, comprising a rear axle with brakes combined with a change speed gear (transmission) and a propeller shaft enclosed in a torsion tube. This system was furnished to builders of cars having 24 hp or less. Several automobile manufacturers used the Stutz transmission-axle system.

By the end of 1910 Stutz was convinced, and with his own enthusiasm was able to convince Campbell, that the time was right to manufacture and merchandise the best car they could build under the Stutz name. Together they made plans to build and sell approximately 500 cars per year. They purchased the real estate at the southwest corner of Capitol Avenue at 10th Street, and started construction of a new four-story factory building on this site. Meanwhile, Harry Stutz made a thunderbolt-fast trip through Europe, visiting many automobile manufacturing plants to learn all he could about automobile manufacture. Supplying rear transaxle assemblies to American automobile manufacturers gave Stutz favorable circumstances under which he could visit their plants and observe their manufacturing methods as well.

While all this was going on in the Stutz organization, something else was happening on 16th Street and Georgetown Road. The Indianapolis Motor Speedway was being paved with fire-glazed paving bricks to make the course faster and safer. On December 17, 1909, Governor Thomas R. Marshall of Indiana placed the Gold Brick (actually a brick molded of brass-bronze alloy) into position at the starting line. This signaled the completion of the Indianapolis Motor Speedway, and it was officially declared open for competition of all racing cars. (See the Allison *et. al.* chapter in Volume 3.)

During May, July, and September of 1910 the Indianapolis Motor Speedway staged nine races of 50, 100, and 200 miles. After the September 5th races, the Speedway announced that it would stage a 500-mile race on the next Memorial Day, May 30, 1911. It had become an accepted fact that in the automobile industry, if you wanted to sell cars you had to prove they could endure the punishment at the race track.

The First Stutz Car

The announcement ignited a spark in Stutz's mind. What a way to introduce a new car! Suppose that his new car, a prototype, were to be entered in the Indianapolis 500, and suppose that it finished in the top ten? What audacity! Even if the car failed to go the whole distance, what better way could you give an untried car a realistic road test? Yet, if the car finished in at least tenth position, there could be no better way of advertising a brand new name in the automobile industry. The more Harry thought about it, the more excited he got about the whole idea. Campbell went along with it, too.

With very little time left to race day, Harry and his men worked feverishly, building the prototype by hand because the factory was still under construction and the machinery was not in place. Stutz decided on a Wisconsin, four-cylinder, T-type engine of 390-cu.-in. displacement, because the Wisconsin engine and the Wisconsin Manufacturing Company had a reputation of building the most powerful and reliable engines of that period. Harry created the finest chassis design that would endure and complement the rugged Wisconsin engine and his own excellent transmission-axle. The final product was a prototype of the cars he would later market under the name Stutz.

Upon completion of the car, it went directly to the Indianapolis Motor Speedway, where he induced his long-time acquaintance and good friend Gil Anderson to test drive the car. After several laps around the track, and after examining the construction of the car, Anderson was pleased and remarked, "I think it will hold together and go the whole distance." Stutz was so sure that his car would finish in at least tenth place that he arranged his application timing so that it became the tenth entry. (In 1911 the racing car numbers were assigned in the order the entries were received.) Thus, his car carried the number 10.

This first Stutz racing car was just like the later Bearcats that would be built for the public, except it had a boat tail rear that covered the fuel tank. In the race this arrangement caused serious delays during refueling due to the awkward position of the filler neck. The first Stutz was not quite as fast as his later cars but it had the stamina and durability to withstand the grind of the 500-mile race. Frequent stops for tire replacements caused the Stutz to lose a lot of time in the pits. Not only was the Stutz plagued with this tire problem, but also Ralph Mulford, who was driving faster than anyone during the race but had to be content with second place due to tire problems, finishing behind Ray Harroun.

Gil Anderson finished in eleventh position in the Stutz #10, going the distance of 500 miles without a single stop for breakdowns, repairs, or adjustments. Thus, the Stutz found its nickname, the Sturdy Stutz, on the spot, and likewise found the slogan that was used in connection with it for a number of years, "The car that made good in a day." Until this time nobody ever heard of a Stutz car, since this was the only Stutz car that had ever been built. When it stood up through the 500 miles of the fiercest kind of competition with the best cars in the world, and came through in eleventh position, it was generally admitted to have "made good."

Ideal Motor Car Company

In order to build and merchandise the Stutz car, the Ideal Motor Car Company was launched. The Stutz Runabout (Bearcat) sold to the public was just like the racing cars except that it had a conventional fuel tank in back of the bucket seats, as well as fenders and Presto-Lite headlamps The Sturdy Stutz racing cars and the Stutz Bearcat were good for each other. The cars that Stutz marketed besides the Bearcat were the Torpedo roadster, a six-passenger touring, and a four-passenger touring car. These cars were also very

1912 Stutz roadster, #1 Gil Anderson in Stutz, #2 Len Zengel in Stutz, #3 Hughes Mercer, #4 DePalma in Mercedes, #6 Hearne Cage.

attractive and high-performing machines. The unequaled success of the Stutz racing cars and the flamboyant Bearcat provided an image that also propelled the other Stutz models to sales successes beyond the production capabilities.

Thus, the magic of the name and the Bearcat itself captured the imagination of all sports-minded Americans. Many who could afford the price ($2,000) bought one, if and where they could find one. To others, who for practical reasons could not afford to buy one, could only sigh and dream about owning such a romantic vehicle. During 1911 Stutz sold all the cars they could build.

For 1912 Stutz announced the new B Series of cars, with little noticeable changes in appearance. However, there were several mechanical improvements, including an improved cooling system, strengthened transaxle, and the addition of a battery, generator, and electric headlamps. On the touring, the tonneau, and the roadster models, the gearshift lever and brake lever were moved to the inside of the car body, to the right of the driver. In 1912 Stutz sales far exceeded production capacities.

More Racing Successes

Stutz racing cars in 1912 made an even better showing, finishing 4th and 6th at the Indianapolis 500, and winning the Tacoma (Washington) 150-mile race and the 203-mile Elgin (Illinois) road race. Charles Merz, Len Zengel, and Earl Cooper were the drivers.

For 1913 the "gleaming white" Stutz racing team fared even better than in 1912, finishing 3rd at the Indianapolis 500 and winning 6 of the 14 championship races held at the major speedways across the United States. It seemed that no one could deny the Stutz team a clean sweep at Tacoma, Santa Monica, Corona (California), and the Elgin races. In 1914 Stutz finished 5th at Indianapolis with Barney Oldfield driving, and Earl Cooper won the Tacoma race again.

The new specifications for the 1915 racing season limited the piston displacement to 300 cu. in. The challenge from the Peugeot racing cars made Stutz go all out with a new four-cylinder racing engine of 295-cu.-in. displacement, with twin overhead camshafts and four valves per cylinder. The greatest glory came to the dazzling white Stutz racing team in 1915. Not only did Howdy Wilcox put a Stutz on the pole at

Len Zengel in 1912 Stutz. (Source: Indianapolis Motor Speedway Corp.)

1913 Stutz racing team: Merz, Anderson, Herr. (Source: Indianapolis Motor Speedway Corp.)

Indianapolis with a new qualifying record of 98.9 mph, but the Stutz racing cars finished in 3rd, 4th, and 7th positions with Gil Anderson, Earl Cooper, and Howdy Wilcox driving.

Success was not limited to Indianapolis. The Stutz racing cars successfully won the major races at San Diego, Elgin (two races), Kalamazoo, Phoenix, and San Francisco. They not only won the races, but finished 1-2 in a clean sweep at both Elgin races and both Twin Cities (Minneapolis-St. Paul) races, with Cooper and Anderson alternately sharing the honors. At the end of the 1915 racing season, when the points were tabulated, Stutz won 8 of the 27 races scheduled for 1915, and finished second and third several times. Stutz was recognized as the winningest racing car. Earl Cooper was crowned 1915 AAA Racing Champion.

Upon completion of the 1915 racing season, Harry Stutz wisely chose to retire the factory-sponsored Stutz racing teams while they were at the peak of their successes. Stutz sold the racing cars to sportsmen, who sponsored them in the 1917 and 1918 races (other than Indianapolis). Indianapolis did not have any races during 1917-18 due to World War I. Stutz racing cars won at the Ascot, Uniontown, Chicago, and Tacoma speedways in 1917, with Earl Cooper, William Taylor, Reeves Dutton, and Cooper, respectively, as drivers. Stutz sold Gil Anderson's #5 Stutz racing car to Cliff Durant, who entered it in the 1919 Indianapolis 500 under the name of Durant Special. Eddie Hearne drove this car and finished in 2nd position, a few seconds behind the winning Peugeot owned by the Speedway and driven by Howdy Wilcox. Cliff Durant won the 250-mile Santa Monica race in another Stutz car earlier that year on March 15, 1919.

Gil Anderson in a Stutz, winner of the inaugural run at Sheepshead Bay, October 1915. (Source: Indianapolis Motor Speedway Corp.)

Stutz Motor Car Company

While racing took up some of Harry Stutz's time, he did not neglect the Stutz cars he was merchandising. For 1913 a new six-cylinder engine, also of T-head design, was added as an option. Thus, for 1913 Stutz B Series cars were offered in five conventional models in addition to the Bearcat models. The body types included four- and six-passenger touring cars, roadsters, and the Bearcats. These were offered with either the 386-cu.-in., four-cylinder, T-head engine, or the 426-cu.-in., six-cylinder, T-head engine. The price range was $2,000 to $2,300. Houk wire-spoke wheels were also offered as an option.

While the 1912 model features included electric headlamps, battery, and generator, a new electric starter of Stutz design and Remy manufacture was featured as standard equipment on the 1913 models. Also, on May 12, 1913, for the purpose of better management control and operation, the Stutz Motor Parts Company and the Ideal Motor Car Company merged to form the Stutz Motor Car Company.

For 1914 the Stutz Series E, referred to as the E4 and E6 models, remained practically the same as the 1913 models, since sales were still outdistancing production capabilities. The four- or six-cylinder engines were fitted in the same chassis wheelbase lengths. The wheelbase on the six-cylinder touring car was 10-in. longer and the touring body was 8-in. longer. A lavishly appointed coupe model was added to the 1914 models on the roadster chassis. This elegant model gracefully bore the Stutz breeding and flair, It was priced at $2,600 for the four-cylinder model, and $2,850 for the six-cylinder model. The coupe was shown in the 1914 Stutz catalog, even though only a few were actually built.

Harry Stutz, impressed by the apparent success of the Bearcat and Roadster models, decided for 1915 to go a step further. He introduced the Stutz Bulldog Special, a smaller, sporty, four-passenger touring car, built on the short Bearcat chassis. It had lower, slender, graceful contours, and gave the exhilaration of driving a Bearcat in a close-coupled four-passenger vehicle. The Bulldog Special remained in the Stutz line for many years.

The other new model was a smaller roadster called the HCS Speedster at $1,475. This new smaller car, while it resembled the Bearcat, had a smaller 3-3/4-in. x 5-in., L-head, four-cylinder engine on a chassis wheelbase of 108 in. The car had an HCS nameplate. The prospective buyers felt that this car in no way could be a Bearcat, and refused to buy it at a price $525 lower than the Bearcat. There were only 125 HCS Speedsters sold in 1915, and the model was dropped for 1916.

The sales momentum created by the successes of the Stutz racing cars and the popularity of the Bearcat continued through 1916. However, there were two new dimensions added in the way Stutz conducted business in 1916. A new 16-valve (4 valves per cylinder) T-head engine replaced the original four-cylinder T-head engine. The new 16-valve engine enjoyed increased performance and could virtually outperform any car on the road. Thus, the Stutz organization was riding on the merits of the performance of their cars, and financial success continued.

Ryan Gains Control of Stutz

The second dimension that had a serious impact on the Stutz organization, and later resulted in Harry Stutz's departure from the company, was triggered by a plan to raise additional capital for plant expansion. Stutz and his corporate officers decided to place an issue of Stutz Motor Car Company stock on the market through the New York Stock Exchange. Everything would have been fine if the stock was sold in small blocks to several hundred small investors. Instead, Stutz's phenomenal success attracted stock trading entrepreneurs, who saw the financial possibilities of getting a "corner" of the Stutz stock. One such entrepreneur was

Allan A. Ryan and his banking associates. Ryan and Company started buying large blocks of Stutz stock, until in a matter of time he gained control of the company. Harry Stutz stayed on as president and general manager during this period.

In the meantime the new R series 1917 model cars were developed and introduced in August 1916. The new R series consisted of the Bulldog Special (touring) with either four- or six-passenger seating, the Stutz roadster, and the Stutz sedan. All body types, except the Bearcat, were on a uniform 130-in. wheelbase chassis with a 16-valve, four-cylinder engine. The 130-in. wheelbase permitted longer bodies and allowed the Bulldog auxiliary seats to fold into the cowl cabinet against the back of the front seats. The Bearcat wheelbase stayed at 120 in.

During 1917 Stutz built several special Bulldog chassis, which were bodied by Parry Manufacturing Company of Indianapolis for ambulances for the Indiana National Guard. Stutz continued to rely on only one engine and chassis, with virtually no changes until 1923. During the war years (1917-18), material was rationed, and a seller's market existed. During this boom period Stutz, along with other car makers, could sell all the cars they could build, without major changes in design. The seller's market lasted until the middle of 1920, when a depression descended on the nation.

Harry Stutz Resigns

Since Harry Stutz created the first Stutz car, established the Stutz Motor Parts Company, the Ideal Motor Car Company, and organized the Stutz Motor Car Company, he felt obligated to stay with the company as long as he could tolerate the agony of Ryan's control. But on June 30, 1919, Harry Stutz resigned as president of the Stutz Motor Car Company, sold the Stutz stock he owned, and departed from the company. Allan Ryan was elected president to succeed Stutz.

Ryan's Downfall

Stutz's departure gave Ryan the chance to continue his plan to own all the Stutz stock, and to manipulate it as he saw fit. This greed finally became the cause of his demise. Ryan and his associates borrowed money and

Stutz plant in Indianapolis, c. 1920. (Source: Indiana Historical Society)

succeeded in buying almost every share of Stutz stock, leaving virtually none to be sold on the market. While the Stutz stock originally sold on the market in the $50-$60 per share range, some of Ryan's purchase of the stock was in the $130-$135 range. Having a corner on the market of Stutz stock allowed Ryan to hold it off the market, and caused the prices to skyrocket upward. By March 31, 1920, the squeeze-play made the stock offer spiral to a record high of $391 per share. The greed became clearly visible when Ryan later quoted his asking price at a reported $750 per share. One important element in the financial transactions that Ryan did not consider was the fact that the financial leaders who were trapped in the bidding on Stutz stock were powerful Wall Street men who had enough friends and associates in the New York Stock Exchange to control its operation.

Now Ryan's squeeze plan would backfire. The Wall Street financiers merely influenced the New York Stock Exchange to discontinue the listing of Stutz stock. When Ryan tried to sell some of the stock, he had no place to sell it. He tried to sell the stock to short traders on the curb at reduced prices but he had no takers. To teach this wheeling-deal stock trader a lesson he would remember, the New York Stock Exchange held the Stutz stock off the market beyond the date that Ryan was to pay back the borrowed money. In desperation Ryan tried to sell the stock at whatever price he could get. It was reported that he had to sell some of his stock for as little as $20 per share. The one-time high roller ended up not only flat broke, but deeply in debt as well.

Schwab Takes Over Stutz

The action by the New York Stock Exchange was possible at that time, before the present-day controls of the United States Securities and Exchange Commission. The buyout of Stutz stock from Ryan on December 1, 1920, by Charles M. Schwab, wrested the control from Ryan, and put it in the hands of Schwab, Eugene R. Thayer, and Charles J. Schmidlapp, prominent industrial and financial giants, but definitely not automobile men. While one could applaud the poetic justice, it did not restore the control of Stutz Motor Car Company to Harry Stutz. Instead, another group of financial speculators owned it.

H.C.S. Motor Car Company

Meanwhile, Harry Stutz, still an automotive engineer and car maker at heart, could not remain idle. Using the funds he had from the sale of his stock, and enlisting the help of his loyal friend and associate Henry Campbell again, a new motor car manufacturing company was organized and incorporated in the state of Indiana on November 3, 1919, under the name of H.C.S. Motor Car Company. The company was capitalized at $1,000,000; $600,000 in common stock, $400,000 in preferred stock. They could not use the name of Stutz. Their aim was to produce a quality car for $1000. Harry C. Stutz was elected president, Samuel T. Murdock vice president, Henry F. Campbell treasurer, and A. Gordon Murdock secretary.

The H.C.S. models were hastily designed and put into production about the middle of 1920. The cars were very attractive, appearing very agile and closely resembling the Stutz. The H.C.S. was available in two body types: a sport roadster and a sport touring. The front fenders circumscribed the front wheels and ended abruptly at the rear of the tire, leaving a space between the front fender and the step-plate on the roadster, and the shortened running board on the touring. The spare wire-spoke wheel and tire were carried in this space.

Neither time nor money permitted Stutz to develop his own engine. He arranged by contract to have Weidely Motors Company of Indianapolis supply engines for his cars. The Weidely four-cylinder valve-in-head engines had a bore and stroke of 3-5/8-in. x 5-in. for 1920-22, and 3-3/4-in. x 5-in. for 1923-24. Stutz

1921 H.C.S. phaeton. (Source: Indianapolis Motor Speedway Corp.)

likewise purchased clutches, transmissions, rear axles, front axles, steering gears, and all other chassis components from reliable original equipment suppliers. While the Weidely was indeed a fine engine, its performance could not match that of the 16-valve engine used in the Stutz.

The untimely introduction of the H.C.S. during the inflationary period, just before the depression that followed, and the purchase of costly machinery and equipment, labor, and material at peak prices, pushed the price of the H.C.S. all the way up to $2,950, when it was intended to sell in the $1,000 price range. Despite the high price tag, the magic of Harry Stutz's name apparently still had some impact, as approximately 800 H.C.S. cars were built and sold in 1920. The depression that followed in 1921 dealt a severe blow to the young company, which was just one year old and could not build up a reserve.

1923 H.C.S. Special racing team: Tommy Milton (left), Harry Stutz (center), Howdy Wilcox (right). (Source: Indianapolis Motor Speedway Corp.)

In 1921 H.C.S. conducted an extensive advertising and promotional campaign. They sponsored a racing car in the 1923 Indianapolis 500, called the H.C.S. Special, with a Miller racing engine. Tommy Milton won the race at an average speed of 90.95 mph. Despite the extra promotional effort, sales dropped sharply during 1921, and H.C.S. suffered sizeable losses for the year. During 1922, when other automobile companies rebounded, H.C.S. did not have the strength nor the resources to stage a comeback. Stutz built the H.C.S. sport roadster and sport touring through 1924. He tried building taxicabs and high-quality fire engines during 1925-27.

Harry Stutz Dies

On February 18, 1929, Harry Stutz and Frank Bellanca organized the Stutz-Bellanca Airplane Corporation to manufacture airplanes and engines in Bridgeport, Connecticut, and Orlando, Florida. On a return trip to Indianapolis in June 1930, Stutz became ill with what was believed to be intestinal flu. A later diagnosis proved to be a burst appendix. Emergency surgery was performed, but it was too late. Harry C. Stutz passed away on June 25, 1930, at the age of 54.

Schwab and the Stutz Motor Car Company

Back at Stutz Motor Car Company, after the buyout of Ryan's stock by Charles M. Schwab and associates, who admittedly knew very little about automobile manufacture and management, they elected William N. Thompson as president, and let the management and operation of the Stutz Motor Car Company ride the crest of the wave created by the Stutz 16-valve engine. One reason for being able to ride the wave was the shortages caused by World War I, when material was rationed by the government and a seller's market existed. After the end of the war, the boom continued for about two years. The all-time sales record for Stutz was in 1920, when almost 4,000 cars were built and sold. The company rode the crest of the wave until 1923, making virtually no changes in the engine or chassis. In fact, Stutz kept the right-hand drive through 1922.

By the end of 1922, when the economic depression had cut deeply into Stutz sales, and the following recovery failed to restore sales, Schwab and his associates realized they needed some changes in their operation. They felt the first change they needed was a new chief engineer who could design and create new engines, new chassis, and new cars. The person they chose for the task was Charles S. Crawford, the chief engineer of Cole Motor Car Company. Crawford certainly had the right credentials, having designed and produced the fine Cole Aero-8, which successfully challenged Cadillac in the V-8 field. (See the Cole chapter in Volume 4.)

It seems that the industry and buying public expected a miracle from Crawford, or at least some magic in the form of a high-performance V-8 engine. Instead, Crawford came up with a valve-in-head six-cylinder engine, designed along conventional industry practices. Weidely Motors Company was chosen to build this engine. It was surprising that Crawford chose a conventional valve-in-head (rocker arm) six, since Weidely had built overhead camshaft six-cylinder engines for Chalmers, and overhead camshaft V-12 engines for the Pathfinder V-12. The reason Crawford compromised on this engine may have been to bow to the wishes of the dominant conservative owners of the Stutz Motor Car Company. Crawford did not enjoy the freedom of decision and operation he would have liked to have.

For 1923 Stutz Motor Car Company offered to the buying public the new Speedway Six with four left-hand drive models: the seven-passenger sport touring, the Sportbrohm, seven-passenger Berline, and seven-passenger

Suburban, plus the four-cylinder T-head-engined Bearcat, Roadster, and Bulldog. In addition Stutz also offered the smaller Special Six models, available in roadster, touring, tourster, and sedan body types. The car bodies were refined and more streamlined in keeping with the trend of the automotive industry. This wide range of models available, ranging in price from $1,995 for the Special Six to $3,800 for the Speedway Six Berline, was very confusing to the car buying public. To make things more complicated, the price range of the model lines overlapped each other. This confusion did not help car sales.

For 1924 the Special Six and the Speedway Six car models were carried over. The T-head-engined cars, Bearcat, Roadster, and Bulldog, were dropped from the Stutz line. Despite the effort and changes, Stutz car sales continued to drop during 1924, ending the year with only approximately 1,000 car sales.

To add to the problems for Stutz, in January 1925 Weidley Motors Company brought suit against the Stutz Motor Car Company. The suit charged breach of contract and an alleged subsequent loss of $750,000 that forced the Weidely Motors Company into receivership. It was claimed that the contracts entered into between the two companies called for 5,000 engines to be delivered to the Stutz Motor Car Company, and that the contract was broken in July 1923 when only 2,191 engines had been delivered to Stutz. A hearing was set for February 9, 1925, and the suit was subsequently settled in court. This caused Charles Schwab and his associates to begin to realize that the operation of an automobile manufacturing company was quite different from operating on Wall Street. Hence, the search for a qualified general manager was initiated, and soon such a person would come calling on Schwab and associates.

Frederick Moskovics

Late in the 19th century a boy was born in Hungary who would become a prominent engineer of a different breed. He would ultimately have a profound effect on the entire automotive industry, and particularly the destiny of the Stutz Motor Car Company. He was christened Frederick E. Moskovics, and was brought to the United States by his parents while he was still a small child. He grew up to be a faithful and patriotic American citizen, but retained strong ties to Europe. He studied engineering at the Armour Institute of Technology, and returned to Europe for post-graduate courses in engineering. Moskovics worked for a short time as an apprentice engineer for Daimler-Benz.

Frederick Moskovics. (Source: Indiana Historical Society)

Upon his return to the United States, he went to work for the American Motors Company of Indianapolis, and later the Remy Electric Company of Anderson, Indiana. He worked his way through the ranks to become the sales manager of Remy. On March 1, 1913, Moskovics resigned his position at Remy and bought an interest in the Jones Electric Starter Company of Chicago. In 1914, Moskovics sold his interest in Jones and joined the Nordyke and Marmon Company in the engineering division. (See the Marmon chapter in Volume 4.)

Moskovics was deeply involved in the design and development of the Marmon 34 which was introduced in 1916. The Marmon 34 featured the famous, all-aluminum, six-cylinder, valve-in-head engine. Aluminum was also used extensively throughout the car, reducing the weight to about 1,000 lb less than competitive cars of the same size. Moskovics likewise spearheaded the building of the famous Liberty aircraft engine by Nordyke and Marmon during 1917-18, gaining for Marmon the award for being the most effective on-schedule producer.

Wartime shortages and the short post-war boom made it possible for Marmon to sell all the Marmon 34s they could build. Moskovics worked his way all the way up to vice president of sales. Being an engineer, Moskovics sensed that the Marmon 34 could not coast along unchanged forever. Prompted by the overwhelming success of the Duesenberg eight-in-line engine with overhead camshaft in the 1921 Duesenberg Model A, Moskovics conceived and developed new and exciting engineering ideas, and came up with a new model that could surpass any fine car being offered to the buying public.

Moskovics presented his ideas to Howard C. Marmon, vice president of engineering, and Walter C. Marmon, president of Nordyke and Marmon Company. He could not convince either of the two brothers, whose conservative philosophy was "If it works, don't fix it." Since the Marmon 34 was still selling well, they saw no reason to spend a lot of money for retooling. As a result, the Marmon 34 stayed on too long. The succeeding Marmon Models 75, 78, 68, Roosevelt, Big Eight, and the Marmon V-16 would never recapture the market for Marmon. The rejection by the Marmon brothers was a frustrating ordeal for Moskovics, which he resolved by resigning from Nordyke and Marmon Company in 1924.

Moskovics joined the H.H. Franklin Manufacturing Company in Syracuse, New York, for a short while in 1924 as vice president of sales (see the Franklin chapter in Volume 4). Herbert H. Franklin was even more conservative than the Marmon brothers, and had a preconceived idea that he would build only air-cooled engines. Moskovics tried his best to sell Franklin his ideas, which led to many disputes regarding Franklin company policies. Moskovics resigned from Franklin in December 1924 and initiated a breach of contract suit against the H.H. Franklin Manufacturing Company. The suit was settled out of court on March 4, 1925, in which the settlement to Moskovics was in excess of $100,000.

In the meantime, in December 1924, Moskovics learned of the woes of Stutz Motor Car Company and their need for a truly advanced motor car with brave imaginative styling. Armed with this knowledge, and the plans for his new car with him, Moskovics called on Charles Schwab and Associates and presented his proposed plans for Stutz's future. Schwab listened intently, and subsequent meetings with Schwab and Eugene V.R. Thayer, chairman of the board of Stutz, resulted in the election of Frederick E. Moskovics as president of the Stutz Motor Car Company on February 25, 1925. William N. Thompson's resignation as president of Stutz was submitted in early January, but was not acted upon by the board until the election of a successor. Moskovics retained Charles Crawford as chief engineer, and assembled the finest engineering team he could muster.

On March 4, 1925, Moskovics announced the appointment of Frederick P. Nehbras as assistant general manager in charge of production. R.A. Rawson was appointed merchandising manager. On October 20, 1925, Moskovics appointed Bert Dingley as Service Manager of Stutz. Dingley was an internationally known race driver from 1904 to 1914, and was one of the leading championship drivers in the days of the Vanderbilt and the Gordon Bennett Cup races.

For the balance of 1925, Stutz merchandised the Speedway Six in the roadster, five- and seven-passenger touring car models, five- and seven-passenger sedan models, and the seven-passenger Suburban. The seven-passenger models had a 130-in. wheelbase, all other models a 120-in. wheelbase. All models were powered by the Crawford-designed, valve-in-head, six-cylinder engine, whose bore was increased to 3-1/2 in. The famous Stutz 16-valve, T-head, four-cylinder engine had been discontinued in 1924. In the meantime all stops were pulled, and it was full steam ahead to develop the new car for 1926. Moskovics cancelled the production of the Speedway Six, even before the new 1926 Stutz went into production.

The Vertical Eight

The new Stutz Vertical Eight with Safety Chassis for 1926 featured a new eight-in-line overhead camshaft engine design. Moskovics (with few exceptions) did not claim to be the inventor or designer of new engineering features; however, he was indeed enough of an engineer and executive leader to comprehend what other men have created, realize their potential, and develop their ideas into practical reality. Moskovics gave credit for the design of the Stutz eight-in-line overhead camshaft engine to Charles R. Greuter, a brilliant American engineer who had worked on the engineering staffs of many automobile companies. Greuter had been designing and building overhead camshaft engines for test purposes since 1902. It was no coincidence that Greuter joined the Stutz engineering team at this time.

The other major design feature of the new car was the inverted worm-drive rear axle, which permitted lowering the entire car by at least 5 in. It was this design feature that James Scripps Booth claimed infringed on his patent of the Da Vinci car design. Booth brought suit against Stutz Motor Car Company, which dragged on in the courts for ten years. The courts finally awarded Booth a monetary settlement, but by this time Stutz Motor Car Company was in receivership and bankruptcy. The settlement barely covered Booth's legal expenses. Apparently Booth's claim must have had some validity, since he was awarded the settlement. (See the Booth chapter in Volume 4.)

The new Stutz for 1926, the AA series, represented a bold departure from conventional passenger car design. The heralded Stutz Vertical Eight embodied numerous new and outstanding engine and chassis features combined with stylish bodies designed under the supervision of Brewster and Company of New York.

1926 Stutz Vertical Eight

Wheelbase (in.)	131
Price	$2,995
No. of Cylinders / Engine	IL-8
Bore x Stroke (in.)	3.37 × 4.50
Horsepower	92 adv, 32.5 SAE
Body Styles	two-passenger speedster (roadster), four-passenger speedster (phaeton), four-passenger Victoria coupe, five-passenger brougham, five-passenger sedan
Other Features	shatterproof windshield

Other features included: nine-main-bearing crankshaft, chain-driven overhead camshaft with automatic chain take-up, 4.8 to 1 compression ratio, final drive by Timken underslung worm and drive gear assembly, Timken hydrostatic four-wheel brakes with six-piece brake shoes, double-drop frame side rails, allowing a 70-in. overall height, floor 20 in. above road, 39 in. headroom, seats 30 in. high. Closed bodies had horizontal glass transoms above the glass that could be tilted for ventilation.

While the new bodies did not constitute a radical change in basic construction, the proportions were altered in such a fashion as to make an attractive and pleasing combination with the chassis design. The net result was a long, low, graceful vehicle, without sacrifice of body space, headroom, or ground clearance. Previous attempts by automobile builders to lower the car silhouette involved reducing wheel diameters, reducing spring jounce distance, and dropping the roof line. All these attempts resulted in poor ride, poor handling, and loss of headroom and road clearance. Aside from the pleasing style attributes, the greatest gain of the Stutz Vertical Eight was the improvement of the handling characteristic on winding roads. The car had little or no tendency to sway, thus the danger of tip-over was practically eliminated.

1926 Stutz models.

The tall engine (due to overhead camshaft) required a higher hoodline, thus the belt line of the doors was considerably lower than that of the hood centerline. To harmonize with the low flowing lines of the car, all cars had a double belt. The area between the two belts (about 4-in. wide) was finished in a lighter shade of the main body color. The entire body was finished in nitro-cellulose lacquer, with the belts in a black color with a bright red hairline stripe.

The front fenders were a one-piece stamping, finished in black baked enamel. The drop at the front of the side rails was hidden by a black enameled, box-shaped dustpan in the area where the hood joined the cowl. The front end of the side rails was hidden by a splash guard between the frame rails and radiator shell. Even those who might have expected a more radical change in body styling were indeed pleased with the functional simplicity of the extremely low-slung chassis and body, the lowest in the automotive industry at that time. The enclosed bodies were equally attractive and also possessed the sporting look.

The excellent performance of the engine was carefully matched by the stable and secure handling characteristics of the car. This was due to the low center of gravity, resulting from the use of the under-mounted worm-drive rear axle and double-drop frame side rails. The low overall height and wide stance of the car gave it a pyramid effect. Aside from its engineering and handling attributes, the New Stutz Vertical Eight with Safety Chassis was indeed an exciting and handsome car; perhaps, the most handsome car built up to that time.

Apparently the dealers and car buyers liked what they saw. It was reported that Stutz dealers submitted their orders for 1,000 new cars that day in December 1925 when the New Stutz Vertical Eight was shown to them at the dealer meeting. Approximately 250 buyers felt the same way at the National Automobile Show in New York, by placing their orders for the new Stutz. Approximately 4,900 Stutz cars were built and sold in 1926.

But in 1927 something happened. Stutz production and sales dropped to about 3,000 cars for the year. On the surface it would appear that someone at Stutz dropped the ball and that the company was floundering, which was not the case. It was a year of retrenchment. The yearly total sales for the industry dropped by approximately 600,000 units, reflecting a sales loss of about 19%. Lincoln sales dropped by 17%, Cadillac lost 27%. Packard with its splendid reputation, fine car lines, and excellent dealer organization gained approximately 6%. With the introduction of the exciting new LaSalle in 1927, there were no less than 13 car makes competing for the buyers in the fine-car field, 16 if you include the Elcar 8-90, the Gardner 8-90, and the Kissel 8-75. The LaSalle sales accounted for 11,258 cars, possibly taking 6,500 car sales from Cadillac. Buyers also encountered a price increase of $155-$200 on the Stutz Standard Line models in 1927.

Product-wise, the Stutz Vertical Eight for 1927 was improved by increasing the bore size of the engine to 3-1/4 in., raising the horsepower to 95, increasing the valve stem length and the guide length for quieter operation and longer life. A 19-1/2-in.-diameter, six-blade fan replaced the 16-1/4-in.-diameter, four-blade fan to provide better cooling under the most severe operating conditions. The engine cooling system was equipped with a Ther-Mo-Cool device that allowed the engine to warm-up quickly to 160°F and maintained an engine operating temperature of 160-185°F summer and winter. A new body model, a touring brougham on a 145-in. wheelbase, was added to the Stutz line in 1927.

1927 Stutz AA convertible coupe.

Stock Car Racing

In promoting the product, the Stutz management entered three stock cars in the Stevens Trophy 24-hour run at the Indianapolis Motor Speedway: a steel-bodied Stutz sedan, a Weymann composition-bodied Stutz brougham, and a two-passenger Stutz speedster. The run was started the evening of April 21, 1927, and the trophy was won by the Stutz sedan at an average speed of 68.44 mph. While the Weymann Brougham weighed some 500 lb less, the engine would miss at high speeds. This problem was cleared up later in the run, but it caused the Weymann brougham to lose valuable time and to finish with a disappointing average speed of 67.17 mph. The Stutz speedster finished with an average speed of 71.35 mph, but it was not eligible for the trophy; they specifically required the winning car to be a production five-passenger sedan.

A 1927 Stutz Roadster was entered in the AAA-sanctioned 75-mile Stock Car Race at the Atlantic City Motor Speedway on May 10, 1927. Tom Rooney, driving the Stutz Roadster, won the race at an average speed of 86.249 mph. On September 5, 1927, three new Stutz Roadsters (Black Hawks) were entered in the AAA Stock Car Race at Atlantic City Motor Speedway. The Stutz Black Hawks came in 1-2-3, with the fastest car averaging 96.305 mph. Also on September 5, 1927, a Stutz-engine-powered racing car won the Pikes Peak Hill Climb with Glenn Schultz driving.

The stock car race at Charlotte, North Carolina, on September 26, 1927, was won by a Stutz sedan averaging 92.22 mph. During the same month a Stutz Black Hawk speedster won the Mexican Road Race from Mexico City to Pueblo. Being the winningest stock car, Stutz won the 1927 stock car championship. Despite the promotional efforts and winning races during 1927, Stutz was able to sell only a dismal 3,413 cars.

On November 21, 1927, the Merz Engineering Company of Indianapolis was organized for the development and production of high-output aircraft engines. The officers of the company included Charles C. Merz, former treasurer of the H.C.S. Cab Company, Frederick Moskovics, and Frank Wheeler, president of the Wheeler-Schebler Carburetor Company. The company bought $140,000 worth of engine manufacturing machinery at a receiver's sale of the former H.C.S. Motor Car Company.

The new 1928 Stutz, known as the Splendid Stutz Series BB, was introduced in December 1927, with numerous changes in the bodies, chassis, and engines. The 1928 Stutz upheld the tradition of custom-built products by setting a policy that every car built would have its own distinct, individual color scheme. The

Glenn Schultz driving Stutz racing car, September 5, 1927.

bodies were lengthened 4 in., and were 2-1/2-in. wider. Full-length horizontal hood side louvres replaced the three rows of shorter horizontal louvres. The roof extension was eliminated at the forward French curve, and a cadet visor replaced the roof extension.

The headlamps were nickel-plated and bowl-shaped, as were the cowl parking lamps mounted on a nickel-plated cowl surcingle. The four-wheel hydraulic brakes were of the internal expanding shoe type. The 1928 Stutz was available in twelve conventional body models, ten new models in the Weymann Custom Line, and three salon custom body types by LeBaron and Fleetwood. Thus, the 1928 Splendid Stutz was available in a total of 30 body types, in a price range of $3,495 to $6,895. Two special chassis of 131-in. and 145-in. wheelbase lengths were available for special one-of-a-kind body builders.

Some Bad Turns

Even with all that the Splendid Stutz had to offer, and some great racing successes, 1928 turned out to be a year of bitter disappointment and terrible tragedy. The racing successes involved two Stutz Black Hawk Speedsters breaking the American Stock Car speed records for 1 mile at Daytona Beach, Florida, on February 23, 1928, when Gil Anderson drove the 1 mile at a speed of 106.52 mph, and Tom Rooney at a speed of 105.54 mph. These two new speed marks exceeded the record of 104.34 mph set by Wade Morton in an Auburn the week before.

The disappointment came as the result of a friendly $25,000 bet between Frederick Moskovics and Charles Weymann (a friend of Moskovics and the supplier of Weymann bodies to Stutz). The bet was that Weymann's (480-cu.-in.) Hispano-Suiza could beat Moskovics' (298-cu.-in.) Stutz Black Hawk in a 24-hour challenge race on the Indianapolis Motor Speedway. The race took place on April 18, 1928. Up to 140 miles the Stutz performed according to plan, keeping pace with the Hispano-Suiza. But then a freak "one in a million" engine failure happened. A valve retainer failed, allowing the valve to drop into the cylinder. Moskovics lost the bet, and Stutz lost prestige along with the race. This failure and disappointment was destructive enough to Stutz's reputation without the tragedy that would follow one week later.

1928 Stutz speedster in the match race with the Hispano-Suiza. (Source: Indianapolis Motor Speedway Corp.)

Frank Lockhart, winner of the 1926 Indianapolis 500, was the product of competition on the race tracks of the Pacific Coast. He came to Indianapolis in 1926 when he was only 23 years old, hoping he could get a chance to drive as a relief driver. He introduced himself and made friends among the drivers, mechanics, and car owners. Opportunity and success came quickly and unexpectedly. Peter Kreis, assigned to drive the Miller racing car #15, came down with the flu and was not in condition to drive. Lockhart substituted for Kreis, and drove with such skill and precision that he set a one-lap record during the race of 120.918 mph and went on to win the race at an average speed of 95.904 mph. In 1926 and 1927 Lockhart finished second in the AAA Championship standings. During 1927 he established a world record of 164.28 mph in a racing car of 91-1/2-cu.-in. engine displacement on the dry lakes of California. These feats established him firmly as one of the most competent and skilled drivers in racing. He was very much in the public view, most respected and admired by nearly everyone.

Frank Lockhart. (Source: Indianapolis Motor Speedway Corp.)

Lockhart's success was due largely to his engineering ability and his ability to make needed changes to make a racing car go faster. The specially built Stutz Blackhawk Special, in which Lockhart was to make the assault on the world speed record, was a product largely of his own creation, and was considered a masterpiece of engineering. World speed records were established by massive cars with two or more aircraft-type

Lockhart in Stutz Black Hawk racing car. (Source: Indianapolis Motor Speedway Corp.)

engines having piston displacements ranging from 3,000 to 4,900 cu. in. The Stutz Blackhawk Special was much smaller, being powered by a 16-cylinder engine (two banks of eight cylinders at 30 degrees), having only 181-cu.-in. displacement. Likewise, the car in every respect was much smaller.

During a trial run at Daytona Beach on February 22, 1928, at a speed of approximately 225 mph, the tires struck an irregularity in the sand, and the car somersaulted end for end into the ocean. Lockhart was pulled from the water by spectators, and was uninjured except for a few bruises and traumatic shock. The damage

Pulling the Black Hawk wreck out of the water. (Source: Indianapolis Motor Speedway Corp.)

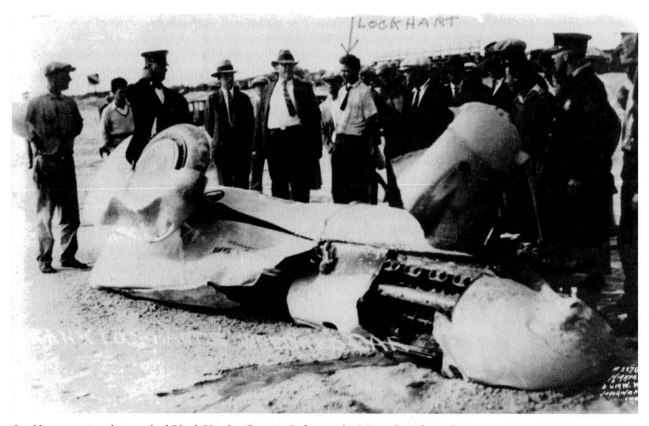
Lockhart viewing the wrecked Black Hawk. (Source: Indianapolis Motor Speedway Corp.)

to the car was mostly a distortion of the frame. On February 27, 1928, the Stutz Motor Car Company announced that Lockhart was recovering satisfactorily, and that Stutz had applied for another sanction by AAA for another assault on the world speed record within a short period of time. The Stutz Black Hawk Special was rebuilt at the Stutz factory in Indianapolis, and returned to Daytona Beach in April for the next try for the world speed record.

After all preparations on Wednesday, April 25th, Lockhart started his first run north at 7:08 a.m. In slowing down at the end of the run, he may have applied the brakes too harshly, locking the right rear wheel for a distance of almost 100 ft. He made the turn around, made whatever adjustments were necessary, and started his run south at 7:32 a.m. He completed his run south, made the turn around, and started the return north run shortly before 7:59 a.m.

Before leaving the south end on his fatal run north, he advised his mechanics that in view of the fact he made a fast run south, he intended to shoot for the record on his north run. About 700 ft. before approaching the timing wire his car went into a skid, with the rear end swinging toward the ocean. After the skid sideways the car seemed to take a course with the beach, when suddenly the tires left the ground completely for a distance of 57 ft. The car left the ground seven times, in jumps through the air, ranging from 33 to 140 ft. in length. The car finally came to rest on the beach, with Lockhart's lifeless body about 51 ft. from where the car stopped.

When Lockhart died, he was trying to break the world's speed record of 207.55 mph for 1 mile established by Ray Keech on April 22, 1928, driving J.M. White's 36-cylinder Triplex. The Triplex power consisted of three 12-cylinder Liberty aircraft engines, totaling 4,950-cu.-in. piston displacement. The tragic death of Frank Lockhart was so crushing to the Stutz Motor Car Company officials, that they declared a halt to all their racing efforts.

Regardless of their plans to quit automobile racing, a racing car named the Stutz Black Hawk Special was already entered in the 1928 Indianapolis 500, with Tony Gulotta as the driver. The racing car was powered by a 90.2-cu.-in. Miller supercharged racing engine. Gulotta finished tenth in the race.

Charles Weymann, who won the $25,000 bet from Moskovics, sponsored and entered a Stutz Black Hawk Speedster in the prestigious 24-hour LeMans Race in France, with a Frenchman Edouard Brisson driving.

1928 Stutz.

The Stutz endured the grueling 24-hour run and finished second behind the winning Bentley. The Bentleys were favored with a four-speed transmission, while the Stutz had a three-speed gear box. This gave the Bentleys a considerable advantage on the twisting course, with zig-zag turns that required shifting into lower speeds. Had the Stutz had a four-speed gear box, it certainly could have been the winner.

Despite the promotional efforts, the Stutz successes at the race courses, and the proliferation of models, the 1928 sales did not improve, but actually dropped to a little over 2,500 units. In the meantime the automobile industry as a whole prospered, almost matching the 1926 production and sales figures. It was indeed a boom, which would carry on through the greater part of 1929, but not for Stutz. Some would want to blame it on the failure to win the match race against the Hispano-Suiza, the tragic loss of the popular racing driver Frank Lockhart, or the comparative loss at LeMans. Stutz's track performance had been commendable, but that alone simply was not enough.

Management Failures

Perhaps it would do more to clarify the situation by taking a closer look at the Stutz organization, its production capabilities, marketing strategy, product line-up, and dealer outlets. Stutz had always been a relatively small producer, building only expensive cars. The selling of only a few expensive cars could not keep a dealer in business. Hence, out of necessity it had to be a dual dealership, also selling lower-priced cars. This divided the selling efforts and resources. During 1928 there were cities with close to 1,000,000 population that would have only one Stutz dealership. Smaller cities had Stutz dealers dualed with other makes. While the Stutz models in 1928 offered 25 body styles, they were all expensive, priced at $3,495 to $6,895.

In an attempt to correct the sales slump after 1927, Stutz management examined their model line-up and price structure. To broaden their market to regain sales in 1929, Stutz decided to introduce a new, less-expensive, companion line of cars to be sold along with the magnificent Stutz Model M. The new Blackhawk was introduced at the National Auto Show in New York in January 1929.

1929 Stutz Blackhawk

Wheelbase (in.)	127.5
Price	$1,995-2,375
No. of Cylinders / Engine	6
Bore x Stroke (in.)	3.37 × 4.75
Horsepower	85 adv, 27.3 SAE
Body Styles	four-passenger coupe, five-passenger sedan, four-passenger speedster
Other Features	eight-in-line L-head engine of 268-cu.-in. displacement optional (discontinued after 1929)

The overhead camshaft engine used the same pistons, connecting rods, valves, timing chains, etc., as the Stutz Model M engine. The Model M engine bore was increased to 3-3/8 in., giving a piston displacement of 322 cu. in. A new four-speed transmission (with internal gear-type silent third speed, and with the "No-back" feature) was offered as standard equipment. Large-diameter, chromium-plated Ryan-Lites were standard, and RA Egyptian Sun-God, symbol of safety, mascot adorned the radiator filler cap.

In the meantime a major shakeup in management was about to occur. Frederick E. Moskovics had come to Stutz in February 1925 and breathed new life into a company that was floundering. He nursed it to good

1929 Stutz Model M speedster. (Source: Indianapolis Motor Speedway Corp.)

1929 Stutz Model M sedan.

1929 Stutz Blackhawk speedster.

health with some of the best-performing cars built in America during 1926-28. The classic styling and handling capabilities were copied by competition in years to come. Somehow, the board of directors had no compassion nor feeling for the merits of the Stutz cars, but were dismayed at the poor sales record and financial performance during 1927 and 1928. Someone had to be blamed, and since he was handy, it was easy to point the finger at Moskovics. At the board of directors meeting in New York on January 24, 1929, Moskovics resigned as president of the Stutz Motor Car Company. Col. Edgar S. Gorrell, former vice president of sales, was elected president, and Edwin B. Jackson was elected chairman of the board. On June 11, 1929, Moskovics was sought out and elected president of the Improved Products Corporation. He was later approached and induced to return to the automotive field, being elected chairman of the board of the Marmon-Herrington Company on March 18, 1931. Marmon-Herrington at that time was pioneering all-wheel drives. (See the Marmon chapter in Volume 4.)

Apparently, the introduction of the new lines of cars for 1929 did not improve sales, nor even stop the slump. According to registration figures, approximately 1,000 Model M Stutz cars and about 1,500 Blackhawk cars were sold in 1929. The Blackhawk sales included about 1,200 overhead cam sixes and about 300 L-head eights. Production figures show 2,100 Blackhawk cars for 1929 and none for 1930; it is possible that the remaining 600 Blackhawk cars built during 1929 were sold in 1930 as carryover models.

Late in October 1929, a new idea in custom bodies was instituted by Stutz with their offering of the new Chateau line of cars with Weymann custom bodies. Expert coach builders were brought from Europe to duplicate in detail the famous individual models owned by European nobility. These new models emerged as the Weymann Chateau series, which included the Versailles and Longchamps on the 134-1/2-in. wheelbase, and the Chaumont and Monte Carlo on the 145-in. wheelbase. The price range was from $3,945 to $4,495 list.

The Depression Marks the End

The timing of the introduction of the Chateau Series could not have been worse, just a few days before "Black Thursday," October 24, 1929, when the stock market crashed on Wall Street. The wealthy and affluent who might have been buyers of the new Stutz line were the first to cut back on their purchases. Consequently, by the end of the year, the expense of developing a new product line created a cash flow problem for Stutz. Late in December three creditors filed suit in the United States District Court, requesting involuntary receivership of the Stutz Motor Car Company. However, by January 6, 1930, Col. Gorrell was able to satisfy the creditors, and had the petition for involuntary receivership dismissed.

While Stutz had withdrawn from racing due to financial squeezes, an enterprising young man named Milton Jones entered a racing car called the Jones Stutz Special in the 1930 Indianapolis 500. The car was a stock Stutz Torpedo Speedster, with the fenders, running boards, top, and windshield removed. L.L. Corum was the driver and Milton Jones was the mechanic and relief driver. The Jones Stutz Special made a good showing, with lap speeds of a reputable 94 mph, finishing in 10th place at an average speed of 85.340 mph. This was a remarkable accomplishment, considering that the Stutz was a completely stock chassis weighing approximately 1,500 lb more than the competing special racing cars. Thus, this proved that the Stutz still had a lot of go-power and could run with the fastest, including Miller and Duesenberg specials.

Despite the introduction of some of the most beautiful and exciting cars in America, and the successful finish at Indianapolis, Stutz did not fare very well in 1930, selling only about 800 cars. That year Stutz suffered financially because of the combination of the merciless Depression, the shrinking of the expensive-car market, and the excessive expenditures of creating new models to keep in step with competition. This resulted in a great financial drain on Stutz resources and a net loss of $1,161,666 for 1930.

For 1931 Stutz became more conservative in their planning. The Blackhawk name was discontinued, and the overhead camshaft six-cylinder model was called the Stutz Model LA. The starting price of the LA was reduced to $1,620. The 1931 Stutz Models MA and MB offered 28 body types, with improved styling and thermostatically controlled radiator shutters. They had a chromium-plated hood center hinge in place of the center panel. The hood side panel hinges were concealed under the belt molding bead. Adjustable door-type vents were used on the Stutz MA and MB hood side panels, while the Stutz LA had vertical hood side louvres.

1931 Stutz town car.

Stutz Motor Car Company operated in a more conservative manner in 1931, but they also made one more attempt to achieve the glory long overdue. In April 1931 Stutz announced that they would build a new Stutz model to be called the DV-32. Production of this new DV-32 engine with Charles (Pop) Greuter's new twin-overhead camshaft and 32-valve train (four valves per cylinder) began on July 6, 1931. By July 28th the Stutz Motor Car Company announced the introduction of the new Stutz Super Bearcat to be built on a shortened 116-in. wheelbase chassis, powered by the DV-32 engine. The body was a specially built convertible roadster with a long hood and a short rear deck. Although advertised with a 100-mph performance capability, only a few Super Bearcats were built and sold during 1931 and 1932.

The Stutz DV-32 was offered with many custom bodies by Weymann, Rollston, Brunn, and the dual-cowl phaeton by LeBaron. The SV-16 was offered with the same body models as the DV-32, except at a price of $600 less. Stutz sales for 1931 dropped to about 530 cars, and net losses for 1931 were $296,270. The losses for 1932 were even worse, since only about 105 cars were sold. Stutz suffered the business slump along with all other car manufacturers; but no one could anticipate that the total U.S. domestic car sales would plunge from 3,880,247 units in 1929 to 1,096,399 cars in 1932.

For 1933 the Stutz Motor Car Company made one desperate attempt to recover its share of the fine-car market. The 1933 models went into production at the end of May, with the formal announcement in July 1932. The model line-up included the Stutz LAA Six, the Stutz SV-16, and the Stutz DV-32, all of which were improved. The Stutz LAA Six was available in four body styles, priced at $1,895. In addition, two new lines of eight-cylinder models were introduced, called the Stutz Challenger Series, CS-16 and CD-32. The Challenger models offered the SV-16 and the DV-32 engines on 134-1/2- and 145-in. wheelbase lengths,

1931-32 Stutz Super Bearcat.

1932 Stutz speedster.

priced at $2,395 to $3,395 for the SV-16 models, and $3,095 to $4,095 for the DV-32 models. The Challenger models differed from the Custom models in trim details and the body builder. While the styling was improved by the slanting windshield and rounded corners, only about 50 new Stutz cars were sold and registered in 1933.

Less than ten Stutz cars were built in 1934, using the residue of 1933 parts. Late one afternoon in 1934, while the finishing operations were being performed on the last precision-built, very beautiful motor car, an important chapter of Indianapolis' automotive history came to a close, as twilight settled and enveloped the beautiful Stutz plant on Capitol Avenue at Tenth Street, which would remain silent in the years that followed.

Harry Ford

Several concurrent events caused the bankruptcy of Saxon Motor Car Company. The failing health of Harry Ford and his untimely death created a void in the Saxon management that could not be filled. The severe post-WWI depression and the subsequent three-digit inflation caused not only Saxon's financial difficulties and bankruptcy, but many large manufacturers' as well. Credit should be given to Saxon Motor Car Corporation for building such a delightful, exciting, modest car during a few years of existence.

* * *

Harry W. Ford (no relation to Henry Ford) was born in the farming community of Knobnoster, Missouri, in 1880. At age 17 he gave up farm work and went to Chicago, Illinois, where he was employed as a messenger-clerk at the Western Union Telegraph Company, working nights from 5:30 p.m. to 1:00 a.m. He practiced telegraphy after hours, and went to school during the days to complete his preparation for college. Entering the University of Chicago in October 1900, he earned his way through the first two years of college by doing extra work for Western Union on Saturdays and Sundays.

During 1902 and 1903 he worked as city editor for the Houghton Mining Gazette in Houghton, Michigan. He returned to Chicago in 1904 to complete his college studies and to do newspaper work for Chicago papers. He moved to Dayton, Ohio, in 1905 and was employed by the National Cash Register Company in the advertising department. In 1908 he became advertising manager for the Sheldon School of Salesmanship in Chicago.

In 1909 Harry Ford moved to Detroit and joined the Chalmers Motor Company as advertising manager. (See the Chalmers chapter in Volume 3.) He made it his particular interest to carefully study the engineering, manufacturing, purchasing, and other departments to prepare himself for the responsibilities of assistant general manager, which he became in 1913. He was well prepared for his next move.

Saxon Motor Car Company

On November 1, 1913, when the formation of the Saxon Motor Car Company was announced, the mention of Harry Ford as president of the new addition to the Detroit automobile manufacturing colony attracted nationwide attention. To the general public, a new "Ford" from Detroit was being introduced, but to the men in the automobile industry, the election of Harry Ford as head of the Saxon organization was but a logical sequence to his previous successes in the industry, and his display of executive leadership ability.

Capitalized at $350,000, the officers of the Saxon Motor Car Company were: Harry W. Ford president and general manager, Percy Owen vice president and sales manager, George W. Dunham vice president, L.R. Scafe secretary and treasurer, and R.E. Cole chief engineer. Lee Counselman was a director. While Hugh Chalmers provided financial support, he was not active in the management of the Saxon Motor Car Company.

The first Saxon plant was located adjacent to the Belt Line Railroad at 1305 Bellvue Avenue at Frederick Street in Detroit, Michigan. The new Saxon car was introduced in November 1913 and deliveries started in January 1914. The Saxon car was a delightful, attractive, agile runabout priced at $395. It was merchandised through Chalmers dealers, and the low price attracted so many buyers immediately that it quickly outsold the production capacity. Approximately 3,000 Saxons were sold in three months.

Thus, in 1914 Saxon moved its production facilities into the former Abbott-Detroit manufacturing plant located on the Belt Line Railroad at Beaufait Avenue at Waterloo Street. In August 1914 Saxon introduced its new 1915 refined Saxon runabout, with a continuous running board, extending the gasoline filler cap at the top of the cowl. It was powered by a four-cylinder, 12-hp, L-head engine, and carried the body on a 96-in. wheelbase chassis, supported by wire-spoke mounted 28 x 3 tires. The price remained at $395, which included a top, lamps, adjustable windshield, baggage container, tools, and tire kit. Sales skyrocketed.

1914 Saxon runabout.

The new 1915 Saxon six-cylinder touring car was not introduced and on general exhibit until after the National Auto Show in New York in September 1914. However, it was shown to dealers at the Saxon factory in Detroit in August. It was built on a 112-in. wheelbase and powered by a six-cylinder, L-head engine built by Continental Motors Corporation. Ignition was by battery and coil, in connection with the charging, lighting, and starting by Atwater-Kent. The clutch was a dry-disc-type, coupled with a three-speed sliding gear transmission, mounted at the three-quarter floating rear axle. The propeller shaft was enclosed in a torque-tube having a double universal joint at the forward end. The touring car was exquisitely styled, with flared-in cowl, giving a neat, streamlined appearance. Standard equipment included artillery wood-spoke wheels mounted with 32 x 32.5-in. tires, windshield, top, electric horn, speedometer, extra rim, tire irons, tools, and headlamps with a dimming device. The Saxon six-cylinder-engine-powered touring car was priced at $785 and the Six touring car with detachable all-season top was priced at $935.

The Saxon Four runabout was called a roadster, and was available with electric ignition, starting, charging, and lighting at $50 extra. The roadster was available with a detachable coupe top and open roadster top included at $455.

Ford Takes Control

On November 1, 1915, Ford bought Hugh Chalmers' stock in the Saxon Motor Company for approximately $450,000. This gave Ford full control of Saxon. In September 1915 Saxon added a roadster model to the six-cylinder line, priced at $785. With the favorable pricing, Saxon sales amounted to approximately 12,000 cars generating a profit of nearly $1,000,000 in 1915.

On November 20, 1915, Harry Ford announced that the Saxon Motor Company had been succeeded by the Saxon Motor Car Corporation, incorporated under the laws of New York, capitalized at $6,000,000, of which much of the stock was purchased by the dealers. The officers were: Harry Ford president and general manager, Percy Owen vice president, and Lincoln R. Scafe secretary-treasurer. Lee Counselman and George Dunham, together with the officers, made up the board of directors.

1915 Saxon.

1915 Saxon Six touring.

The 1916 Saxon models were announced in September 1915. They were basically the same designs as the 1915 models except for the improved styling of the Four roadster, and the addition of a roadster body type to the Six models. The prices remained the same, encouraging the buyers, which gave Saxon a production and sales volume of approximately 24,000 units. The popularity of the Saxon cars made it possible for Saxon to earn a record profit of $1,316,273. The demand for Saxon cars indicated a need for larger manufacturing facilities.

New added features and attractions marked both the Six and the roadster for 1917. The body of the Six was built with entirely new lines, making it roomier, and a slanted windshield enhanced the grace of the overall lines. The inclusion of the Wagner Electric two-unit starting, charging, and lighting system as standard equipment was a dominating feature of the roadster. Another point of interest was the fitting of 30 x 3 Goodyear tires as standard. All this was included in the new price of $495.

However, with the added cost of the new features, and the increased cost of all materials, even though Saxon had record production and a sales record of almost 28,000 units, the net profit was $663,768 in 1917.

The Success is Short-lived

During 1917 several events occurred that had a profound effect on the future of Saxon Motor Car Corporation. Among those was the declaration of war against Germany in April 1917, which created a shortage of materials, with the accompanying inflation of the costs of all materials. And the over-optimism of the demand for new cars resulted in the building of the excessively large production plant at Wyoming and McGraw Avenues in west Detroit. The plant construction started in late 1917, and the inflationary costs and cost overruns resulted in the depletion of Saxon Motor Car Corporation's operating capital.

Concurrently, Harry Ford became extremely ill, causing him to retire from the presidency. The board of directors prevailed upon Benjamin Gotfredson to accept the presidency in December 1917. With the retirement of Harry Ford, Lincoln R. Scafe resigned and was succeeded by E.E. VonRosen as secretary-treasurer.

1915 Saxon roadster.

While the plant at Wyoming and McGraw Avenues was under construction, Saxon Motor Car Corporation occupied additional factory space at 621-59 Bellvue Avenue near Jefferson Avenue in Detroit. In retrospect, the additional space was not really needed because the wartime restrictions resulted in producing and selling fewer cars in 1918. To compound Saxon's problems, Harry Ford, in his weakened condition, succumbed to influenza on December 23, 1918. The morale at Saxon Motor Car Corporation dropped to its lowest level.

During 1919 not only did inflation plague Saxon, but there was a full-blown depression in the making. C.A. Pfeffer, a long-time Chalmers and Saxon executive, became comptroller-secretary, and C.L. Fox was named sales manager. Less than 3,500 Saxon cars were built and sold during 1919. On August 30, 1919, the Saxon Motor Car Corporation's new plant at Wyoming and McGraw Avenues was sold to the General Motors Corporation to pay off the creditors. Saxon was still building cars at the Beaufait Avenue plant.

On November 5, 1919, the Saxon shareholders approved the plan for reorganization and financing. Under this plan the creditors would receive payment in full. The Saxon Motor Car Corporation would start its new course with a new working capital of $1,200,000, by disposing of additional common stock through a banking syndicate set up for that purpose.

Considering his duties in connection with the reorganization of the Saxon Motor Car Corporation completed, Benjamin Gotfredson resigned as president of the company on December 13, 1919, to devote more of his time and attention to his own business. He did this despite the strong pleas of both the creditors' committee and the directors, who wanted him to stay in the post which he had filled so acceptably during the period of Saxon's re-establishment. But, believing the company's future was assured, and with his own body-trimming business growing to large proportions, Gotfredson felt that he could be spared to manage his own enterprises more closely. Gotfredson had worked in close cooperation with Pfeffer, and lent the Saxon comptroller every support in the measures that had been taken toward rehabilitating the company. He strongly supported Pfeffer, who was elected to succeed Gotfredson as president of Saxon Motor Car Corporation.

During 1919 and 1920 only the Saxon Six was built, with roadster and touring car models available, priced at $1,195 in 1919 and $1,295 in 1920. Regardless of what effort was taken, the severe depression had taken its toll. For 1921 Saxon tried to resurrect the four-cylinder model, calling it the Duplex, available in roadster, touring, and sedan body styles, priced at $1,675 to $2,475; but the Duplex could not save Saxon. In April 1921, Saxon sold its service parts to Puritan Auto Parts Company for a reported $500,000 to pay off its creditors. In 1922 Pfeffer resigned as president and there was no successor. Saxon Motor Car Corporation went out of business.

The Wyoming-Avenue plant was subsequently used by General Motors for additional production of Buick from 1924 to 1926. It was used for the building of the new LaSalle from 1927 until 1936. In 1936 the plant was sold to the Chrysler Corporation. DeSoto cars were built at this plant in 1937 and 1938, when DeSoto production was moved to the Chrysler Jefferson Avenue plant. In 1959 the plant was used by the Chrysler Export Corporation, and in 1960 it became the Chrysler Glass Plant.

Robert, Joseph, and Ray Graham. (Source: David B. Graham)

Graham Brothers

From the beginning it was always "My brothers and I" prefacing a conclusion or announcement of any kind. One would not make a decision until all three had consulted together. Each was an emissary of the other two. The tribune was not only thoroughly organized, but was completely protected against incursion of any kind. Outsiders never participated in family council meetings. The fraternalism as a family trait was probably responsible for the many rapid attainments of the Graham brothers.

* * *

The three Graham brothers were sons of Ziba Foote Graham and Margaret Cabel Graham of Washington, Indiana. Joseph B., the mechanical genius, was born on September 12, 1882; Robert C., the natural salesman, was born on August 21, 1885; and Raymond A., born on May 28, 1887, possessed great financial and administrative ability. The talent distribution was excellent, natural, and complementary to each other, marked by a bond that was firm and unique. Internal differences that any brethren may have had were never visible; to the world they were invariably united.

Ray graduated from the University of Illinois in 1908 with a degree in agriculture. He then assumed the management of the family's many farming and business ventures. Robert enrolled at Cornell University to study agriculture; but, attracted by the bright lights of New York City, he continued his education at Fordham University, graduating in 1906 with a degree in chemistry.

After completing college, Joseph chose to come home to his grandfather's business in Washington, Indiana. He then went into business for himself at age 23, and later, with his father, Ziba, he established a glass manufacturing business in Loogootee, Indiana, where natural gas had been discovered.

The Start in Glass

The greater part of glass manufacture was in bottles for beer, soda pop, and mineral water. Their first glass contract was with a mineral bottler in West Baden Springs, Indiana. The bottles of that period were very fragile: the necks were too thin, they easily cracked, and were frequently broken. This was an industry-wide problem, solved only when Joseph Graham developed a process for forming the bottles upside down. In this

Graham automatic bottle-making machine. (Source: David B. Graham)

position the molten glass easily flowed down the side and formed a thicker and stronger shoulder and neck. The Graham process is still being used today.

As the glass business prospered, Joseph had a new glass plant built in Evansville, Indiana. At about this time Robert joined Joseph in the glass business, handling and selling the products. Later the Grahams established two more glass plants at Okmulgee and Checotah, Oklahoma. In 1911 Joseph and Robert convinced Ray to join them as secretary and treasurer of the Graham Glass Company. By 1914 the combined output of the Graham Glass plants climbed to 144,000 bottles daily. In 1916 the Graham Glass Company merged with the Owens Bottling Company of Toledo, Ohio, which later became the Owens-Illinois Glass Company.

The Conversion Truck

Meanwhile, on the farms the Grahams were not satisfied with the small light trucks of the time, and had a lot of experience in developing mobility for the farm equipment and products. Out of this experience, and out of the fertile mind of Ray, came the idea for a light 1-ton truck made by converting a Ford passenger car. By splicing and extending the Ford frame and building a sturdy special rear axle, they were able to make this inexpensive truck conversion capable of hauling 1-ton loads. They developed a cab and body, introduced the conversion outfit in October 1917, and built it in a plant they acquired in Evansville, Indiana.

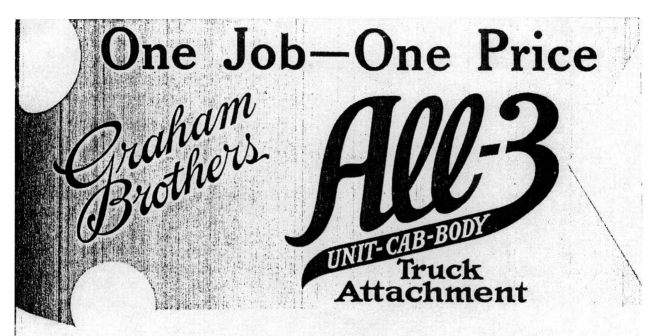

One Job—One Price

Graham Brothers All-3 UNIT-CAB-BODY Truck Attachment

Ready to attach to a Ford chassis and make a **complete** truck, ready for work—not merely a stripped unit without body or cab.

EASY TO APPLY—Axle shafts not cut off; differential not dis-assembled. Standard specifications, including Hess Springs and Axles, Bock Bearings, Cullman Sprockets, Diamond Chains.

Frames, springs and axles assembled before shipment—less work for you. Every "All-3" body is set up and fitted before being knocked down and shipped. Every job is a perfect fit.

There is a tremendous satisfaction in **now** being able to buy the **complete** job.

Write us today for booklet and full details.

GRAHAM BROTHERS
Dept. B, Evansville, Ind.

BRANCHES: New York, Chicago

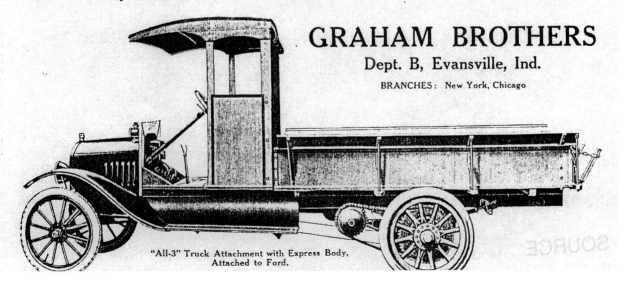

"All-3" Truck Attachment with Express Body, Attached to Ford.

The unique feature of the Graham brothers' conversion package was that it was complete with chassis converting components, cab, and body. It was available in an express body type, with a loading area of 45.5 x 106 in., and a stake body type with stakes 36 in. above the floor, and a loading floor area of 55 x 112 in. The chassis was supported on two, parallel, semi-elliptical springs on a forged steel axle, 1.75 x 2.25 section of Hess manufacture. The drive was through a bevel ring and pinion gear, driveshafts, and Cullman sprockets. Final drive to the Prudden artillery-type wheels was by side chains of 1.25-in. pitch. These truck conversion outfits were priced at $395 and were distributed through Ford dealers.

The Original Truck

By mid-1917 Ford Motor Company was producing its own version of a 1-ton worm-drive truck chassis. Thus, in 1919 the Graham brothers developed a 1.5-ton farm truck totally of their own design. They also developed a 15- and 30-hp tractor, and a kerosene-burning-engined cultivator. Although the Grahams were well known for their truck units, the new farm truck was their first attempt at building a complete truck. Reputable components and units were used in the manufacture of the truck including Continental engine, Torbenson internal-gear rear axle, Eisemann magneto, Stromberg carburetor, Monarch governor, Lavigne steering gear, and Disteel wheels. For the Graham brothers, building the trucks was one thing, but trying to sell them during the post-war depression (1919-20) was another.

Graham Brothers Incorporated

The Dodge brothers died in 1920: John F. Dodge in January and Horace E. Dodge in December (see their story in Volume 1). In January 1921, Frederick J. Haynes was elected president and general manager of Dodge Brothers, and during the next five years the business flourished just as well under the capable management of Haynes as it would have for the Dodges.

In 1921 Graham Brothers Incorporated entered into a contract with Dodge Brothers, whereby the Graham Brothers Incorporated would design and build trucks of 1- to 3-ton capacity using Dodge Brothers engines and transmissions. The Graham Brothers trucks were marketed through Dodge Brothers dealerships. Additional manufacturing facilities were provided in Stockton, California, and Toronto, Ontario, Canada. Graham Brothers established a reputation for dependability, performance, and economy, rivaled only by Dodge Brothers itself. By 1925 Graham Brothers was manufacturing buses, too, and became one of the major truck producers.

Real Estate

Robert's son, David B. Graham, was born in Miami Beach, Florida, in 1925, during the height of the Florida land boom. (He was the first boy to be baptized in a chapel converted from a polo pony stable.) While awaiting the birth of his son, Robert purchased 250 acres of undeveloped land at the northern end of the beach. Robert organized a company with Joseph, Ray, Walter O. Briggs, and Charles T. Fisher as partners. They had fun building roads, pumping in sand and soil for landfill, and planting coconuts to grow the trees.

The partners held the dormant land through the boom, bust, and World War II, when it was rented to the United States Government for $1.00 to be used as a training site and firing range. Robert started development of the property after World War II, and shortly afterward, the lots were marketed by the original builders and developers, with Robert as president of the selling company. The area emerged as Bal Harbour, Florida, and

Bal Harbour, Florida.
(Source: David B. Graham)

now has some of the finest homes, estates, fashionable apartments, glamorous shops, luxury hotels, and finest beaches in the world. Uninhabited when purchased by the Grahams, Bal Harbour now has over 3,000 permanent, and 12,000 seasonal residents.

New Relationship with Dodge

In April 1925, Dillon Read and Company negotiated a deal to purchase the assets, name, goodwill, and plant facilities of Dodge Brothers. The deal was consummated on May 1, 1925, by the issuance of a check in the amount of $146,000,000. Immediately after the purchase, Clarence Dillon began the controversial restructuring of the corporate finances and organization. The new company was named Dodge Brothers Incorporated.

On November 23, 1925, Dodge Brothers Incorporated purchased for cash a majority interest in Graham Brothers Incorporated. The Grahams bought huge blocks of Dodge Brothers Incorporated common stock and became the largest Dodge stockholders. All three of the Graham brothers were elected to the board of directors. Ray was named general manager, Joseph was elected vice president in charge of manufacturing, and Robert was elected vice president in charge of sales. The Graham Brothers truck production for the first ten months of 1925 was 16,500 units compared to 1,085 produced in 1921. By 1926 Graham Brothers truck sales exceeded 37,000 units yearly, which made them the largest exclusive truck manufacturer.

Graham-Paige Motors Corporation

On April 30, 1926, the Grahams disposed of their stock holdings in Dodge Brothers Incorporated, and withdrew from the automobile industry until May 5, 1927, when they purchased Harry M. Jewett's interest in the Paige-Detroit Motor Car Company for a reported $4,000,000. (See the chapter on Jewett in Volume 3.) After the purchase, the Grahams concentrated on strengthening the sales organization. At a special stockholders meeting on January 5, 1928, the stockholders voted to change the name of the company to Graham-Paige Motors Corporation. In this organizational move, Ray Graham was the prime mover, carrying the heavy responsibilities of negotiation up to the point where the triumvirate was able to take executive action.

17 ACRES DEVOTED TO MANUFACTURING TRUCK-BUILDERS

Announcing the Complete Line of
TRUCK-BUILDERS

1 -- Ton Truck-Builders for Fords. (*Chain Drive*)

1½-2 Ton } Truck-Builders for Dodge Brothers.
2½-3 Ton } (*Torbensen Drive*)

1½-2 Ton } Truck-Builders for Other Cars.
2½-3 Ton } (*Torbensen Drive*)

3-5 - Ton } Traction Truck-Builders with Fifth Wheel
5-7 - Ton } and Semi-Trailers. For Fords, Dodge Brothers and other makes of cars. (*Torbensen Drive*)

Truck Bodies: Nine distinctive types of truck bodies for individual requirements—prices on request.

1926 Graham Brothers stock rack truck.

On opening day of the National Auto Show in New York in January 1928, Graham-Paige introduced their new line of cars with five wheelbase lengths. Models 610, 614, 619, and 629 had new six-cylinder engines, and Model 835 had a new eight-cylinder engine. The body styling was very attractive, conforming to the beauty standards of the time.

	1928 Model 610	**1928 Model 614**	**1928 Model 619**
Wheelbase (in.)	111	114	119
Price	$875	$1,295	$1,795
No. of Cylinders / Engine	L-6	L-6	L-6
Bore x Stroke (in.)	2.87 × 4.50	3.12 × 4.50	3.50 × 5.00
Horsepower	52 adv, 19.84 SAE	71 adv, 23.44 SAE	97 adv, 29.4 SAE
Body Styles	coupe, cabriolet, five- and seven-passenger sedans, seven-passenger limousine		
Other Features	North East starting, charging, ignition 6V electrical systems, hydraulic four-wheel brakes		

	1928 Model 629	**1928 Model 835**
Wheelbase (in.)	129	135
Price	$1,985-2,185	$2,285-2,485
No. of Cylinders / Engine	L-6	L-8
Bore x Stroke (in.)	3.50 × 5.00	3.37 × 4.50
Horsepower	97 adv, 29.4 SAE	120 adv, 36.45 SAE
Body Styles	coupe, cabriolet, five- and seven-passenger sedans, seven-passenger limousine	
Other Features	North East starting, charging, ignition 6V electrical systems, hydraulic four-wheel brakes	

The Graham-Paige shield-emblem emblazoned on the radiator shell portrayed the profile of three helmeted knight images of the Graham brothers in solid bronze relief from an original creation by Lorado Taft. The name GRAHAM PAIGE in pyramid-shaped letters appeared below the images. The radiator, a tube and fin type, was V-shaped, with an attractive nickel-plated radiator shell and sloping flat top panels meeting in a wide obtuse angle in the middle. The hood had sloping flat top panels, hinged in the middle, with vertical louvres in the hood side panels. The large headlamps with tilting beam filaments had depressed beam lenses in shallow, bullet-shaped bodies, and were mounted on vertical supports and the fender tie bar.

The car bodies, styled and built by the Murray Corporation, had graceful lines. The high belt line encompassed the body along the upper edge of the hood side panels, terminating at the radiator shell. The rear

The Graham-Paige shield-emblem. Note from designer Lorado Taft at left states: "My studio has designed this symbol of the integrity and unity of purpose back of the Graham-Paige."

corners of the body were rounded, and the windshield header featured a cadet-type sun visor. The heavily crowned fenders had a flowing clamshell shape, with wells in the front fenders for the spare wheels and tires on the 629 and 835 models. Wood-spoke artillery wheels were standard on all models, with six-disc or wire-spoke wheels optional on the 629 and 835 models at no extra cost. Nickel-plated front and rear bumpers were standard on all models. For 1928 only closed-body types were available.

All engines were the L-head type, featuring forged counterbalanced crankshafts, supported by seven main bearings on the six-cylinder engines, and by five main bearings on the eight. All had force-fed pressure lubrication throughout. Pistons were of aluminum-alloy, with Invar struts to control expansion. All models except the 610 featured the Warner Hy-flex four-speed transmission with a silent internal-gear third speed.

Graham-Paige enjoyed record sales for 1928 of 58,523 new car registrations, the highest recorded for any new model introduction year.

For 1929 the Graham-Paige wheelbase lengths were increased, and the model designations were changed to 612, 615, 621, 827, and 837. A new eight-cylinder Model 827 replaced the former 629. Chromium-plating was used instead of nickel on all exposed bright metal parts, including the bumpers. Otherwise the 1929 cars were the same as the 1928 models, except for the use of Delco-Remy electrical equipment, and the four-speed was used on all models except the 612. Graham-Paige enjoyed a good sales year, reflected by 60,487 domestic new car sales. Then came "Black Thursday" on October 24, 1929, which affected all businesses.

	1929 Model 612	1929 Model 615	1929 Model 621
Wheelbase (in.)	112	115	121
Price	$955	$1,195	$1,595
No. of Cylinders / Engine	L-6	L-6	L-6
Bore x Stroke (in.)	3.00 × 4.50	3.25 × 4.50	3.50 × 5.00
Horsepower	62 adv, 21.6 SAE	76 adv, 25.35 SAE	97 adv, 29.4 SAE
Body Styles	coupe, cabriolet, five- and seven-passenger sedans, seven-passenger limousine, touring, roadster		
Other Features			

	1929 Model 827	1929 Model 837
Wheelbase (in.)	127	137
Price	$1,925	$2,355
No. of Cylinders / Engine	L-8	L-8
Bore x Stroke (in.)	3.37 × 4.50	3.37 × 4.50
Horsepower	120 adv, 36.45 SAE	120 adv, 36.45 SAE
Body Styles	coupe, cabriolet, five- and seven-passenger sedans, seven-passenger limousine, touring, roadster	coupe, cabriolet, five- and seven-passenger sedans, seven-passenger limousine
Other Features		

The "first series" 1930 Graham-Paige models had no changes through December 31, 1929. The "second series" 1930 models were introduced at the National Auto Show in New York on January 4, 1930. Paige was dropped from the car name at that time. Double chrome-plated headlamp tie bars carried the bronze Graham medallion in the center, with the name GRAHAM in chrome-plated block letters below the medallion.

	1930 Standard Six	1930 Special Six
Wheelbase (in.)	115	115
Price	$995	$1,225
No. of Cylinders / Engine	6	6
Bore x Stroke (in.)	3.12 × 4.50	3.25 × 4.50
Horsepower	66 adv, 23.43 SAE	76 adv, 25.35 SAE
Body Styles		
Other Features	chromium-plated vertical vanes pressed integral with V-type radiator shell in front of the radiator	

	1930 Standard Eight	1930 Special Eight
Wheelbase (in.)	122-134	122-134
Price	$1,445	$1,595
No. of Cylinders / Engine	8	8
Bore x Stroke (in.)	3.37 × 4.50	3.37 × 4.50
Horsepower	100 adv, 36.45 SAE	100 adv, 36.45 SAE
Body Styles		
Other Features	chromium-plated vertical vanes pressed integral with V-type radiator shell in front of the radiator vertical automatic radiator shutters	

	1930 Custom Eight	1930 Custom Eight
Wheelbase (in.)	127	137
Price	$2,025	$2,455
No. of Cylinders / Engine	L-8	L-8
Bore x Stroke (in.)	3.37 × 4.50	3.37 × 4.50
Horsepower	120 adv, 36.45 SAE	120 adv, 36.45 SAE
Body Styles		
Other Features		vertical automatic radiator shutters

At the 1930 National Auto Show in New York, Graham-Paige Motors Corporation introduced two Paige commercial models, a panel delivery and a screen delivery, both priced at $1,095 f.o.b. Detroit. These delivery units were mounted on the Special Six chassis with minor modifications. The appearance, comfort, accessibility, and attention to detail were the outstanding attributes of these new commercial vehicles.

The 1930 Graham models continued unchanged through December 31, 1930. The cars produced from August 1 through December 31, 1930, were registered as "first series" 1931 models. Due to the depressed economic conditions throughout the United States, Graham registrations for 1930 dropped to 30,140 vehicles, or less than one-half of the 1929 sales. Graham-Paige Motors Corporation sustained a loss of $880,960 for the first six months of 1930.

In January 1931 the "second series" Graham 1931 models were introduced.

	1931 Standard Six	1931 Special 820	1931 Custom Eight
Wheelbase (in.)	115	120	134
Price	$995	$1,195	$1,845-2,095
No. of Cylinders / Engine	L-6	L-8	L-8
Bore x Stroke (in.)	3.25 × 4.50	3.12 × 4.00	3.25 × 4.50
Horsepower	76 adv, 25.35 SAE	86 adv, 31.25 SAE	100 adv, 33.8 SAE
Body Styles			
Other Features	vertical louvres, chrome-plated vanes stamped integral with radiator shell	rectangular hood ventilating doors, chrome-plated vanes stamped integral with radiator shell	

The Ghost of Dodge

On January 13, 1931, Chrysler Corporation filed suit against Graham-Paige Motors Corporation and the Graham brothers to enjoin them from ever using the "Graham" name in the manufacture and sale of trucks and buses, and to enjoin them for a period of five years subsequent to April 30, 1926, from the manufacture and sale of any trucks or buses in competition with Chrysler Corporation.

The Graham brothers filed their answer on March 25, 1931, asserting that the Paige commercial vehicles manufactured by Graham-Paige Motors Corporation, comprising passenger car chassis and light delivery bodies, did not fall within the meaning of the Grahams' agreement with Dodge Brothers Incorporated, which referred to trucks for heavy duty. The Graham brothers declared that Chrysler's allegations as to the value of the Graham name were "a gross exaggeration." They asserted that Chrysler Corporation, after acquiring Dodge Brothers Incorporated, undertook to undermine the value of the Graham name, and substituted the name "Dodge" on the trucks and buses formerly known under the name of "Graham Brothers," types of vehicles that were never manufactured by Dodge Brothers Incorporated. The suit was subsequently dismissed in court. While the litigation did not help the image of the Grahams, it certainly did not seriously hurt the sales of Paige commercial vehicles because the sales of all automotive vehicles skidded downward in 1931. After 1931 Paige commercial vehicles were discontinued.

On April 20, 1931, Matilda Wilson (former Mrs. John D. Dodge) was elected to the board of directors of Graham-Paige Motors Corporation. Wilson's election to the board revived the relationship of the Graham brothers with the Dodge family. As far as it was known, Wilson was the first woman to serve on the board of directors of a large automobile manufacturing firm.

On May 10, 1931, Graham-Paige Motors Corporation entered into competition in the low-priced six-cylinder car field with the introduction of the Graham Prosperity Six. Style-wise it resembled the other Graham models, except it was slightly smaller and lighter. The second series 1931 models continued through December 31, 1931. The cars registered between July 1 and December 31, 1931, were considered 1932 models. The sales of 1931 Graham vehicles dropped to 19,209 units, with a resulting loss of $833,056 for the first six months of 1931.

1931 Prosperity Six

Wheelbase (in.)	113
Price	$785-825
No. of Cylinders / Engine	6
Bore x Stroke (in.)	3.12 × 4.50
Horsepower	70 adv, 23.44 SAE
Body Styles	two-passenger coupe, four-passenger coupe, two sedans
Other Features	

On August 10, 1931, Charles Matheson, sales manager of Graham-Paige Motors Corporation, notified their dealers that Graham-Paige would participate in the compensation of automobile salesmen. The plan was believed to be the first of its kind in the industry, enabling the Graham-Paige dealers in North America to employ additional salesmen. The plan provided a regular salary plus commission on the sales, and it was hoped that it would spur new Graham sales.

Graham Blue Streak

On December 9, 1931, the Detroit Graham-Paige distributor had preview showings of the new 1932 Graham Eight cars, which became known as the Graham "Blue Streak" models. The formal presentation and debut was at the National Auto Show in New York on January 1, 1932. The boldly new and fresh styling was a design masterpiece by Amos E. Northup, and was meticulously crafted by the Murray Corporation. It would set a styling trend that would be copied by the industry for the next several years.

The new Graham Blue Streak was distinguished by its graceful appearance, in which all basic lines were carried in a streamlined curve to the rear. It featured a rearward-sloping, V-shaped radiator grille of chromium-plated, vertical louvres. The radiator shell was omitted, with the hood forward edges meeting the grille outline. The radiator filler neck was concealed under the hood. The sweeping, one-piece, clamshell-shaped, front fenders had a deep outer flange which followed the tire outline, forming a fender skirt.

The body had a rearward-sloping windshield pillar with rounded corners. The car roof outline had a gentle convex arc, with the rear quarter uppers leaning gently forward. The inside adjustable sun visor replaced the outer cadet visor. The Blue Streak introduced the "Pearl-escence" (mother of pearl) exterior lacquer finishes on a production basis. Chromium-plating was sparingly used for the bumpers, grille, and body hardware. Demountable wire-spoke or wood-spoke wheels with 17 x 6.00-in. balloons were standard equipment.

The most innovative mechanical feature of the Blue Streak was the chassis with new frame design. The frame was cradled between widely spaced springs supporting the outrigger spring and shackle brackets. The front springs were spaced 8 in. farther apart, and the rear springs were spaced 3 in. farther apart than the previous 1931 models. The car tread was increased to 60-5/8 in. in front, and 61 in. in the rear. The use of outrigger spring hangers allowed the reduction of frame kick-up and lowering of the entire car. The frame rear kick-up was reduced by increasing the depth of the side rail web (at the rear axle area), and stamping a parabolic opening in the frame side rail web for the rear axle tube.

With this construction, the frame cross-members were located in a plane of the lower flange of the frame side rail. Two of the frame cross-members extended transversely beyond the frame side rails to provide rigid mounting for the running boards and sheet metal. Aside from the increased stability of the low center of gravity, the spring arrangement and the shock absorbers controllable from the driver's compartment provided a smooth, cushioned ride. Driving control was greatly enhanced by shackling the front springs at the front. Thus, the front end of the longitudinal steering drag link followed practically the same arc as the center of the front axle.

On June 15, 1932, the Graham Blue Streak Six was introduced, with features identical to the Eight.

	1932 Blue Streak Six	1932 Blue Streak Eight
Wheelbase (in.)	118	125
Price	$875-1,075	$1,095-1,270
No. of Cylinders / Engine	6	8
Bore x Stroke (in.)	3.12 × 4.50	3.12 × 4.00
Horsepower	70 adv, 25.35 SAE	90 adv, 31.25 SAE
Body Styles	two-passenger coupe, four-passenger coupe, four-passenger cabriolet, five-passenger sedan	five passenger sedan, two-passenger coupe, four-passenger coupe, two-passenger Deluxe coupe, four-passenger Deluxe coupe, four-passenger cabriolet, five-passenger Deluxe sedan
Other Features	Synchro-Silent three-speed transmission with dash-controlled free-wheeling, Super hydraulic brakes equipped with "centri-fuse" brake drums	

Graham Race Car

In May 1932 a Graham racing car, #72 "Folly Farm Special," was entered in the Indianapolis 500 by E.D. Stair, Jr., driven by Ray Campbell, with Howard Dauphin (who later himself became famous as a driver on one-half- and one-mile dirt race tracks in the Midwest) as the riding mechanic. The car had a stock Graham chassis and engine of 245-cu.-in. displacement. Campbell qualified at 108.969 mph for the 34th starting position of the 40-car field. Campbell was forced out of competition with a broken crankshaft after completing 60 laps. While a Lincoln was the official pace car for the start of the race, the AAA officials used a Graham Blue Streak sedan and convertible as the official cars for track inspection and other needs. The Graham racing car was again entered in the 1933 Indianapolis 500, but failed to qualify.

Ray Graham Dies

Concurrent with the decline of business in 1932, Ray Graham suffered from a severe sinus condition and, as the result of stress, had a nervous breakdown in late spring. He was treated in hospitals in Detroit and New York. In June 1932 he entered the hospital at Battle Creek, Michigan, for convalescence. Somewhat

Graham race car at 1932 Indianapolis 500. (Source: Indianapolis Motor Speedway Corp.)

improved, he went to his brother Robert's summer home in Conway, Michigan, in late July. To provide more rest, the brothers agreed that Ray would go to the Loyola House of Retreat in Morristown, New Jersey. They decided to travel in easy stages, covering short distances each day. On August 13, 1932, they stopped in Chatham, Ontario, Canada, for overnight lodging. Ray drowned that night. The exact details of his passing were not known at the time, since there were no witnesses, and the subsequent medical examination was inconclusive. It is believed by the family that Ray was in such excruciating pain at the time, that he broke away from his brother Robert and ran. It is very probable that Ray accidentally fell into the stream in the nearby park, sustaining injuries that caused his death. It was very cruel and unfair that the news media and journalists conjectured that Ray's death was a suicide. Ray's deep and sincere Catholic faith would not allow him to take his own life. Further, if a person was contemplating suicide, he certainly would not have selected a stream with only a few inches of water. The news media made no effort to correct the erroneous impressions it left by hasty and irresponsible reporting.

The whole automotive industry mourned the loss of a great automotive pioneer and a beautiful, virtuous human being. Ray was buried in Washington, Indiana, on August 17, 1932, while all business firms of the city were closed during the day of his internment. Ray had a profound religious faith, and was a Knight of Saint Gregory, as were his brothers.

Always the Innovators

Despite the beautiful styling, advanced innovative features in design, precise manufacturing, and reasonable price, the buying public did not respond to the Graham Blue Streak. Graham-Paige was able to sell and deliver only 12,858 cars in 1932. The irony of all this was that the copycat competitors later flourished, proclaiming design features pioneered by Graham. Although Graham-Paige Motors Corporation, along with the rest of the automobile industry, sustained losses in 1932 due to the economic depression, at no time were the Graham families ever in financial trouble, nor was there ever a threat of receivership of the Graham-Paige Motors Corp.

For 1933 Graham-Paige Motors Corporation offered three different chassis models: two eights and a six. Again, as in 1932 with the Blue Streak, Graham was the innovator: The new, two-piece, V-shaped bumper revealed the full depth of the radiator grille; the central instrument cluster contained temperature gauge, fuel level gauge, oil pressure gauge, and ammeter, and was encircled by the speedometer dial that was a translucent ring; the three-position headlamp control, in addition to having "both-up" and "both-down" positions, had a third position which provided a "down beam" on the left headlamp and an "up beam" on the right headlamp for roadside illumination when passing.

The frame construction was the same as the 1932 Blue Streak models, except for a K member which was added for rigidity in the engine front-mounting area. On the exterior, a chrome-plated molding extended along the running board outer edge, and forward along the lower edge of the front fender skirt. Competitors had a field day the following year, picking up Graham innovations to catch up.

	1933 Standard Six	1933 Standard Eight	1933 Custom Eight
Wheelbase (in.)	118	118	123
Price	$745-835	$845-895	$1,045-1,095
No. of Cylinders / Engine	6	8	8
Bore x Stroke (in.)	3.25 × 4.00	3.12 × 4.00	3.12 × 4.00
Horsepower	85 adv, 25.35 SAE	95 adv, 31.25 SAE	95 adv, 31.25 SAE
Body Styles	two-passenger coupe, rumble-seat coupe, five-passenger four-door sedan		
Other Features	V-shaped front bumper, central instrument cluster, three-position headlamp control		

Despite the great value it represented, with many engineering and styling innovations, Graham domestic sales for the 1933 models were a mere 10,128 cars. This was perhaps due to the depressed national economy in 1933. Even so, Graham-Paige Motors Corporation showed a modest profit of $66,997 in 1933, against a loss of $2,810,852 for 1932.

For 1934 the new Graham models were refined, but proliferated into six separate car lines: three with six-cylinder engines and three with eight-cylinder engines. A centrifugal supercharger designed by Floyd F. Kishline and built by Schwitzer-Cummins Company of Indianapolis was introduced on the 1934 Graham Custom Eight at the National Auto Show in New York in January 1934. The 1934 Grahams featured pressed-steel, artillery-type wheels standard, and wire-spoke wheels optional. A built-in trunk was available on the Graham Standard Six, Special Eight, and Custom Eight. Available in 25 body styles, priced at $745-$1,730, the 1934 Graham domestic sales amounted to 12,887 cars.

Two Graham racing cars were entered in the 1934 Indianapolis 500. Both cars had a stock Graham eight-cylinder engine: car #24 "Lucenti Special," driven by Herbert Ardinger with 281-cu.-in. displacement, qualified at 111.722 mph to start in the 14th position; car # 72 "Economy Gas Special," driven by G. Tramison with a 245-cu.-in. displacement engine, failed to qualify. Despite the "Lucenti Special" weighing 2,642 lb, Ardinger successfully completed the race without a tire change, finishing in 11th place at an average speed of 95.936 mph.

In August 1934 a Graham Custom Eight with supercharger won the Mexico City to Pueblo race on a hazardous mountain course. The distance of 135 km over the terrain with 57 dangerous curves was covered in 1 hour, 5 minutes, 42.2 seconds. The previous record set by Jose Estrade Menocal was over a distance of 119 km in 1 hour, 8 minutes, and 31 seconds. The record was broken not only with a time shorter by 2 minutes and 49 seconds, but while traveling 16 km farther. The Championship prize and trophy were awarded to Graham by the Mexican National Automobile Association.

Graham race car at 1934 Indianapolis 500. (Source: Indianapolis Motor Speedway Corp.)

1934 Graham supercharged Custom Eight.

The 1935 Graham models were introduced on January 2, 1935, consisting of four car lines.

	1935 Series Six	**1935 Special Six**	**1935 Eight**	**1935 Supercharged Eight**
Wheelbase (in.)	111	116	123	123
Price	$595-685	$695-810	$765-825	$865-1,170
No. of Cylinders / Engine	L-6	L-6	L-8	L-8
Bore x Stroke (in.)	3.00 × 4.00	3.25 × 4.50	3.12 × 4.00	3.25 × 4.00
Horsepower	60 adv, 21.6 SAE	85 adv, 25.35 SAE	95 adv, 31.25 SAE	140 adv, 33.8 SAE
Body Styles	4	5	4	5
Other Features				

The 1935 Graham domestic sales amounted to 15,965 cars. On August 10, 1935, Floyd Kishline was promoted to chief engineer of Graham-Paige Motors Corp.

The 1936 Graham models were introduced on January 2, 1935, consisting of three basic series. The styling was new with a two-piece, safety, V windshield, flat body side panels, semi-pontoon-shaped fenders, and the rear doors hinged at the B pillar. The doors extended down to the running boards, eliminating the need for splash panels. The sloping front grilles were made of narrow vertical blades, and the hood side panels were adorned with five die-cast, scallop-shell-shaped louvres. In order to contain material costs, the body panels were shared with the Reo Flying Cloud models. The 1936 Graham sales amounted to 16,439 cars.

	1936 Crusader 80	**1936 Cavalier 90**	**1936 Supercharger 110**
Wheelbase (in.)	111	115	115
Price	$635-695	$765-825	$865-1,170
No. of Cylinders / Engine	6	6	6
Bore x Stroke (in.)	3.00 × 4.00	3.25 × 4.37	3.25 × 4.37
Horsepower	70 adv, 21.6 SAE	85 adv, 25.3 SAE	112 adv, 25.3 SAE
Body Styles	4	6	7
Other Features			

The 1937 Graham models introduced on October 1, 1936, were in fact a carryover of the 1936 models, with the addition of the Graham Custom Supercharger Series 120. The carryover models consisted of the Graham Crusader Series 85, Cavalier Series 95, and the Supercharger Series 116. The engine output ranged from 70 hp on the Series 85 and 95 to 112 hp in the Series 116 and 120. Offered in 24 body-engine-chassis combinations, priced at $690-$1,195, the 1937 Graham domestic sales accounted for 13,984 cars.

Spirit of Motion

The 1938 Graham models announced in October 1937 were indeed new with the "Spirit of Motion" styling that featured a new front-end design and hood side panel louvre arrangement. The headlamps were truly fared into the leading area of the front fenders, and the body pillars were narrow to improve visibility. The new styling was in fact so far advanced that that the competitors chose not to copy it. It was not until the late 1980s and '90s that the Formula I Grand Prix racing cars used the "Spirit of Motion" motif.

Floyd F. Kishline. (Source: Floyd Kishline)

For 1938 the Graham car lines were confined to two series with six-cylinder, naturally aspirated and supercharged engines. Domestic sales for 1938 amounted to 4,139 cars.

	1938 Series 95	1938 Series 96
Wheelbase (in.)	120	120
Price	$940-965	$1,025-1,075
No. of Cylinders / Engine	6	6
Bore x Stroke (in.)	3.25 × 4.37	3.25 × 4.37
Horsepower	90 adv, 25.3 SAE	116 adv, 25.3 SAE
Body Styles	2	2
Other Features		

Graham-Bradley Tractor

In January 1938 Graham-Paige Motors Corporation started production of the Graham-Bradley farm tractor, merchandised through Sears, Roebuck and Company. Graham had an initial order from Sears for 2,000 tractors. The main reason for the building of the Graham-Bradley tractor was to provide full employment for Graham's loyal employees. In May 1938 Floyd Kishline resigned from Graham-Paige Motors Corporation to become chief engineer of Willys-Overland Motors Incorporated.

Due to the crippling labor strikes and economic recession, the entire automotive industry suffered great losses. The plunge of sales industry-wide was about 50%, aptly shown by the volume of 1,891,021 units in 1938 against 3,483,752 units in 1937. Graham's dismal 1938 sales, along with just 1,532 Graham-Bradley tractor sales, resulted in a net loss of $1,920,186 for 1938.

Partnering with Hupp

Coincident with Graham's enterprise of building the "Spirit of Motion" cars and the Graham-Bradley farm tractors, another seemingly unrelated turn of events took place in 1938. Norman DeVaux, general manager of Hupp Motor Car Company (former builder of the DeVaux car in 1932), managed to acquire the patent rights, drawings, dies, jigs, and tooling for the 1936-37 Cord 810 and 812 for a reported $45,000. DeVaux engaged John T'jarda (famed stylist) to restyle and redesign the Cord front end to conform to the proposed Hupp Skylark 115-in. wheelbase and rear-wheel drive. The redesign was excellent and a restored Hupp Skylark (if one could be found) is still handsome even today. That part was fine, and the first pilot model was built in 1939, but Hupp Motor Car Company had no operating plant, nor the money to finance the building of the car.

DeVaux was able to interest Robert Graham in the project. An agreement was reached between DeVaux and Graham, by which Graham-Paige Motors would build the cars for Hupp Motor Car Company in exchange for the licensed right to build the Graham Hollywood models on the same assembly line. Graham used its six-cylinder, naturally aspirated as well as the supercharged engines. The Hupp Skylark was powered by Hupp's own 101-hp, six-cylinder engine.

For 1939 Graham-Paige Motors Corporation wisely chose not to spend money for redesign and tooling, considering the depressed national economy. The 1939 Graham models were a carryover of the 1938 models, and the same technical specifications were used. Available in twelve body and chassis combinations, priced at $940-$1,225, the 1939 Graham domestic sales accounted for 3,660 cars.

On October 2, 1939, the 1940 Graham model line-up was announced, consisting of five series. The four carryover (Spirit of Motion) models were designated as the Senior models, with applicable 1939 technical specifications. The new Graham Supercharged Hollywood model had entirely new styling. The 1940 Graham domestic sales amounted to 1,856 cars, of which about 1,000 were Senior models.

	1940 DeLuxe Six	1940 DeLuxe Supercharger Six	1940 Custom Six
Wheelbase (in.)	120	120	120
Price	$995	$1,135	$1,160
No. of Cylinders / Engine	6	6	6
Bore x Stroke (in.)	3.25 × 4.37	3.25 × 4.37	3.25 × 4.37
Horsepower	92 adv, 25.3 SAE	120 adv, 25.3 SAE	92 adv, 25.3 SAE
Body Styles	3	3	3
Other Features			

	1940 Custom Supercharger Six	1940 Supercharged Hollywood
Wheelbase (in.)	120	115
Price	$1,295	$1,250
No. of Cylinders / Engine	6	6
Bore x Stroke (in.)	3.25 × 4.37	3.25 × 4.37
Horsepower	120 adv, 25.3 SAE	120 adv, 25.3 SAE
Body Styles	3	1
Other Features		

The production of the Graham Senior (Spirit of Motion) model was discontinued on June 10, 1940. Graham continued to build the Hupp Skylark, the Graham Hollywood, and the Graham-Bradley tractor through September 1940, to keep Graham employees working.

For 1941 only the Graham Supercharged Hollywood and naturally aspirated Hollywood models were offered. Priced at $1,045 and $895, respectively, 544 Graham cars were sold domestically. During 1940 and 1941, Graham-Paige Motors Corporation built and sold 314 Rupp Skylark cars. Graham sold the cars directly to the Hupp dealers because Hupp had no distribution organization.

When the production of civilian vehicles ended in September 1940, Graham-Paige Motors Corporation was engaged in the building of amphibious military vehicles and Army ordnance through 1945.

Postscript

On October 1, 1943, Joseph W. Frazer was elected president of Graham-Paige Motors Corporation. By August 1944, Frazer held controlling interest in Graham-Paige Motors Corp.

In February 1947 all the plants, equipment, and assets of the Graham-Paige Motors Corporation were sold to Frazer and Henry J. Kaiser, who formed the Kaiser-Frazer Corporation. In 1950 the Graham-Paige Corporation moved their investment enterprises to New York City, where it was renamed the Madison Square Gardens Corporation in 1960. The Madison Square Gardens is now a Gulf and Western subsidiary.

The Graham families remained very active in sports, particularly in owning and sponsoring athletic teams such as the New York Knickerbockers (basketball) and the New York Rangers (hockey), and many others. Robert died in 1967 and Joseph died in 1970.

The Family Farm

The history of the Graham Family in Indiana dates back to 1825, when James Graham, great-grandfather of Joseph, Robert, and Ray, migrated from Kentucky and purchased 121 acres of fertile land in Davies County, Indiana. Thomas B. Graham, James' son, added considerably to the land holdings of the Graham family. Ziba F. Graham, Thomas' son (and father of the three brothers), also made purchases of land that added to the acreage owned by the Grahams. It is interesting to note that some of the 5,000 acres of land now owned by the Graham families was never owned by anyone but the Grahams.

Some of the land was purchased from the United States Government, some from the state of Indiana, and some from the former Wabash and Erie Canal Company. The canal was located along the present right-of-way of the former New York Central (Penn Central) Railroad that serves the Graham Farms elevator of 225,000-bushel capacity, and the feed mill. The Graham Farm enterprises were so successful that they expanded in subsequent years to the point that today the Graham Farms encompass more than 5,000 acres of the best farm land in mid-America.

At the turn of the century, each harvest season the farm product surplus would go to the Graham Elevators, of 60,000-bushel wheat capacity, and 55,000-bushel corn capacity. The farm supported 200 head of prize dairy cattle. By contract, the Graham Farms supplied all the milk, cream, butter, and egg requirements of the New York Central and Baltimore and Ohio (B&O) Railroad Systems dining cars. The dairy surpluses along with the fresh milk collected from the surrounding dairy farms went to the Graham Cheese factory.

The Graham Farms and Graham Cheese Corporation are still family-owned and operated by sons of Robert C. Graham: David B. Graham president, Ziba F. Graham III, Robert C. Graham, Jr., Thomas E. Graham, and their children. Approximately 3,000 acres of grain including corn, wheat, rye, and grain sorghums are harvested each year. Over 250,000 turkeys and 2,500,000 chickens are raised each year on the Graham Farms feeding program. The livestock raised includes 1,000 head of beef cattle, 5,000 hogs, 1,000 sheep, and 200 Holstein dairy cattle each year.

Officers of Graham Farms, Inc., Graham Cheese Corp., and Graham Brothers, Inc., in 1960, l. to r.: David B. Graham, secretary; Robert C. Graham, chairman; Ziba F. Graham, treasurer; Robert C. Graham, Jr., president; and Thomas E. Graham, vice president. (Source: David B. Graham)

Charles Nash

Through his intelligent business management, fair labor policies, and technical experience, success followed Charles W. Nash as president of General Motors Company and then Nash Motors Company, which would become one of the few pioneering automobile companies to weather the Great Depression. Automotive leaders of today could learn a lot from him about labor relations, communication, cooperation, and prosperity.

* * *

Early Lessons

Charles Warren Nash was born in DeKalb County, Illinois, on January 28, 1864. Because of the separation of his parents, at age seven he was "bound out" by court order to a farmer named Lapworth of Flushing, Michigan. At age twelve he ran away to Grand Blanc, Michigan, about 10 miles southwest of Flint, to become a carpenter's helper. The carpenter paid him $7 for the first month's work, $8 for the second, and $9 for the third. After three months, when harvest time came, he was hired by Alexander in Mt. Morris, 10 miles northeast of Flint. Alexander paid him $12 for the month. That brought Charles Nash's total earnings for the four months to $36 of which he saved $25. With his savings he purchased ten sheep. In five years his flock numbered 80 sheep. This was his first and lasting lesson in economics, as he continued working and saving.

Nash learned his skill at carpentry from John Shelben, maintaining farm buildings. He acquired his mechanical skills by operating and maintaining farm machinery, including a hay baler. In 1883, while baling hay at a neighbor's farm, he met the farmer's daughter, Jessie Halleck, whom he married on April 23, 1884. Nash continued working on the farm, saving his money, and he was also hired out as a member of a cherry-picking crew.

In 1889, while picking cherries on the farm of Josiah Dallas Dort, he met Dort who was very impressed with Nash's ideals, work, and enthusiasm. This led to a job as a cushion stuffer at the Durant-Dort Carriage factory in Flint, Michigan. His progress in the manufacturing industry was steady from then on. After working several years as a semi-skilled worker, then as a skilled mechanic, he became the shop foreman. He then became the superintendent of the carriage factory. It was here, in 1906, that Charles Nash met James T. Wilson, who was his secretary. Wilson subsequently married one of Nash's daughters. (Nash's other daughter married C. Hascall Bliss, future Nash sales manager.)

In 1907, after 16 years with the company, Nash became vice president and general manager of the Durant-Dort Carriage Company. In 1910 he resigned his position to become the president and general manager of the Buick Motor Company. Wilson followed Nash to Buick.

Leading General Motors

Because of Nash's tireless energy, management skills, and organizational ability, Buick prospered. Production and sales increased to 28,000 units in 1912, realizing a profit of over $1,500,000. On November 19, 1912, Nash was elected president of General Motors Corporation, succeeding Thomas W. Neal, who had retired. After lengthy interviews with Charles Nash and James J. Storrow, Walter P. Chrysler, a former American Locomotive Company superintendent, was chosen to succeed Nash as general manager of Buick.

During this period General Motors was comprised of Cadillac, Oldsmobile, Buick, and Oakland car companies, Northway Motor Manufacturing, Weston-Mott Company, and General Motors Truck. Success followed Nash, and he was one of the best presidents General Motors ever had. His management skills and direction enabled General Motors to enjoy a prosperity to which it was not accustomed. Within three-and-a-half years, General Motors' profits quadrupled from $7,460,000 in 1913 to $29,150,000 in 1916. It was during this period that James Storrow developed an unshakable respect and confidence in Nash's judgment and ability, an important factor in their subsequent business relationship. Nash remained president of General Motors until 1916, when William C. Durant regained control of the organization. Nash submitted his resignation on April 18, 1916, accepted by the board on June 1st, effective August 1, 1916. Storrow, who was the chairman of the finance committee and a member of the board of directors, also resigned on June 27, 1916. Durant and Nash were friends for many years dating back to the Durant-Dort Carriage days. Durant urged Nash to stay on as president, and offered him an outlandish salary; but Nash refused Durant's offer and decided to leave General Motors, realizing that his business policies of conservatism, frugality, and economy were directly opposed to those of Durant's free-spending attitudes. (See the Durant chapter in Volume 1.)

Asked what he intended to do, Nash replied, "I'm going fishing." In truth Nash was an avid fisherman and big game hunter. He never lost his taste for the great outdoors and the simple life. However, his fishing expedition was of a different kind this time. Storrow and Nash contacted Henry B. Joy and asked him to consider selling Packard Motor Car Company to Nash. While Joy was somewhat receptive to the idea, no one else on the Packard board of directors was. Joy resigned as Packard president shortly thereafter and Alvan Macauley was named president (see the Packard chapter in this volume).

Nash Buys Jeffery

Hence, the next move of Storrow and Nash was to find another reputable automobile manufacturing company that might be bought. Nash found such a company in the Thomas B. Jeffery Company. The transaction took place on a scorching hot July 13, 1916, in a room in the Blackstone Hotel in Chicago, and took very little time to complete because of two circumstances. First, Nash was anxious to buy in order to get back into the automobile manufacturing business since he was only 52 years young, and second, Charles T. Jeffery was anxious to sell because of two traumatic experiences in his life: the sudden death of his father (Thomas B. Jeffery) in Pompeii, Italy, on April 3, 1910, and his own miraculous survival of the torpedoed Lusitania on May 7, 1915. (See the Jeffery chapter in Volume 4.)

Nash Motors Company

The price was agreed upon, reportedly between $9,000,000 and $10,000,000, whereupon Nash wrote out a check for $500,000 as a deposit to bind the deal. Nash and Jeffery shook hands, and the purchase of the Thomas B. Jeffery Company by Charles W. Nash was completed. Nash then made arrangements with James Storrow in conjunction with Lee Higginson and Company of Boston. Nash Motors Company was incorporated in Maryland on July 29, 1916, with a capitalization of $25,000,000. With Nash and his associates in possession of the Jeffery properties, Nash took a delayed vacation during August, and returned to take charge on September 1, 1916.

Nash returned to find his organization in place and working. A letter to the dealers and a press release was sent out on Thomas B. Jeffery Company stationery, outlining the plans. "The purchase of the Jeffery Company by Nash seems essentially fitting. Throughout his career he stood for that which is right and proper. He has always been considered of having a high degree of honesty and square dealing. He is the man who will easily slip into the management of a successful business to his liking. Also having the name and reputation of one of the oldest and best known automobile manufacturers in the business. The Jeffery Company is in every way poised for an uninterrupted period of progress and prosperity in the continuation of the Jeffery line."

Nash did not purge the Jeffery organization. Charles T. Jeffery, who succeeded his father in administration responsibilities in 1910, agreed to stay on the board directorate, along with his brother Harold W. Jeffery, only as long as was necessary to enable Nash to pick up the operation of the company. Eventually both retired from the board of directors in 1917. During the latter half of 1916 and 1917, Nash spent most of his time setting up his company organization, making production studies, and arranging plant facilities for maximum efficiency and profitability.

James Storrow was named chairman of the board without fanfare, Nash the president, James Wilson assistant to the president, Charles B. Voorhis sales manager, C. Hascall Bliss assistant sales manager, and Nils Erik Wahlberg director of engineering. Voorhis, Bliss. and Wahlberg had had similar positions at Oakland Motor Car Company. Walter H. Alford, who had been associated with Nash since 1910, was named comptroller and treasurer. John A. Rose, who handled export for Jeffery, stayed on as export manager for Nash until 1924. All key men were tops in their business.

By mid-1917 the new Nash for 1918 was shown to dealers at Kenosha, by invitation from Charles Nash. The new Nash was formally introduced to the public on September 1, 1917. A five-passenger touring car

1918 Nash four-passenger coupe Model 695. (Source: Nash Motors Co.)

priced at $1,275, it was totally new, incorporating the finest design, engineering, and styling of the period. The exterior lines were smooth and graceful. The chassis was scrupulously clean in design, with no unnecessary rods, tubes, or plates. Yet the frame was surprisingly light, rugged, and stiff due to the proper placement of cross-members. The Nash was powered by a perfected valve-in-head, six-cylinder engine.

The engine design was a masterpiece in itself in performance, efficiency, silence, and flawless appearance. The most unique feature was the placement of the generator and fan, mounted at the front of the engine block, driven by a pulley on the front end of the crankshaft and a short V-type drive belt. The fan and generator pulley was cast integral with the four-blade aluminum fan, attached to the front end of the generator armature shaft. All rotating parts of the engine were fully enclosed, except for the fan, belt, and crankshaft pulley, and they were adequately lubricated with practically no oil leaks in operation. The engine was truly simple, clean, quiet, and reliable.

The new Nash five-passenger sedan for 1918 was introduced on September 10, 1917, a four-passenger chummy-roadster on November 17, 1917, a seven-passenger touring car on April 18, 1918, and a four-

1918 Nash touring car Model 681. (Source: Nash Motors Co.)

*1919 Nash Model 681.
(Source: Nash Motors Co.)*

passenger coupe on May 21, 1918. The chummy-roadster was priced at $1,295, the sedan at $1,985, the seven-passenger touring at $1,545, and the four-passenger coupe at $2,085. All prices were f.o.b. Kenosha, Wisconsin. The Jeffery trucks and Jeffery Quad became the Nash trucks and Nash Quad. By September 1918 the whole plant changeover was made, and Nash Motors Company enjoyed a profit of over $2,000,000 in the first 15 months of operation.

With the introduction of the new Nash, production was slow for several months because there were the Jeffery cars still to be sold, and Nash Quad trucks to be built for the United States Army Transportation Corps. However, a number of Nash cars and trucks were built for domestic sales. After the Armistice, from December 1, 1918, approximately two months were spent in completing and terminating U.S. Government contracts and readjusting the Kenosha plant. The production totals for 1918 were 10,283 domestic cars, 2,500 domestic trucks, and 11,494 U.S. Army contract trucks. While many politicians were decrying huge war profits for war materiel producers, it certainly was not the case at Nash Motors Company. The average profit for all manufacturers of war materiel was approximately 5%; Nash Motors Company earned a modest $2,790,331 for the entire year of 1918.

The 1919 Nash models had no significant changes except that two new body types were added: a two-passenger roadster and a four-passenger sport touring model with wire-spoke wheels, nickel-plated radiator shell, and nickel-plated headlamps. The sport touring was priced at $1,595 f.o.b. Kenosha, $200 more than the five-passenger touring car. Sales were good, but limited by production and material handicaps from strikes in the coal and steel industries. Despite those hindrances to production, 25,114 passenger cars and 4,727 domestic trucks were produced in 1919, producing earnings of $6,711,367.

During 1919 one-half interest in the Seaman Body Corporation was acquired by Nash Motors Company at a cost of $255,031.

With the opening of the National Auto Show in New York on January 3, 1920, the automotive trade and the public got its first glimpse of the new Nash Four. It had been rumored for a long time that a new car would be built in the new assembly plant in Milwaukee as the plant neared completion. The new Nash Four featured a four-cylinder, perfected valve-in-head engine, identical in many respects to the Nash Six engine. The appearance of the new Nash Four closely resembled that of the Nash Six, except for a shorter (112-in.) wheelbase and a shorter hood. Only one body type, a touring car, was completed in time for the show. Deliveries of the new car were not made until the late summer of 1920.

LaFayette Motors Company

A short distance from the Auto Show, at the Grand Central Palace, another new car was making its debut in the Grand Promenade lobby of the Commodore Hotel. The car was the LaFayette, built by the LaFayette Motors Company of Indianapolis, Indiana. The LaFayette Motors Company was incorporated in the state of Maryland in November 1919. James J. Storrow was elected board chairman, Charles W. Nash president, D. McCall White vice president in charge of engineering and production, Earl C. Howard vice president of sales, and M.J. Moore secretary and treasurer.

After viewing the LaFayette at both the New York and Chicago exhibits, many potential automobile buyers, in their eagerness for LaFayette ownership, asked that their names be entered on the waiting list even before the price was announced. But, like the introduction of Leland's Lincoln, Nash couldn't have picked a worse time to introduce an expensive new car. The nation was rapidly plunging into a severe depression.

The new LaFayette had many new and innovative features, reflecting the influence of aircraft engine design on White's design of the new car. The LaFayette was for customers who could afford the best of everything. All the features LaFayette offered were truly the best available at that time. Being an entirely new design, it was unfettered by conventional habits, and a great number of new innovative features were incorporated into the design. The LaFayette was truly a very attractive and exciting car, rendered in good taste.

1920 LaFayette

Wheelbase (in.)	132
Price	$5,025-7,500
No. of Cylinders / Engine	V-8
Bore x Stroke (in.)	3.25 × 5.25
Horsepower	90 adv, 33.8 NACC
Body Styles	five-passenger touring, four-passenger Torpedo model, four-passenger coupe, touring sedan, seven-passenger limousine
Other Features	

The touring and Torpedo models were distinguished by lines suggesting high speed, resembling the expensive, swagger sport models of Europe. Technically it was a very advanced design. The new V-8 engine had 348-cu.-in. displacement. The valve stem angle was 9 degrees from parallel of the cylinder bore, providing better engine breathing and increased engine power output. The exhaust-jacketed intake provided complete vaporization of the fuel mixture. The exhaust manifolds were integrally cast in the cylinder blocks, permitting a single exhaust pipe connection to each block. The cooling of the engine was controlled by thermostatically operated vertical shutters in front of the radiator. The engine featured a forged counterbalanced crankshaft, supported on five unusually large main bearings. Lubrication was by force-feed to all main bearings, connecting rod bearings, and camshaft bearings by an oil pump driven by the crankshaft.

Ignition, starting, and lighting was by a modified Delco dual system. The drive line was through a large-diameter tubular shaft to a full-floating rear axle. Semi-elliptical springs were used at all four locations. The rear springs were 60 in. in length. The tires were 33 x 5-in. cord casings, with a ribbed front tread design, and a non-skid pattern on the rear tire tread.

The post-war shortages and strikes that affected most automotive supplier plants seriously delayed the production of the LaFayette car, which finally got going slowly in August 1920. As the post-war depression

*1921 Nash Model 687.
(Source: Nash Motors Co.)*

deepened, many potential expensive-car buyers, who may have had a desire to purchase a LaFayette, had second thoughts about spending $5,000 for a new car at this time. LaFayette could muster only a dismal 274 registrations through the end of 1921. Lincoln Motor Company, which got started in production earlier than LaFayette, fared slightly better with 1,089 registrations for 1920 and 1921. By late 1921, LaFayette was represented by only 30 direct distributorships, many of whom were also Nash distributors.

During 1920 Nash had a total output of 40,820 cars, a few trucks, and earned a net profit of $7,007,471. For 1921 the output dropped to 20,737 cars, and the profit plummeted to $2,226,078. The losses sustained by LaFayette Motors Company reduced the combined profits of Nash Motors Company.

In 1921 James T. Wilson was elected vice president in charge of production, and in 1922 Earl H. McCarty resigned from Studebaker to become director of sales for the Nash Motors Company.

During early 1922 a merger was attempted between the LaFayette Motors Company and the Pierce Arrow Motor Car Company. The two companies could not come to terms and the plan was abandoned. On June 20, 1922, at a meeting of the LaFayette stockholders, the holders of the stock were asked to authorize the sale of the LaFayette Motors Company property to the LaFayette Motors Corporation, a Maryland corporation. The Nash Motors Company had advanced LaFayette Motors Company $500,000 in addition to endorsing bank loans of about $1,000,000. Because of this assistance, about 80% of the original LaFayette Motors Company stock was transferred to the Nash Motors Company.

Despite the financial burden which had used up almost $2,000,000 of Nash capital, Nash Motors Company was able to build about 40,350 cars and earn a profit of $7,613,246. The Nash Four Carriole, a five-passenger, two-door, sport model priced at $1,350, was introduced in 1922. The Carriole, LaFayette, and Nash Six Sport Touring were continued through 1924, and then discontinued along with the Nash Truck and Nash Quad.

Nash Growth and Change

Nash Motors Company produced 56,569 cars and earned $9,280,032 in 1923. Nash could have sold 100,000 cars in 1923, and was asked by his sales people to build that many. To this Charles Nash replied, "I'd rather build 60,000, and have them all sold, with none left on the dealer's floor at the end of 1923."

During the years 1920-23 Nash made minor changes and improvements on all the cars. In 1922 the headlamp bodies were changed to the popular drum shape. In 1923 wood-spoke artillery wheels were standard, with Budd-Michelin disc wheels as an option.

In 1923 Nash introduced a Sport Roadster model in the Nash Six line featuring a Burbank cloth top, nickel-plated radiator shell, disc wheels, with the two spare wheels carried one at each side of the cowl. The Sport Roadster also had a convenient cloth-covered trunk at the rear. Nash also introduced a four-door coupe model (on the longer 127-in. wheelbase chassis) in the Nash Six line. This body type in effect was a handsomely styled, close-coupled, luxurious closed car with a sizeable trunk at the rear. For years afterward the four-door coupe was one of the most sought-after closed models of the Nash line. The Victoria model was also introduced in 1923. The Victoria was a two-door closed model with seating for four adults and room for a child behind the driver's seat. The car featured a built-in trunk in the deck lid area.

The refined 1924 Nash models were introduced on July 20, 1923, retaining the price structure of the 1923 models. However, on April 5, 1924, the prices on open models were increased by an average of $35.

On January 28, 1924, the Nash Motors Company purchased the real estate and buildings of 2,900,000 sq. ft. area from the Mitchell Motors Company Incorporated of Racine, Wisconsin, for a reported $450,000. This commodious plant, conveniently located between the Kenosha and Milwaukee plants, was later used to manufacture the Ajax car, introduced on June 1, 1925.

In February 1924, James T. Wilson was elected to the board of directors of the Nash Motors Company. During 1924 Nash Motors Company put on a drive to obtain better dealer representation, and to close many open points in smaller cities. This action reflected the selling skill of Earl McCarty, which resulted in Nash building and selling 53,135 cars in 1924. The net income for 1924 was $9,280,541, an average profit of $175 per car.

The Milwaukee plant, which produced the LaFayette and Nash Four, was retooled for the new Nash Special Six models for 1925.

The new 1925 Nash models were introduced on August 1, 1924, featuring many innovations and very attractive styling. The Advanced Six replaced the former Nash Six, and featured new body styling with a double belt line molding, heavily crowned fenders, with triple crease lines to accentuate and reinforce the

1924 Nash Six roadster Model 696. (Source: Nash Motors Co.)

fender sheet metal. The radiator shell was slenderized and nickel-plated. The hood sides had a greater number of slender vertical louvres. All closed-body models featured a one-piece windshield with automatic wipers. The windshield on the open models had a gentle rearward slope. The improved Nash Advanced Six engine was refined for quieter operation and oil-tight construction.

The Nash Advanced Six chassis featured test-proven, four-wheel mechanical brakes, with external contracting bands on the rear brakes, and internal expanding shoes for the front brakes. During 1924 Charles W. Nash observed that 75% of the car sales included the optional disc wheels at an additional cost of $25. Thus, for 1925, Nash made Budd-Michelin disc wheels finished in matching body colors standard equipment at no extra cost. Also standard were 33 x 6.00-in. balloon tires with ribbed tread on the front wheels, and non-skid tread on the rear and spare wheels.

The new Nash Special Six had all the features of the Advanced Six, except scaled down in size. The styling was similar to the Advanced Six, except it did not have the double belt line or the crease in the fenders. The radiator shell and headlamp doors were nickel-plated. Budd-Michelin disc wheels finished in matching body colors were standard, as were 31 x 5.25-in. balloon tires with ribbed tread for the front, and non-skid tread on the rear and spare wheels. All closed models were attractively styled with rounded corners. The open models featured a rakish folding top and a gracefully sloping windshield. A permanent top with glass enclosures was available as an extra-cost option on all open models. All 1925 Nash models were finished with nitro-cellulose lacquer.

	1925 Nash Advanced Six	**1925 Nash Special Six**
Wheelbase (in.)	121/127	113
Price	$1,375-2,290	$1,095-1,545
No. of Cylinders / Engine	6	6
Bore x Stroke (in.)	3.25 × 5.00	3.12 × 4.50
Horsepower	60 adv, 25.3 SAE	46 adv, 23.44 SAE
Body Styles	12	five-passenger touring, two-passenger roadster, five-passenger two-door sedan, five-passenger four-door sedan
Other Features	three-oval-cluster instrument panel: an ignition switch and lighting switch in the left cluster; a speedometer, odometer and an eight-day Waltham clock in the center cluster; ammeter, oil pressure gauge, and a K-S Telegage fuel gauge in the right cluster	fuel gauge was at fuel tank, and instrument clusters were round instead of oval

The August and September 1924 Nash sales exceeded production capacity by 24,000. Thus, the second week of October 1924 was observed and celebrated by Nash dealers as National Oversold Week.

On May 26, 1925, the first Nash-built Ajax Six car was shipped to the dealers, and was officially introduced to the public on June 1, 1925. The Ajax Six was similar in appearance to the Nash Advanced Six by having a double belt molding, disc wheels, and balloon tires as standard equipment. The engine had a piston displacement of 169.65 cu. in., and featured a seven-main-bearing crankshaft with full-pressure lubrication to all vital moving parts. This was the first car below $1,000 to offer these features. The electrical equipment was manufactured by the Electric Auto-Lite Company, with the distributor mounted vertically above the cylinder head.

1925 Nash Special Six two-door sedan Model 133. (Source: Nash Motors Co.)

1925 Ajax Six

Wheelbase (in.)	108
Price	$865-995
No. of Cylinders / Engine	L-6
Bore x Stroke (in.)	3.00 × 4.00
Horsepower	40 adv, 21.6 SAE
Body Styles	five-passenger touring and four-door sedan
Other Features	mechanical front-wheel brakes, selective three-speed transmission with Borg and Beek single-disc dry-plate clutch, disc wheels with 30 x 4.95-in. balloon tires

The bodies were attractively styled with a one-piece windshield on both body types. The sedan roof had an integral sun visor, extending about 7 in. ahead of the windshield. The touring model had a rakish top and a gently sloping windshield. The fenders were heavily crowned with a concave flute following the crown line. The sedan upholstery was trimmed in velour, while the touring had Morroco leather. The hood had gently sloping flat top panels and narrow vertical louvres in the side panels. The nickel-plated radiator shell had gently sloping flat top panels, resembling the radiator shell of some of the very expensive cars. The headlamps were a modified drum shape, with nickel-plated headlamp doors,

1925 Nash Advanced Six Victoria Model 165. (Source: Nash Motors Co.)

While the Ajax Six was modestly priced, handsomely styled, and performed and handled well, representing an outstanding value, the buying public did not respond to the Ajax name. In the meantime Nash sales climbed to a record level. During the 1925 model year 73,384 new Nash cars and 4,157 new Ajax cars were registered in the United States. Nash and Ajax had a total output of 93,397 cars (including export shipments) and an earned profit of $16,256,216. On June 3, 1925, the Nash Motors Company common stock soared to an all-time high on Wall Street, closing at $445, culminating an advance of about 200 points in about two months.

John A. Rose, who single-handedly managed export sales during the early days of Rambler and Jeffery, stayed on with Nash Motors Company after the purchase of the Jeffery Company in July 1916. Under the Nash Motors administration he developed and expanded the export market to the extent that, by 1925, Nash Motors was enjoying export sales equivalent to 10% of their total production. Competitive automobile manufacturers recognized Rose's export management ability, and on October 15, 1925, the Willys-Overland Company succeeded in hiring Rose to head up their Export Division.

The new 1926 Nash Advanced Six models were introduced on June 1, 1925. The 1926 Nash Special Six models were introduced a month later, sporting a double belt molding, closely resembling the Advanced Six. The Special Six also offered a two- to four-passenger rumble-seat roadster. While the styling was generally improved, the most notable feature of the 1926 Advanced Six and Special Six models was the placement of the parking brake handle to the cowl left side, and the two-piece gearshift lever was designed so that the lower end was about 2-in. farther forward, allowing for easy entrance and exit through either front door. Sales of the 1926 models were brisk in late 1925. On November 1, 1925, the prices on some Nash models were cut by $35-$200, which must have helped sales.

On December 1, 1925, Nash introduced a new, larger engine for the Advanced Six. This new engine was called the Enclosed Car Motor. It featured a forged seven-main-bearing crankshaft. The cylinder bore size was increased to 3-7/16 in., retaining the 5-in. stroke, for an engine displacement of 278.4 cu. in. The Enclosed Car Motor was fully protected against dirt and moisture by an oil filter, fuel filter, and air cleaner at the carburetor air inlet.

The Nash Advanced Six also offered a two- to four-passenger rumble seat roadster. On January 1, 1926, Nash Motors Company announced a new, lower-priced, five-passenger, four-door sedan for $1,525. The United States registrations for 1926 showed that 98,804 Nash cars and 17,965 Ajax cars were delivered during the 1926 model year. Total Nash output for 1926 (including export sales) amounted to 135,520 cars.

1926 Nash Light Six Model 21 (former Ajax). (Source: Nash Motors Co.)

1927 Nash Advanced Six roadster Model 266. (Source: Nash Motors Co.)

Nash Motors earned a record profit of $23,346,306 in 1926, equivalent to earning approximately 56% of the company's net worth.

In analyzing the 1925 and 1926 sales figures, Charles Nash and Earl McCarty were convinced that the Ajax name did not add anything to Nash's stature and was not attracting customers. Even though the car was attractively styled, a quality product, and competitively priced, on June 1, 1926, they decided to change the Ajax name to Nash Light Six.

Sadly, James Storrow, who championed Charles Nash in all his business activities since the days at General Motors, died in March 1926. On July 1, 1926, Earl H. McCarty was elected to the board of directors of Nash Motors Company.

The 1927 Nash model introductions were staggered: the Nash Light Six was announced in early July 1926, the Nash Advanced Six on August 1, 1926, and the Nash Special Six on September 1, 1926. The Advanced Six and the Special Six featured a "Circassian" inlaid walnut instrument panel, as well as the crowned

1927 Nash Advanced Six seven-passenger sedan Model 264. (Source: Nash Motors Co.)

deflector panel above. The steering wheel was also of inlaid Circassian walnut. The instruments were grouped in a single panel under glass, with the Advanced Six featuring a hydrostatic K-S (King-Seeley) fuel gauge. Specially silversmith-crafted and silver-plated interior body handles and controls gave an air of elegance.

At the National Automobile Show in New York in January 1927, the new Nash Ambassador model on the Advanced Six chassis and the new Nash Cavalier model on the Special Six chassis made their debut. The contour of these bodies was patterned after European practice, particularly in the rear where the roof corners were inclined forward. The abrupt front corners were inclined slightly to the rear. A new convertible cabriolet on the Special Six chassis, priced at $1,290, and a new Sport Touring on the 127-in. wheelbase Advanced Six chassis, priced at $1,540, also made their debut at the National Automobile Show. Production and sales of the 1927 Nash models amounted to 109,979 cars in domestic sales, and calendar year production including export sales amounted to 122,606 units. The sales and production performance provided Nash Motors Company with earnings of $22,670,744.

Three new series of 1928 Nash cars offering 27 body types on four different chassis lengths were announced by Nash Motors Company on June 29, 1927. The former Nash Light Six had been superseded by the Nash Standard Six, powered by a six-cylinder L-head engine, whose bore was increased to 3-1/8 in. The prices were reduced on two body types of the Standard Six and five body types on the Advanced Six chassis. The Advanced Six and the Special Six closed models featured the new cadet-type sun visor, while the Standard Six continued with the protruding extended roof visor. The Standard Six radiator shell shape was similar to that of the Advanced Six and Special Six shapes. The nickel-plated radiator filler cap was adorned by a pair of small "angel wings." The road wheel diameter of all models was reduced to 20 in.

On December 3, 1927, Nash Motors Company introduced a new seven-passenger Imperial Sedan model on the 127-in. wheelbase Advanced Six chassis. This luxury model doubled as a seven-passenger sedan and a limousine by a sliding glass partition separating the driver's compartment from the auxiliary and rear seat passengers. This attractive seven-passenger model priced at $2,165 provided the Nash dealers with a competitive product to enter the profitable (chauffeur-driven) car market.

Nash Motors certainly had a profitable and prosperous 1928, producing and selling a record number 115,172 domestic cars, with a calendar year production of 138,137 units. This performance resulted in a net income of $20,820,085. While this represented a slight decrease from the record income of $22,670,744 reported for 1927, it was due to the expenditures for the additional buildings, equipment, tooling, and factory changes needed for the completely redesigned 1929 models.

1928 Nash Standard Six coupe Model 325. (Source: Nash Motors Co.)

1928 Nash Special Six Landau sedan Model 338. (Source: Nash Motors Co.)

The 1929 Nash 400 Series models were introduced on June 21, 1928, consisting of the Standard Six, Special Six, and Advanced Six. Without changing cylinder dimensions, the top speed capabilities of the three model lines was increased to 70, 75, and 85 mph, respectively, by increasing the compression ratio to 5 to 1. The Advanced Six and the Special Six valve-in-head engine performance was aided by a twin-ignition system consisting of two sets of ignition contacts, two ignition coils, two sets of spark plug cables, and twelve metric spark plugs. The two spark plugs per cylinder were located one in the cylinder head and one in the cylinder combustion chambers. Auto-Lite electrical and ignition systems were used on all 1929 Nash models.

The chassis wheelbases were 112.25 in. for the Standard Six, 116 and 122 in. for the Special Six, and 121 and 130 in. for the Advanced Six. The longer wheelbases on the Special Six and the Advanced Six were to accommodate the seven-passenger body types. The Advanced Six featured Houdaille shock absorbers and the BiJur one-shot chassis lubrication. The Special Six and Standard Six had Lovejoy shock absorbers and Alemite chassis lubrication. The Advanced Six and Special Six employed a new transmission with a close ratio (faster speed) in second gear operation, which was also considerably quieter. The gearshift was relocated to the center of the toeboard, mounted in the sloping portion of the clutch housing, to allow for more room in the driver's compartment.

The Nash 400 Series styling was distinctively new, with long, sweeping, clamshell front fenders, and a new, higher, restyled, chromium-plated radiator shell meeting a new center peaked hood with vertical-louvred hood side panels. The single belt molding was of the applied style, extending from the radiator shell, along the hood, the cowl side panels, and under the window reveals to around the rear of the body. A new cadet-type sun visor was used on all closed models.

1929 Nash Special Six roadster Model 436. (Source: Nash Motors Co.)

The chromium-plated headlamps were bullet-shaped, as were the cowl parking lamps (chromium-plated) mounted on a hoop molding across the cowl. The chromium-plated combination stop and tail lamp were mounted on a bracket extending rearward from the rear fender. The chromium-plated twin-bar spring steel bumpers were shaped to conform to the styling of car. The Advanced Six and Standard Six had a single row of slim vertical louvres on the hood side panels; the Special Six sported two rows. The Advanced Six and Special Six closed models were upholstered in pleated, tufted, velvet mohair cushion covers, while the roadster and phaeton model seats were upholstered in genuine leather. Twenty-four body types were available in the 1929 Nash 400 Series, priced from $885 for the Standard Six coupe to $2,190 for the Advanced Six seven-passenger limousine. The 1929 domestic sales were 105,146 cars, while the calendar year production was 116,622 units. Sales for the 1929 model year might have been better had it not been for the stock market crash. However, in the final accounting, Nash Motors Company showed a profit of $18,013,781 for 1929.

McCarty Retires

On February 4, 1929, Nash Motors Company announced the retirement of Earl H. McCarty, director of sales and member of the board of directors. While the news came as a surprise to the automotive industry, it had been known for some time in the Nash organization that McCarty had been planning his retirement since 1927. In contemplation of McCarty's retirement, the sales personnel had been strongly augmented. Thus, the position was not filled and the broadened sales activity was carried by C. Hascall Bliss, sales manager. After retirement, McCarty made his home in the Northwest.

The 1930s

Three new lines of 1930 Nash cars were announced on October 4, 1929.

	1930 Nash Single Six	1930 Nash Twin-Ignition Six	1930 Nash Twin-Ignition Eight
Wheelbase (in.)	114.25	118/128.5	124/133
Price	$985	$1,385-1,695	$1,695-2,085
No. of Cylinders / Engine	L-6	6	IL-8
Bore x Stroke (in.)	3.12 × 4.37	3.37 × 4.50	3.25 × 4.50
Horsepower	60 adv, 23.4 SAE	74.5 adv, 27.3 SAE	100 adv, 33.8 SAE
Body Styles	9	9	9
Other Features		BiJur centralized chassis lubrication system, non-shatterable glass in all windows, and starter control button on instrument panel	

The most notable feature of the new straight-eight engine was the nine-main-bearing counterbalanced crankshaft. Lynite forged-aluminum connecting rods were used, and an engine oiling heat exchanger (oil cooler) was incorporated. The performance of the Nash Twin-Ignition Eight was so phenomenal that Nash Motors could proudly boast "0 to 80 mph in eight blocks." The Twin-Ignition Six was powered by a six-cylinder valve-in-head engine with twin ignition and a seven-main-bearing crankshaft. The Single Six engine had a seven-main-bearing crankshaft. While the general appearance of the 1930 Nash models was similar to the 1929 Nash 400, there were many refinements. The Twin-Ignition Eight and Twin-Ignition Six featured a narrower rim radiator shell to accommodate the thermostatically controlled radiator shutters.

1930 Nash Twin-Ignition Six sedan, with Tom Mix. (Source: Nash Motors Co.)

On February 5, 1930, Walter H. Alford, vice president and comptroller of Nash Motors Company, died at his home after an illness of only two days. Alford, affectionately known as the Judge, pursued his industrial career in St. Louis, Missouri, and in 1910 joined the General Motors Company as comptroller. When Charles Nash left General Motors and formed the Nash Motors Company in 1916, Alford was provided with the opportunity to direct the financial destiny of Nash Motors Company as comptroller and treasurer. He discharged his responsibility so well that he had been credited with the success and financial stability of Nash Motors Company.

While "Black Thursday" (October 24, 1929) and the subsequent Depression took its toll on all businesses, Nash Motors Company was able to weather the storm. Through the fruitful years of operation, by prudent management, frugality, and wise spending within reasonable limits, Nash Motors Company was able to accumulate a sizeable amount of cash surplus.

Thus, on April 14, 1930, Nash Motors Company was able to declare a $1.50 per share dividend, payable on May 1, 1930. Charles Nash stated that the accumulation of cash surplus rightfully belonged to the stockholders and in his judgment this was the right time to distribute it. During the first quarter of 1930 (December 1929, January and February 1930) Nash Motors Company earned $1,782,512. These earnings, plus the cash surplus, made the dividend possible. At this time Nash Motors Company had approximately 3,000 dealers.

McCarty Returns

By earnestly soliciting, Charles Nash was able to persuade Earl McCarty to return to Nash Motors Company. At the board of directors meeting on April 14, 1930, McCarty was elected vice president, general manager, and director of the Nash Motors Company. Grinning, Nash quipped, "I knew he (McCarty) could not stay retired very long." During 1930 Nash Motors Company built and sold 51,086 cars domestically, and the calendar year production was 54,605 units, resulting in a net profit of $7,691,164 for 1930.

On October 9, 1930, Nash Motors Company announced the four 1931 model lines.

	1931 Nash 6-60	1931 Nash 8-70	1931 Nash 8-80	1931 Nash 8-90
Wheelbase (in.)	114	116	121	124/133
Price	$845	$995	$1,295	$1,925
No. of Cylinders / Engine	L-6	L-8	IL-8	IL-8
Bore x Stroke (in.)	3.12 × 4.37	2.87 × 4.37	3.00 × 4.25	3.25 × 4.50
Horsepower	65 adv, 23.4 SAE	75 adv, 26.4 SAE	87.5 adv, 28.8 SAE	115 adv, 33.8 SAE
Body Styles	5	5	5	9
Other Features	mechanical four-wheel brakes, and 19-in.-diameter road wheels			

Mechanically the refined twin-ignition valve-in-head 8-90 engine continued the use of the Lynite forged-aluminum-alloy connecting rods. The engines of the other models used forged-steel connecting rods, rifle-drilled for piston pin lubrication. Marvel downdraft carburetors were used on the 8-70 and 8-80 engines, while a Stromberg dual-throat (two-bore) updraft carburetor was used on the 8-90 engine. For 1931 a mechanical fuel pump was used on all models replacing the vacuum tank on previous models. BiJur one-shot chassis lubrication was continued on the 8-70, 8-80, and 8-90 models.

The sales of 1931 Nash models dipped to 39,366 units domestically, a drop of about 23% below the 1930 sales. Despite the depressed economy and fewer sales, by continued strict cost control, Nash Motors Company was able to post a profit of $4,807,680 for 1931. Only two other automobile manufacturers showed a profit in 1931. In October 1931, Nash Motors Company elevated chief engineer Nils E. Wahlberg to vice president of engineering.

Production of the 1932 Ninth Series Nash models began on June 28, 1931, even though the public announcement was somewhat earlier. The model designations of 9-60, 9-70, 9-80, and 9-90 were carried over until February 27, 1932. The appearance of the 9-70, 9-80, and 9-90 models were characterized by the new V-shaped radiator shell and the new V-shaped grille of vertical louvres. The radiator shells were finished in body color lacquer, accentuated by chromium molding. The 9-60 radiator shell was chromium-plated, with a vertical chrome trim bar down the front center.

The bullet-shaped headlamps were chromium-plated, as were the parking lamps now mounted atop the front fenders. Chromium-plated single-bar bumpers were used, front and rear. The attaching bolts of the demountable wood-spoke wheel mounting were concealed by cover plates. The rear end body panel was concave at the lower edge, giving a very streamlined beavertail effect.

1931 Nash Model 881 sedan.
(Source: Nash Motors Co.)

First 1932 Nash Big Six sedan Model 1060 ever built, with: James Wilson (second from left); Nils Wahlberg (third from left); Meade Moore (by driver's door); Earl McCarty (second from driver's door); and Charles Nash at right. (Source: Nash Motors Co.)

Technically, the engine dimensions remained the same as the 1931 models, except the power output was increased on the 9-70 and 9-80 engines through the use of a Stromberg dual throat (two-barrel) downdraft carburetor. The new transmissions used throughout the entire Nash line were of the synchro-mesh silent-second type, employing the use of helical cut gears. The use of the oil temperature regulator (heat exchanger) had been extended across all lines of Nash cars.

On February 27, 1932, further important changes and great improvements were made on the Tenth Series Nash models. The model designations were changed to: 1060 - Big Six, 1070 - Standard Eight, 1080 - Special Eight, and 1090 -Advanced Eight and Ambassador Eight. The 1090 Advanced Eight wheelbase was 133 in., while the 1090 Ambassador Eight wheelbase was 142 in. The Nash Models 1070, 1080, and 1090, in addition to having the Ninth Series improvements, had the following additional features: synchro-mesh three-speed transmission with free-wheeling, underslung silent worm drive rear axle, rubber isolated spring

Charles Nash and Mayor Anton Cermak with a 1932 Nash Model 994. (Source: Nash Motors Co.)

shackles, dual resonator exhaust mufflers, full range ride regulator, remote-control hood latches, inside sun visors, aircraft-style instruments, twin glove compartments in instrument panel, twin windshield wipers, twin chromium-plated trumpet-type horns, and five adjustable ventilating doors in the hood side panels. The Tenth Series models were available in 27 body types, priced from $777 for the 1060 Big Six coupe to $2,065 for the 1090 Ambassador seven-passenger limousine.

Despite all the fine features and ultimate quality, Nash domestic sales amounted to a dismal 20,233 cars and the calendar year production was only 17,696 units. But Nash Motors Company made the best of a poor market: a 1932 second quarter net income of $534,208, a net profit of $183,981 for the third quarter, and a 1932 fiscal year earning of $1,029,500. At a time of severe business conditions caused by the cruel Depression, Nash Motors Company was the only other automobile manufacturer besides General Motors to operate on a profitable basis in 1932.

The 1933 Nash models were announced on December 1, 1932, and production started on December 15, 1932. A new model was identified as the Nash Standard Eight Model 1130.

1933 Nash Standard Eight Model 1130

Wheelbase (in.)	116
Price	$830-900
No. of Cylinders / Engine	L-8
Bore x Stroke (in.)	3.00 × 4.37
Horsepower	85 adv, 28.8 SAE
Body Styles	5
Other Features	

This model addition gave Nash five model lines: Big Six, Standard Eight, Special Eight, Advanced Twin-Ignition Eight, and Ambassador Twin-Ignition Eight, identified by model numbers 1120, 1130, 1170, 1180, and 1190, respectively, with 31 body types in a price range of $695 to $2,095.

First Operating Loss

Even with the ultimate in styling and motor car value, and the proliferation of body styles and models, Nash's 1933 domestic sales amounted to a dismal 11,353 cars, with a calendar production of only 14,493 units. As a matter of business interest, according to the Federal Reserve Board Indexes of 1933, the industries operating under the NRA Blue Eagle sustained an employment increase of 4% during September 1933, with an accompanying 17% decline in production. Needless to say Nash Motors Company sustained the first operating loss in its history.

The new 1934 Nash models, introduced in October 1933, consisted of three model lines: Big Six, Advanced Eight, and Ambassador Eight, on wheelbase lengths of 116, 121, and 133/142 in., respectively. The most significant change was the reinstatement of the Big Six valve-in-head, twin-ignition, six-cylinder engine of 3-3/8-in. bore and 4-3/8-in. stroke. Welded, pressed-steel, artillery-type wheels were standard equipment. The 1934 Nash models were styled by the famous designer, Count Alexis De Sakhnoffsky.

Stylewise, Nash was not caught up in the sudden maelstrom of streamlined styling trends. They fortunately did not give up the classical beauty they attained in the 1932 and 1933 models, but improved on it. The significant changes were the skirted front fenders that extended downward in the front to the bumper level to conceal exposed chassis components.

The longitudinal, embossed speed ribbons on the hood front carried out the "Speedstream" styling motif. The hood sides had six, horizontal, adjustable, ventilating doors on the Nash Eight models and four on the Big Six models. The parking lamps were fared into the front fenders as were the stop and tail lamps into the rear fenders. The beavertail rear body panels were continued and extended farther rearward.

The buying public reacted very favorably to the 1934 Nash model lines. However, Nash Motors Company was forced to shut down the Kenosha assembly operations due to a walkout of assembly workers. This put 3000 people out of work for an indefinite period. After successful negotiations, the workers returned to their jobs.

At the National Automobile Show in New York in January 1934, Nash Motors Company introduced the new LaFayette. The LaFayette resembled the larger Nash models except for the front grille and the LaFayette badge in place of the Nash crest.

	1934 LaFayette
Wheelbase (in.)	113
Price	$635-695
No. of Cylinders / Engine	L-6
Bore x Stroke (in.)	3.25 × 4.37
Horsepower	75 adv, 25.3 SAE
Body Styles	5
Other Features	

In April 1934 Nash Motors Company supplemented the LaFayette with two lower-priced body types: a two-door sedan at $595 and a four-door sedan at $645. These two models were intended to compete head-on with Chevrolet, Ford, and Plymouth.

Charles and Jessie Nash on their 50th wedding anniversary, April 23, 1934. (Source: J.A. Conde)

On April 23, 1934, Charles and Jessie Nash celebrated their Golden Wedding Anniversary, and on April 27, 1934, Charles W. Nash drove the 1,000,000th Nash car off the assembly line in Kenosha. The historic vehicle was a Nash Twin-Ignition Big Six sedan.

Despite the favorable buyer response, with sales of 23,616 cars domestically (double that of 1933) and a calendar year production of 28,664 units, the economic conditions and the work stoppages caused Nash Motors Company to sustain a loss of $1,625,078 on a dollar volume sales of $19,679,777.

Charles Nash with the 1,000,000th Nash car in 1934. (Source: Nash Motors Co.)

Charles Nash and Earl McCarty with the 1,000,000th Nash car. (Source: Nash Motors Co.)

1935 LaFayette sedan. (Source: Nash-Kelvinator Co.)

The 1935 Nash and LaFayette models were introduced on January 1, 1935.

	1935 Nash Advanced Six	**1935 Nash Advanced Eight and Ambassador Eight**	**1935 LaFayette**
Wheelbase (in.)	120	125	113
Price	$945	$1,165-1,290	$585-750
No. of Cylinders / Engine	6	IL-8	L-6
Bore x Stroke (in.)	3.37 × 4.37	3.12 × 4.25	3.25 × 4.37
Horsepower	88 adv, 27.3 SAE	100 adv, 31.25 SAE	80 adv, 25.3 SAE
Body Styles	four-door sedan and two-door Victoria		5
Other Features	hydraulic four-wheel brakes		mechanical four-wheel brakes

The LaFayette's conventional radiator shell was eliminated, and the hood extended forward to blend with the sloping radiator grille. The hood sides had three horizontal louvres on each side. All Nash and LaFayette had Startix automatic starting, controlled by clutch pedal operation.

In May 1935 the new Nash 400 model was introduced. The unusual styling design was that the front grille, hood sides, and hood top panel formed a single assembly; hinged at the rear it gave easy access to the engine.

	1935 Nash 400
Wheelbase (in.)	117
Price	$675-790
No. of Cylinders / Engine	L-6
Bore x Stroke (in.)	3.37 × 4.37
Horsepower	90 adv, 27.3 SAE
Body Styles	6
Other Features	all-steel construction, with one-piece steel-stamped roof

Nash and LaFayette domestic sales amounted to 35,184 cars, with a calendar year production of approximately 44,600 units. This resulted in a reported loss of $610,227 for fiscal year 1935. The Nash 400 was the forerunner of the Nash and LaFayette models for 1936 which were announced on October 15, 1935.

1936 Nash eight-cylinder touring sedan. (Source: Nash-Kelvinator Co.)

Generally, the 1936 specifications were the same as for the corresponding models in 1935. The Nash Ambassador Six chassis wheelbase was increased to 125 in. in order to be able to use common chassis components with the Nash Ambassador Eight. The styling for 1936 was in keeping with the trend of the automotive industry, including chromium-plated die-cast grilles and hood louvres, and hinging the hood at the rear. The LaFayette adopted hydraulic four-wheel brakes.

The purchasers of 1936 Nash sedans had a choice of a sloping fastback body rear end or a built-in trunk. The Nash 400 and LaFayette sedan offered a full-size convertible bed arrangement, utilizing the rear seat and trunk space. Domestic sales for 1936 amounted to 43,070 cars and the calendar year production was 53,038 units, resulting in a net profit of $1,020,707.

The balance of Seaman Body Corporation interests was purchased by Nash Motors Company in 1936. On August 1, 1936, Nash Motors Company celebrated its 20th anniversary. Over 1,100,000 Nash cars had been built and sold since the founding of Nash Motors Company in 1916.

In July 1936 Earl McCarty announced his plans for retirement again. The effective date was set for November 30, 1936, to give Charles Nash a chance to locate and select a successor.

Mason and Nash-Kelvinator

George W. Mason was recommended to Nash by Walter P. Chrysler. Meetings and negotiations between Nash and Mason went on during 1936, and the final arrangement, resulting in the merger of Nash Motors Company and Kelvinator Corporation, was announced on October 27, 1936. The Nash-Kelvinator Corporation was formed on January 4, 1937, with George W. Mason president and Charles W. Nash chairman of the board.

George Mason was born on March 12, 1891, in Valley City, North Dakota. He earned his B.S. degree at the University of Michigan in 1913, and entered the automobile industry at Studebaker in South Bend, Indiana. He served in a management position at

George Mason. (Source: Nash-Kelvinator Co.)

1937 Nash Ambassador Eight sedan. (Source: Nash-Kelvinator Co.)

Dodge Brothers from 1915 through 1918, and at the Irving National Bank in New York in 1919 and 1920. He was with Walter Chrysler from 1921 through 1925. He was president of Copeland from 1926 to 1928 and president of Kelvinator Corporation from 1926 through 1936.

As president of Nash-Kelvinator Corporation, Mason assigned Nash engineers to work with the Budd Manufacturing Company engineers to develop a new lightweight car, the Nash 600, which would be introduced in the fall of 1940. The new 1937 Nash models were announced on October 1, 1936, consisting of three lines: the Nash Ambassador Eight, Nash Ambassador Six, and the Nash-LaFayette 400, a combination of the LaFayette and the Nash 400.

The styling of the 1937 models was similar to that of 1936 with die-cast grilles. The headlamps were now mounted on the radiator cradle in the forward area of the hood side panel. A new three- to five-passenger cabriolet was added to each series line. Fifteen body styles were offered priced at $650 to $960. Nash sold 70,571 cars domestically with a calendar year production of 85,949 units. A net profit of $3,640,747 was reported for the fiscal year of 1937. Nils E. Wahlberg, vice president of engineering, passed away on October 23, 1937.

The new 1938 Nash models were shown on October 15, 1937, consisting of three lines: Nash Ambassador Eight, Nash Ambassador Six, and the LaFayette. The Nash 400 designation was dropped from the LaFayette models. The styling and specifications were similar to the 1936 models. New aircraft-type telescoping hydraulic shock absorbers were used, arranged in a "sea leg" fashion. A vacuum-powered gearshift control, with the short lever located on the instrument panel, was offered as optional equipment. Eighteen body types were offered, priced at $770 to $1,240.

Because of the business downturn as the result of a full-blown recession, only 31,814 cars were sold domestically during 1938, less than one-half of the 1937 sales. The calendar year production was only 32,017 units as the result of work stoppages. Thus, Nash-Kelvinator Corporation sustained a loss of $7,655,138 for fiscal 1938.

The new 1939 Nash models styled by George W. Walker were introduced on October 15, 1938. Walker certainly possessed the necessary credentials, having previously designed the famous Weymann and Walker LeGrande custom bodies for the finest American cars. The new 1939 Nash models featured a very narrow radiator grille, with zinc-chromium-plated die-cast horizontal grille bars. There was a wide catwalk between the hood sides and the front fenders, with an opening at the forward area embellished by chromium-plated

1938 Nash Ambassador Eight cabriolet. (Source: Nash-Kelvinator Co.)

vertical grille bars. The headlamps that mounted flush at the forward area of the massive front fenders were blended into the fender contour. The hood was hinged at the rear, tapering to meet the narrow front grille.

The rear end of the bodies was available with the conventional trunk or the streamlined fastback. The full-size double bed arrangement was optional in the sedan body types. Nash pioneered the new, improved, conditioned air heating system, including an automatic temperature control known as the "Weather Eye." Column-mounted gearshift lever was standard on all models. Nineteen body types were available priced at $770 to $1,290.

Nash domestic sales amounted to 54,050 cars, and the calendar year production was 65,662 units, nearly double that of 1938. While the final shipments exceeded those of a year earlier, labor interruptions delayed the introduction of the 1939 models to the extent that practically no cars were shipped during October 1938. The result was a reported net loss of $1,573,524 for the fiscal year ending September 30, 1939.

The 1940s

The 1940 Nash was announced on September 15, 1939, consisting of three series: the Nash Ambassador Eight, Nash Ambassador Six, and the LaFayette Six. Available in 18 body styles, the price range was from $795 to $1,295. The styling and technical features were basically a carryover of the 1939 models. The most noticeable change was the finer horizontal grille bars in the narrow grille center. The forward portion of the catwalks had a larger, flatter opening supporting a grille of slender chromium-plated vertical louvres. The sealed beam headlamps introduced on the 1940 Nash models were fared into the forward area of the front fenders.

The 1940 Nash models were well-received by the buying public, resulting in domestic sales of 52,853 cars, and a calendar year production of 63,617 units. At the end of the 1940 fiscal year, September 30, 1940, Nash-Kelvinator reported a net profit of $1,505,151.

The 1941 Nash models were introduced on October 1, 1940, offering three lines of cars: Nash Ambassador Eight, Nash Ambassador Six, and the Nash Ambassador 600. The Ambassador 600 marked Nash's re-entry into the low-priced field. The 600 designation represented the distance the fully loaded car could travel on a tank (20 gal) of gasoline. The LaFayette model designation was dropped.

1941 Nash Ambassador 600 touring sedan. (Source: Nash-Kelvinator Co.)

Stylewise, the 1941 Nash was a carryover of the 1940 models, except the narrow front center grille was omitted, replaced with a narrow chromium-plated molding. The new catwalk grilles were continued, and two horizontal grilles, having five stainless-steel louvres each, were located around the front, below the catwalks, above the bumper, and extending around the front fender to the wheel opening.

The technical specifications were the same as the corresponding 1940 models. However, the Nash Ambassador 600 pioneered the unitized frame and body construction, which eventually was adopted by all automobile manufacturers. The 600 also featured four-coil spring suspension.

During 1941 Nash-Kelvinator Corporation received a number of United States Government contracts for the production of defense materiel. The Racine, Wisconsin, plant produced tens of thousands of cargo trailers. Hamilton-Standard aircraft propellers were built by Nash-Kelvinator in the former Reo truck plant at Lansing, Michigan.

Nash-Kelvinator experienced a very productive and profitable 1941. In addition to the modest profit on the production of defense materiel, Nash domestic sales of 77,824 cars, and a calendar year production of 80,408 units, resulted in a net profit of $4,617,052 for fiscal year 1941.

The 1942 Nash cars were announced on October 1, 1941, with the same models and similar styling as 1941, except that the catwalk grilles were omitted, replaced by a single broader grille of six stainless-steel horizontal louvres. The four, long, horizontal, stainless-steel moldings extended across the entire front, just above the bumper to the wheel openings. Fifteen body types were available priced at $918 to $1,209.

While the 1942 Nash cars were gaining public acceptance rapidly, the attack on Pearl Harbor on December 7, 1941, signaled the end of car production on January 31, 1942. Thus, approximately 31,700 Nash cars were sold during the model run, but only 5,428 Nash cars were built during January 1942. By this time all Nash-Kelvinator plants in Kenosha, Racine, Milwaukee, Detroit, Lansing, and Grand Rapids were totally producing war materiel. Aircraft engines, propellers, helicopters, flying boat sub-assemblies, cargo trailers, bomb fuses, and other war-related items, valued at approximately $600,000,000, were manufactured by Nash-Kelvinator Corporation from 1941 to 1945.

1946 Nash Ambassador. (Source: Nash-Kelvinator Co.)

After V-J Day in the fall of 1945, Nash-Kelvinator Corporation was able to convert quickly from production of war materiel back to car production. Despite the material shortages, Nash-Kelvinator was able to build 6,148 cars during 1945. Stylewise, the Nash Ambassador and the Nash 600 models were a carryover of the 1942 models. The main change included a wider center grille, and the parking lamps were moved to the front of the catwalk next to the headlamps.

Priced at $1,295 to $1,930, Nash was able to sell 85,169 cars domestically with a calendar year production of 98,769 units. During the 1946 model year, Nash introduced a four-door suburban model with wood panels on the door sides and quarter panels. About 1,000 suburbans were built and sold during 1946-48.

In May 1946 the American automotive industry celebrated its Golden Anniversary in Detroit, Michigan, where Charles W. Nash and eleven other automotive pioneers were honored.

The 1947 and 1948 Nash models were carried over from 1946 with minor changes and a slight price increase. Ranging from $1,415 to $2,225 in 1947, the prices were raised to $1,475 to $2,240 in 1948. Nash-Kelvinator Corporation established assembly plants in El Segundo, California, and Toronto, Ontario, Canada, in 1947. The 1947 Nash 600 had half-closed rear fenders, partially concealing the rear wheels. A 1947 Nash Ambassador sedan was chosen as the pace car for the 1947 Indianapolis 500, with George W. Mason driving. Domestic sales in 1947 were 102,808 cars and 104,156 cars in 1948. Calendar year production was 113,315 units in 1947 and 118,621 units in 1948.

Charles Nash Dies

Sadly, on June 6, 1948, Charles W. Nash passed away at age 84. George W. Mason succeeded him as chairman of the board while remaining president. George W. Romney was elevated to executive vice president. Romney joined the Nash-Kelvinator Corporation as assistant to president Mason after a successful career as managing director of the Automotive Council for War Production, and later as director of the Automobile Manufacturers Association of America.

In October 1948 the new 1949 Nash Airflyte models were introduced, successfully styled toward aerodynamics. In wind tunnel tests of 60 mph, the new car had only 113 pounds of surface drag. Unitized single-unit

R.E. Olds and Charles Nash in 1946. (Source: P.M.C. Co.)

1948 Nash Ambassador Super 4860. (Source: Nash-Kelvinator Co.)

frame and body construction was used on all models, as well as a one-piece curved safety glass windshield. The Uniscope single instrument cluster was mounted on the steering column. The half-closed rear fender wheel opening was extended to the front fenders, where the upper half of the wheel and tire were concealed. The Nash 600 and Nash Ambassador models were offered in the Super, Super Special, and Custom lines, with three body styles in each line. The price range was from $1,810 to $2,365. In 1949, 104,156 cars were sold domestically on a calendar year production of 142,592 units.

1949 Nash Super 600 four-door. (Source: Nash-Kelvinator Co.)

The 1950s

The production of the 1950 Nash models began in September 1949. The Nash 600 was renamed the Nash Statesman, while the Nash Ambassador series remained the same as the 1949 models.

	1950 Nash Statesman	1950 Nash Ambassador	1950 Nash Rambler
Wheelbase (in.)	112	121	100
Price	$1,738	$2,046	$1,808
No. of Cylinders / Engine	L-6	6	L-6
Bore x Stroke (in.)	3.12 × 3.75	3.37 × 4.37	3.12 × 3.75
Horsepower	82 adv, 23.4 SAE	112 adv, 27.3 SAE	82 adv, 23.4 SAE
Body Styles	7	6	6
Other Features	optional reclining front seat equipped with safety seat belts		

A Super line and a Custom line were available, offering three body styles: a five-passenger two-door brougham, a six-passenger two-door sedan, and a six-passenger four-door sedan in each series. The Nash Ambassador offered the Hydramatic Drive automatic transmission as an extra-cost option. The new compact Nash Rambler convertible made its debut on April 14, 1950. Nash domestic sales for 1950 amounted to 175,722 cars, with a calendar production of 189,543 units.

The 1951 Nash models introduced on September 22, 1950, featured the return of the sloping fastback styling on the Statesman and Ambassador four-door sedan models. On June 28, 1951, Nash added a third body type to the Rambler line, a five-passenger two-door hardtop called the Country Club Coupe. The price range of the entire Nash model line-up was from $1,840 to $2,500.

In 1951 Nash-Kelvinator Corporation obtained a defense contract from the United States Air Force to build Pratt and Whitney aircraft engines. The Korean War effort placed many restrictions on critical materials such as nickel and other vital metals. Despite the material shortages Nash-Kelvinator Corporation was able to build 161,209 cars. Domestic sales amounted to 140,025 units, resulting in a profit of $16,220,173 for 1951.

March 14, 1952, marked the introduction of the totally new 1952 Nash models, celebrating the 50th anniversary of the first Rambler car built by the Thomas B. Jeffery Company, predecessor of the Nash Motors Company.

	1952 Nash Statesman	1952 Nash Ambassador	1952 Nash Rambler
Wheelbase (in.)	114.25	121.25	100
Price	$2,178	$2,557	$2,003
No. of Cylinders / Engine	L-6	6	L-6
Bore x Stroke (in.)	3.12 × 4.25	3.50 × 4.37	3.12 × 3.75
Horsepower	88 adv, 23.4 SAE	120 adv, 29.4 SAE	82 adv, 23.4 SAE
Body Styles	7	5	5 including Country Club Coupe
Other Features	rear seat width increased by 12-1/2 in.		

The new Nash Statesman and Nash Ambassador models were beautifully styled by the famous Italian designer, Pinin Farina. The sheet metal was all new. The belt line was moved up, and the window lower edge was moved down to meet the belt line. This provided a pleasing appearance by eliminating the roll or bulge on the body side panels. The half wheel openings were continued at the front fenders and the rear quarters. The hood was lowered, providing improved eye level vision and creating a unique design of the windshield lower corners. The broad rectangular grille consisted of 20 chromium-plated die-cast vertical bars, and the outer frame had rounded corners. The front fenders extended further forward, placing the headlamps ahead of the grille. Solex tinted safety glass was available as an option on the Ambassador models. The Nash Rambler Country Club Coupe sported as an option the Continental spare tire carrier in the rear. Sales amounted to 142,442 cars, with a 1952 calendar year production of 152,141 units.

The handsome 1952 Nash Healy Sports Roadster made its debut in late 1951. Built in England by the Donald Healy shops, it was powered by a Nash valve-in-head six-cylinder engine, having a bore of 3-3/8 in. and a stroke of 4-3/8 in., developing 125 hp. The fuel induction system included two Dual Jet Fire side draft carburetors. Among the unique styling features was the location of the headlamps within the front grille frame. There were about 100 Nash Healy Sports Roadsters built in 1951, and about 150 in 1952.

The 1952 Nash Pinin Farina styling was continued through 1953 unchanged. The Nash Healy LeMans Dual Jetfire engine was provided as an option on the 1953 Nash Ambassador models. The Pinin Farina styling was also extended to the Nash Rambler, except the front grille had a single, horizontal, chromium-plated bar. Five body styles were available in each of the series of the Nash Rambler, Nash Statesman, and Nash Ambassador models. Priced at $2,005 to $2,830, Nash domestic sales amounted to 137,350 cars in 1953, on a calendar year production of 153,753 units. Thus, Nash-Kelvinator Corporation reported a net profit of $14,123,026 for fiscal year 1953.

The 1954 Nash Ambassador and Statesman models announced in the fall of 1953 were in fact a carryover of the Pinin-Farina-styled Golden Anniversary models of 1952 and 1953. The 1954 grille had concave, vertical, chromium-plated bars. The bodies featured new interiors and instrument panels. All Custom models had the Continental spare tire carrier as standard equipment. The price range of the Nash Statesman and Nash Ambassador models was $2,110 to $2,425.

The new 1954 Nash Rambler models had all-new styling similar to that of the Nash Statesman and Nash Ambassador models. Ten body styles on 100- and 108-in. wheelbases were priced from $1,550 to $2,050. The engine and technical specifications were the same as those for the 1953 Nash Rambler models. The Statesman and Ambassador technical specifications were the same as those for the 1953 models, except the

Statesman engine output was increased to 110 hp, and the Ambassador engine developed 130 hp. While the sales of the Nash Statesman and Nash Ambassador dropped, the Nash Rambler sales increased to over 36,000 cars.

The new Nash Metropolitan, a two-passenger, sub-compact roadster model built in Great Britain by Austin and Fisher-Ludlow, was introduced in March 1954. It had an 85-in. wheelbase, powered by a 42-hp four-cylinder-engine, averaging 40 mpg. A two-door hardtop model was also offered.

American Motors Corporation

In early spring of 1954, negotiations were in progress between Abraham E. Barit, president of Hudson Motor Car Company, and George W. Mason, president of Nash-Kelvinator Corporation. As the result of these meetings, on May 1, 1954, the Hudson Motor Car Company and Nash-Kelvinator Corporation merged to form the American Motors Corporation.

On June 22, 1954, after months of meetings and negotiations, an agreement to merge the Packard Motor Car Company and Studebaker Corporation was signed by James J. Nance of Packard and Harold C. Vance of Studebaker. The effective date of the merger was October 1, 1954. (See the Hudson, Packard, and Studebaker chapters in this volume.)

Between May 1 and October 1, 1954, Mason negotiated with Packard Motor Car Company and the Studebaker Corporation, exploring the possibility of bringing them into the American Motors combine. An unexpected blow came to the American Motors Corporation on October 8th with the sudden death of George W. Mason after an illness of only five days. Mason was 63 years of age. George W. Romney was elected president and chairman of the board of American Motors Corporation on October 12, 1954.

(l to r) Abraham Barit, George Mason, and George Romney. (Source: American Motors Corp.)

Romney had different plans for American Motors Corporation, and those were to build small economical cars. In view of the Rambler's excellent sales, Romney saw no need for large cars, nor Packard, nor Studebaker. Nash and Hudson sales and dealer organizations were consolidated.

Romney organized an experienced, aggressive management team. In November 1954, Roy Abernathy, with years of successful merchandising experience, was elected vice president and general sales manager. Fred W. Adams was appointed director of advertising and public relations. The domestic sales of Nash Statesman and Nash Ambassador models amounted to 41,116 cars, while Nash Rambler domestic sales were 35,613 cars in 1954.

The 1955 models of the Nash and Rambler cars were introduced in October 1954. Beginning in 1955 and through 1956, the Rambler could be purchased as a Nash or a Hudson, the only difference being the identifying badge (nameplate). The single, horizontal grille bar was replaced by six vertical and three horizontal grille bars, arranged in a pigeonhole or egg crate pattern. The Nash Statesman was offered with the popular Nash valve-in-head six-cylinder engine. The Nash Ambassador, in addition to the valve-in-head six-cylinder engine, offered a 208-hp valve-in-head V-8 engine equipped with Ultramatic Drive transmission purchased from the Packard Motor Car Company.

For 1955 the production of Hudson cars was terminated in Detroit and transferred to Kenosha, Wisconsin, and they were built using Nash unitized body components and styling. The Hudson Wasp and Hudson Hornet models were powered by the famous Hudson Hornet L-head six-cylinder engine. The 1955 Nash models were updated Pinin Farina styles, with a new wraparound curved windshield with offset vertical A-pillars. The rear quarters were restyled and the front fenders extended farther forward. The headlamps were placed within the radiator grille frame area, similar to the previous Nash Healey arrangement. Two body styles, a six-passenger four-door sedan and a six-passenger two-door hardtop, were available in each of the series: the Nash Statesman Six Series 40, the Nash Ambassador Six Series 60, and the Nash Ambassador V-8 Series 80, priced from $2,215 to $3,095. The Nash Rambler was available on 100-in. and 108-in. wheelbase lengths, with a total of seven body types offered. The price range was from $1,770 to $2,095.

A total of 37,192 Nash Statesman and Nash Ambassador models, and 72,227 Nash Rambler models were sold domestically during 1955. The calendar year production amounted to 96,156 units, of which about 56,000 cars were Nash Rambler models. American Motors Corporation reported an operating loss of approximately $6,900,000 for 1955.

The 1956 Nash models were presented in October 1955. The V-8 engine and Ultramatic Drive automatic transmission were built for Nash by the Packard Division of the Studebaker-Packard Corporation.

	1956 Nash Statesman	1956 Nash Ambassador Series 60	1956 Nash Ambassador Series 80	1956 Nash Rambler
Wheelbase (in.)	114.25	121.25	121.25	110
Price	$2,345	$2,644	$2,956	$1,795
No. of Cylinders / Engine	L-6	6	V-8	L-6
Bore x Stroke (in.)	3.12 × 4.25	3.50 × 4.37	4.00 × 3.50	3.12 × 4.25
Horsepower	130 adv, 23.4 SAE	135 adv, 29.4 SAE	220 adv, 51.2 SAE	120 adv, 23.4 SAE
Body Styles	four-door sedan	four-door sedan	six-passenger Super four-door sedan, six-passenger Custom four-door sedan, six-passenger Custom two-door hardtop model	7

On April 9, 1957, the American Motors Corporation announced its new British-built Nash Metropolitan, available as fabric-top or hardtop convertible models. Also in April, Nash announced the new Nash Ambassador Special, Series 50, powered by a new valve-in-head V-8 engine of 250-cu.-in. displacement, developing 190 hp. This new engine was of Nash design and manufacture. The 1956 Nash Rambler introduced a new six-passenger four-door hardtop sedan and a four-door hardtop station wagon. On March 27, 1956, the 2,000,000th car was built at the Kenosha, Wisconsin, plant.

Even with the proliferation of Nash models, the 1956 sales dropped to 25,271 cars domestically, while the Nash Rambler sales dropped only slightly to 70,867 cars. American Motors Corporation calendar year production amounted to 17,842 Nash cars, 79,166 Nash Rambler cars, and 7,182 Hudson cars. American Motors Corporation reported a loss of $19,746,244 for fiscal year 1956.

During 1957 American Motors Corporation discontinued merchandising the Rambler as a Nash or Hudson car. Instead, the Rambler came out on its own as a separate make. Only the Ambassador V-8 sedan was offered as a Nash model, powered by a 327-cu.-in. displacement V-8 engine of American Motors manufacture. On June 25, 1957, the last Hudson Hornet model was built at the Kenosha, Wisconsin, plant. After that, all sales and manufacturing effort was aimed at Rambler.

American sold 9,474 Nash cars, 4,596 Hudsons, and 91,469 Rambler cars domestically in 1957. The Nash and Hudson model lines were discontinued after 1957. Despite sales of over 105,500 cars, representing a dollar volume of over $362,000,000, the American Motors Corporation reported a loss of $11,833,200 for fiscal 1957.

The decision to build only the Rambler cars must have been a good one; because of Rambler's success, American Motors Corporation was able to turn around the long sustained losses of several years to a record net profit of $60,341,823 for the fiscal year ending September 30, 1959. A profit of approximately $26,000,000 had been reported for 1958.

Postscript

George W. Romney resigned as president and board chairman to campaign as a candidate for governor of Michigan. He was elected on November 6, 1962, and inaugurated on January 1, 1963. Roy Abernathy became president of American Motors Corporation and Roy D. Chapin chairman of the board. Abernathy retired in 1978 and was succeeded by Gerald Meyers as president and chief operating officer.

Through a series of acquisitions and mergers, the Willys Overland plants and facilities were purchased by Henry J. Kaiser in 1953 and renamed Willys Motors Incorporated. In March 1963 the Willys name was dropped and it became the Kaiser Jeep Corporation. The Kaiser Jeep Corporation was purchased by American Motors Corporation on February 5, 1970. Then, on August 5, 1987, American Motors Corporation was merged with the Chrysler Corporation, becoming the Jeep-Eagle Division of the Chrysler Corporation.

Index

Page numbers followed by *p* indicate a photo.

Abernathy, Roy, 194, 196, 373, 458, 459
Adams, Fred W., 196, 458
Aerodyne models by Hupp, 325*p*, 325–326, 326*p*
Aircraft/marine engines
    by Duesenberg, 277–278, 279*p*, 280, 280*p*
    Merlin engines, 183
    by Packard, 154*p*, 154–156, 162*p*, 162–163, 180, 183, 184
    by Studebaker, 48, 48*p*, 50
Alborn, F.G., 245
Alford, Walter H., 429, 442
Alger, Russell A., 162
Alley, Tom, 277, 277*p*
Allison, William, 193
American Gear and Manufacturing Company, 312
American Motor Car Company, 376
American Motors Corporation, 373, 457–458, 459
Ames Manufacturing Company, 4–5
Anderson, Gilbert, 376, 377, 378*p*, 380, 380*p*, 392
Andrews, Archie M., 82, 304, 327
Anibal, Benjamin J., 261, 262
Annual Gardner Trophy Race, 91
Anthony, Walter M., 344
Auburn-Cord-Duesenberg Company, 296
Auto Poise torsion bar sway eliminator, 369
Auto races
    Duesenberg's interest in, 274–276, 277
    Gray Wolf, 135
    Hudson cars, 352
    Maxwell cars, 343
    Panther by Packard, 193
    Peerless cars, 251*p*, 252–253
    Race Around the World, 204–206, 205–207*p*
    Thomas Flyer in, 199, 200*p*, 203, 204–206
    Times-Herald race, 5–7
    see also Indianapolis Motor Speedway races
Autocar Company
    auto design details and prices (1901-04), 95–97, 96*p*, 97*p*
    auto design details and prices (1905-07), 97–100, 98–100*p*
    commercial vehicles business, 100–101, 101*p*, 104, 104–106*p*, 106
    early autos, 94*p*95*p*
    incorporation and officers, 94–95
    military business, WWII, 104
    ownership changes, 101–103
    service agreement for buyers, 96
    truck photos, 102*p*, 103*p*
Automotive ads, 219–220
Avanti Motors Corporation, 52
Ayers, F.W., 84

Baits, Stuart G., 367–368, 368*p*
Baker, Rauch and Lang, 15
Ball, Frederick and Thomas, 259
Bangham, A.W., 327
Bare, Erwin L., 172
Barit, Abraham Edward, 355, 367, 367*p*, 373, 457, 457*p*
Barno, Peter S., 194
Bates, Donald E., 243
Batten, Norman, 288
Bean, Ashton G., 39
Belden, E.H., 144*p*
Bellanca, Frank, 385
Benny, Jack, 175*p*
Berger, Cliff, 40
Berry, George M., 150
Bethune, John, 242
Bezner, Frederick O., 218, 220
Biggers, H.A., 34
Bliss, C. Hascall, 429, 441
Blue Streak by Graham, 417–418
Bohannon, John A., 268
Booth, James Scripps, 388
Boyce Moto-Meter, 352
Boyer, Joe Jr., 144*p*, 150, 150*p*, 283, 286, 286*p*
Bradley, Thomas M., 329, 330
Brady, James J., 216, 218
Brandt, Arthur J., 327
Bremer, Rodger, 193
Briggs, Clarence E., 194, 196
Briggs Manufacturing Company, 185
Briggs, Walter O., 218
Briscoe, Benjamin and Frank, 218
Brodie, E.H., 193
Brodie, George H., 163, 194
Bronner, H.C., 343
Brown, Neil S., 194
Brownell, Wayne, 194
Budd, B.C., 194
Budd Manufacturing Company, 450
Bugatti engines, 278
Bugatti, Ettore, 278
Buggyaut by Duryea, 9
Buick, David Dunbar, 218
Buick Motor Company, 428
Bulldog Special by Stutz, 381
Bundy, G.W., 60
Burke, D.A., 261, 262
Burman, Bob, 277*p*
Burnside, Frederick W., 327
Burst, Carl, 69, 82, 83
Butterfield, H.K., 297

Cagney, James, 291p
Campbell, Henry, 376, 383
Campbell, Malcolm, 366
Campbell, Ray, 418
Caribbean sport convertible by Packard, 193
Carlson, Willie, 343, 343p, 344
Cermak, Anton, 444p
Chalfant, E.L., 209
Chalfee, Robert M., 267
Chalmers-Detroit Company, 202
Chalmers, Hugh, 202, 220, 344
Chalmers Motor Company, 202, 344
Chamberlain, R.E., 148, 175
Champion by Studebaker, 45
Chandler-Cleveland Motors Corporation, 319
Chanter, Arthur J., 40–41, 119, 125
Chapin, Roy Dikemam, 218
    American Motors chairmanship, 459
    background, 219–221
    death, 367
    Hudson Motor presidency, 355, 364
    photos, 219p, 350p, 355p
    start of Thomas-Detroit company, 200
Chase, Herbert, 144p
Chedru, Gustave, 203
Chevrolet, 87
Chevrolet, Gaston, 282
Chevrolet, Louis, 285p
Christopher, George T., 172, 174, 180p, 180–181
Chrysler Corporation, 416
Churchill, Harold, 52
Clark, Frank G., 213
Clarke, Louis S., 93p
    Autocar start, 93–95
    death, 106
Clifton, Charles, 107p
    background, 107–108
    death, 119
Club Sedan by Packard, 153
Coffin, Howard E., 218
    government appointment, 352
    Hudson Motor vice-presidency, 355
    start of Thomas-Detroit company, 200
    Thomas Flyer redesign, 203
    work at Olds, 220
Cole, R.E., 402
Cole, Rufus S., 324, 327, 328
Collins, Richard, 259–262
Columbia Motors Company, 341
Connolly, E.J., 245
Cooper, Earl, 379, 380
Corbett, James, 144p
Cord Corporation, 288
Cord, E.L., 288, 295
Corum, L.L., 398
Cost of vehicles. *See under individual companies*
Counselman, Lee, 402, 403
Crawford, Charles S., 385
Cugnot, Nicholas, 1
Cunliffe, C.R., 261

Cunningham, E.P.J., 192
Cunningham, Harry, 135
Curtiss-Wright Corporation, 52, 195

Darrin, Howard (Dutch), 78
Dauphin, Howard, 418
Dawley, Herbert M., 115
De Sakhnoffsky, Alexis, 445
DeBoutteville, DeLamarre, 2
Deering, Ray S., 15
DeHautefeville, Jean, 1
DeLorean, John Z., 192, 193, 196
Dension, W.H., 60
DePalma, Ralph, 36, 150, 283
DePaolo, Pete, 287, 287p
Depression's impact on auto companies
    Elkhart, 65
    Gardner Motor, 91
    Hudson Motor, 360–361
    Hupp Motor, 320–321, 330
    LaFayette Motors, 433
    Moon Motor, 82
    Nash Motors, 442
    Peerless, 268–269
    Pierce Arrow, 120–121
    Reo Motor, 237
    Stutz Motor, 398
DeRivaz, Isaac, 1
DeVaux, Norman, 330, 331, 423
DeVlieg, Ray A., 243
Diamond T Motor Car Company, 246–247
Diana Motors Company, 78, 81p
Dictator Six by Studebaker, 37, 38p, 42
Dietrich, Raymond, 35
Dingley, Bert, 387
Dirigibles, 154
Dort, Josiah Dallas, 428
Douglas, Earl M., 194
Drake, J. Walter, 308p
    background, 308
    Chandler-Cleveland and, 319
    government appointment, 316
    Hupp chairmanship, 313
    litigation against Andrews, 328
    purchase of Hupp, 311
    return to Hupp (1927), 318
Drum, Vern R., 328
Duesenberg, August, 274, 285p, 287p
Duesenberg Automobile and Motors Company, 281
Duesenberg, Denny, 294p, 296
Duesenberg, Fred
    background, 274
    death, 293–294
    photos, 273p, 277p, 285p, 287p, 294p
    presidency of Duesenberg, 286
Duesenberg, Fritz, 296, 296p
Duesenberg, Gertie, 276, 276p

Duesenberg Incorporated
    car photos, 290*p*, 291*p*, 292*p*
    celebrity car owners, 290
    death of Fred, 293–294
    model Js, 288–290
    Mormon Meter, 294–295, 295*p*
    organization under Cord, 288
    racing results, 290, 291
    sale of company, 295–296
    supercharged SJ cars, 293
Duesenberg Motors Corporation
    aircraft/marine engines, 277–278, 279*p*, 280, 280*p*
    auto racing interest, 277
    car photos, 285*p*
    financial problems (1923-24), 284–285
    name change, 288
    race car design (1920), 281
    racing successes, 283–284, 286, 287
    speed record set, 282
    straight-eight engines, 280–281, 282, 283*p*
Dunham, George W., 402, 403
Dunn, James H., 286
Durant, Cliff, 380
Durant-Dort Carriage Company, 428
Duray, Leon, 287
Duryea, Charles, 3*p*, 6*p*
    background, 3–4
    post-auto pursuits, 16
    powered buggy prototype, 4*p*, 4–5, 5*p*
    powered vehicle business attempts, 8–9
    Times-Herald race win, 6–7
Duryea, J. Frank, 3*p*, 6*p*
    background, 3–4
    death, 16
    powered buggy prototype, 4*p*, 4–5, 5*p*
    retirement, 13
    Stevens-Duryea and, 9–10
    Times-Herald race win, 6–7
Duryea Motor Wagon Company, 7, 7*p*
Duryea Power Company, 8*p*, 8–9, 11*p*

E-M-F Company. *See* Everitt-Metzger-Flanders Company
Earhart, Amelia, 173, 363*p*
Earle C. Anthony Incorporated, 136
Egbert, Sherwood, 52
Eight-in-line engine cars by Packard, 151, 151*p*
Eldridge, Clarence, 237
Electric Hand gearshift, 366
Electronic fuel injection systems, 259
Elkhart Carriage and Harness Manufacturing Company, 55*p*
    catalog selling initiatives, 54–55
    Elcar model (1916-19), 55–56, 57–59*p*
    horseless carriage prototype, 55
    sale and name change, 60
Elkhart Motor Car Company
    Elcar model (1925-29), 60, 61*p*, 62–63, 62*p*, 64*p*
    financial impact of Depression, 65

Endurance runs
    in Peerless cars, 253
    in Reos, 224, 227
E.R. Thomas Motor Company
    car photos, 208*p*, 209*p*
    end of company, 209
    first powered vehicles, 197–198, 198*p*
    Flyer in auto races, 203, 204–206
    Flyer models (1904-06), 198–200, 199*p*
    product line and prices (1907-08), 202–203, 203*p*, 204*p*
    product line and prices (1909), 207
    reorganization (1911), 207
    taxicab production, 203
Erskine, Albert R., 28, 29, 29*p*, 40
Erskine car by Studebaker, 35, 37–38
Essex Challenge by Hudson, 358
Essex Motor Company, 352
Eugene Mayer and Company, 207, 209
Evans, Dave, 288
Evans, Oliver, 1
Evans, Robert, 275
Everitt, Byron P. (Barney), 218, 334*p*
    background, 334
    exit from EMF, 338
    new partnership with Flanders and Metzger, 27, 340
    qualifications, 335
    retirement, 346
Everitt-Metzger-Flanders Company (E-M-F Company), 335*p*
    exit of Everitt and Metzger, 338
    incorporation and officers, 334–335
    incorporation into Studebaker, 27
    manufacturing set up, 337
    production facilities, 337
    qualifications of principals, 335–336
    touring cars, 336*p*, 338*p*
Ewald, M.C., 125

Fassett, D.F., 224
Faulkner, Roy, 39, 41, 123, 125
Feltes, Nicholas R., 344
Fenn, Francis H., 328
Ferguson, J.R., 183
Fergusson, David, 108, 116
Ferry, Hugh J., 158, 175, 191
Fetch, Thomas, 132
Finley, George G., 209
Fish, Frederick S., 28, 28*p*, 29, 42
Fisher, Carl Graham, 144*p*, 218
Fitness, R.J., 243
Fitzsimons, ?, 209
Flanders Manufacturing Company, 27
Flanders, Walter, 333*p*
    auto company start, 340
    background, 333
    contract with Studebaker, 338
    death, 346
    manufacturing company start, 340

Flanders, Walter *(cont.)*
    Maxwell Motor role, 341–342
    qualifications, 336
    retirement, 345
    takeover of Chalmers, 344
    work at Studebaker, 27
Flint, John F., 344
Fly-ball governor, 251
Flying Cloud by Reo, 233–234, 234*p*
Folly Farm Special by Graham, 418
Forbes, Myron E., 116, 117
Ford, Harry W., 401*p*
    background, 401–402
    presidency of Saxon, 402
Ford, Henry, 404, 405
Fox, C.L., 405
Frazer, Joseph W., 424
Frech, Theodore, 256, 257
Front-Drive car by Gardner, 91
Fry, Theodore I., 246
Fuller, Alvan T., 134

Gable, Clark, 294*p*
Gardner Motor Company
    car photos (1930), 92*p*
    financial impact of Depression, 91
    incorporation and officers, 87
    liquidation, 92
    product line and prices (1920-30), 88–90, 89–91*p*
Gardner, Russell, 85*p*
    auto company start, 87, 88*p*
    background, 85–86
    buggy company, 86, 86*p*
    carriage company start, 87
    Chevrolet business, 87
Gaskin, D.C., 194
Gasoline-engine-powered vehicles, 2
General Electric Company, 257
General Motors Company, 428
General Parts Corporation, 270
General Vehicle Company, 257
George N. Pierce Company
    name change to Pierce Arrow, 112
    Pierce Great Arrow models (1904-09), 109–111, 111*p*
    Pierce motorette, 108–109
    start of company, 107
German, Leon R., 264, 267, 268
Gilman, Max M., 171, 178, 180
Gleason, Jimmy, 289*p*
Gleason, W.L., 209
Glover, Frederick, 245
Goethals and Company, G.W., 116, 117
Gold Bug by Kissel, 300, 302*p*
Golden Hawk by Studebaker, 51, 51*p*
Gordon, Al, 366
Gordon, Charles, 28
Gorrell, Edgar S., 397
Gotfredson, Benjamin, 404

Graffis, A. Michael, 60
Graham Brothers, 407*p*, 425*p*
    auto company start, 411
    death of Ray, 418–419
    deaths of Robert and Joseph, 424
    family farm, 425
    glass business, 407
    truck conversion package, 408, 409*p*, 410
Graham Brothers Incorporated
    contract with Dodge, 410
    investment in Dodge, 411
    real estate holdings, 410–411
Graham, George M., 38
Graham-Paige Motors Corporation
    Blue Streak models, 417–418
    financial status (1930s), 416, 420, 423
    incorporation and officers, 411
    litigation with Chrysler, 416
    new commission on sales, 417
    product line and prices (1928-31), 413–416, 417
    product line and prices (1933-34), 420, 421*p*
    product line and prices (1935-38), 422, 423
    product line and prices (1939-41), 423–424
    racing cars, 418, 419*p*, 420, 421*p*
    Spirit of Motion, 422, 423, 424
Graham, Robert, 331
Graham, W.S., 324
Grant, W.R., 193, 194
Graves, William H., 196
Gray, Gilda, 156*p*
Gray Wolf by Packard, 135*p*, 135
Greater Hudson by Hudson, 358
Greiner, Karl, 187
Greuter, Charles R., 388
Groves, Wallace, 102–103
Gunn, E.G., 148
Gunn, J.N., 28
G.W. Goethals and Company, 116, 117

H. and C. Studebaker, 19–21
Hall, Elbert J., 147, 245
Hall, Ira, 289*p*
Hammond, Eugene, 228*p*
Hampden Automobile and Launch Company, 9–10
Harding, F.I., 256
Harlow, Jean, 167*p*
Harroun, Ray, 343
Hastings, Charles D., 218, 309*p*
    background, 308
    Chandler-Cleveland and, 319
    Hupp chairmanship, 324, 327
    purchase of Hupp, 311
    return to Hupp (1916), 313
Hatch, Darwin, 144*p*
Hatcher, William A., 130, 134
Haupt, Willie, 200*p*, 275
Hayes, Albert B., 242
Hayes, Clarence, 218

H.C.S. Motor Car Company, 383–385, 384*p*
Headlamps by Pierce Arrow, 115
Hearne, Eddie, 282, 380
Heaslet, James G., 29
Henry, Guy, 35
Hershey, Franklin, 192
Highway construction, 143, 220
Hines, H.L., 19, 21
Historical overview of auto industry, 1–2
Hitchcock, Hugh, 178–179, 178*p*, 191
Hoelzle, E.C., 193
Hoffman, Paul G., 34–35, 35*p*, 122
Hoge, W.B., 193
Holley, George and Earl, 218
Howard, Earl C., 432
Hudson, Joseph L., 348, 351
Hudson Motor Car Company, 349*p*
    auto racing, 352, 363, 366
    business success (1924-26), 355–356
    Chapin's return, 364
    end of company, 374
    Essex car (1919-23), 353, 354*p*, 355
    financial impact of Depression, 360–361
    financial status (1920s), 355, 357
    financial status (1930s), 362, 364, 369
    financial status (1940s), 370
    financial status (1950s), 372
    incorporation and officers, 349–350
    military business, Korean War, 372
    military business, WWI, 352
    military business, WWII, 370–371
    new capital (1935), 365
    officers (1923), 355
    partnership with American Motors, 373, 457
    personnel changes (1936), 367–368
    product line and prices (1913-17), 351–352
    product line and prices (1919-23), 353, 354*p*, 355
    product line and prices (1930-31), 359, 360–361, 361*p*
    product line and prices (1932), 361*p*, 361–363, 363*p*
    product line and prices (1935-36), 366*p*, 366–367
    product line and prices (1937-41), 368–370, 370*p*
    product line and prices (1948-51), 371–372
    product line and prices (1954-57), 373*p*, 373–374
    sales declines (1952-54), 372
    Super Sixes (1927-28), 356–358, 357*p*
Huff, Russell, 135, 143
Humpage, F.R., 209
Hunt, Ormond E., 139, 143, 148
Hupmobiles. *See* Hupp Motor Car Corporation
Hupp Comet, 324, 324*p*
Hupp Motor Car Company
    early business success (1909-12), 309–311, 310*p*, 311*p*
    exit of Hupp, 311
    incorporation and officers, 309
    name change, 313
    product line and prices (1915), 312
Hupp Motor Car Corporation
    Aerodyne models (1934), 325*p*, 325–326, 326*p*
    Century Six series (1929), 319–320
    Chandler-Cleveland purchase, 319
    end of auto business, 331
    financial impact of Depression, 320–321, 330
    financial status (1928), 319
    financial status (1932), 324
    Hupp Comet, 324, 324*p*
    incorporation, 313
    litigation and officer changes (1933-35), 327–328
    officers (1932), 324
    product line and prices (1916-25), 313–315, 313–316*p*
    product line and prices (1928), 318–319, 319*p*
    product line and prices (1930-31), 320–321
    product line and prices (1932-33), 321–325, 321–325*p*
    product line and prices (1936-38), 329–330, 330*p*
    racing cars, 324
    sale of body plant, 316–317
    six-cylinder engine cars, 317, 317*p*
    Skylark production, 330–331
    Steeldraulic brakes, 318
Hupp, Robert C., 218, 308*p*
    background, 307
    death, 321
    exit from Hupp, 311
    exit from R.C.H., 312
    start of company, 309
Hurlbut, William R., 328
Hutchinson, R.A., 194
Huygens, Christian, 1
Hydraulic brake systems, 367

Ideal Motor Car Company, 378
Indianapolis Motor Speedway races
    Duesenberg cars in, 283
    Packard cars in, 150
    photos, 275*p*, 276*p*
    Studebaker cars, 39, 40*p*
    Stutz car debut, 377–378
    see also Auto races
Internal-combustion engine, 2
Inverted worm-drive rear axle, 388

J. Stevens Arms and Tool Company, 9–10
Jackson, Edwin B., 397
Jackson, Roscoe B., 218
    background, 347–348
    death, 359
    work at Olds, 220
James, William S., 36
Janney, Walter, 103
Jay, John C., 116
Jeffery, Charles T., 429
Jenkins, Ab, 125, 126, 295
Jewett, Harry Mulford, 411
Johnson, Orlin, 155*p*
Jones, Milton, 398
Jones Stutz Special, 398

Joy, Henry B., 133p
    background, 133
    death, 176
    exit from Packard, 428
    highway association work, 143
    work at Packard, 134
Joy, Richard P., 158
Judge, Arlene, 171p

Kahn, Albert, 134
Kaiser-Frazer Corporation, 424
Keech, Ray, 395
Kelly, William E., 335, 336, 337
Kelsey, John, 218
King, Charles Brady, 215–216, 218, 278
Kishline, Floyd F., 422p, 423
Kissel, Adolph P., 297
Kissel, George A., 297
Kissel Motor Car Company
    body style changes (1928), 304
    Double-Six (1917-18), 299–300
    end of company, 305
    financial status (1927), 303
    growth and expansion, 299
    incorporation and officers, 297
    New Era deal, 304–305
    product line and prices (1908-12), 297–298, 298p
    product line and prices (1913-16), 299
    product line and prices (1917-18), 299–300
    product line and prices (1920-25), 301–303, 302p
Kissel, Otto P., 297
Kissel, William L., 297
Kittredge, Lewis H., 249, 256, 257, 259
Klein, Art, 277p
Klenke, William H., 193
Kohlsaat, H.H., 5
Kollins, Michael, 192p, 276p
Krarup, Marcus, 132
Kreis, Peter, 287
Kummer, Felix A., 163, 185

LaFayette Motors Company
    financial impact of Depression, 433
    first auto (1920), 432–433
    incorporation and officers, 432
Land Cruiser by Studebaker, 42
Laying, George H., 261
Leland, Henry Martyn and Wilfred, 218
Lenoir, Etienne, 2
Liberty aircraft engines, 145p, 147–148
Lichten, Adolph, 329
Lincoln Highway Association, 143
Lindbergh, Charles, 164p
Lockhart, Frank, 393p, 393–395, 394p
Loewy, Raymond, 321, 326, 329
Longaker, V.A., 376

Longenecker, E.D., 187
Lyons, R.P., 324

Mabley, Christopher, 348
Macauley, Alvan, 220
    background, 139–140
    chairmanship, 178
    photos, 129p, 140p, 163p
    work at Packard, 144, 174
Macauley, Edward R., 163, 163p, 194
MacDonald, Stewart, 78, 82
MacDonald, W.A., 329
Macke, W.E., 194
Mahan, Harry B., 286
Manning, Lucius B., 295
Marcus, Siegried, 2
Marine engines. *See* Aircraft/marine engines
Marion Motor Car Company, 376
Marmon-Harrington Company, 296
Mason, Edward R., 274
Mason, George W., 373, 449p, 449–450, 453, 457, 457p
Mason Motor Car Company, 274
Maxon Incorporated, 192
Maxwell, Jonathon D., 218
Maxwell Motor Company Incorporated
    financial status (1918-19), 345
    Flanders' retirement, 345
    Flanders' role in, 341–342
    military business, WWI, 344
    product line and prices (1913-15), 342p, 342–343, 343p
    product line and prices (1918), 344
    racing cars, 343–344, 343p
    takeover of Chalmers, 344
May, Henry, 117
Mayo, William B., 245
Maytag, Frederick I., 274
Maytag-Mason Motor Company, 274
McAdam, George, 205p
McAneeny, William J., 328p, 348p
    background, 348–349
    hiring by Hupp, 328
    Hudson chairmanship, 364
    Hudson presidency, 359
    Hudson vice-presidency, 355
McCarty, Earl H., 441, 442, 444p, 449
McClain, George L., 148
McCormack, O.H., 355
McCulla, W.R., 144p
McFarland, Forest, 193, 196
Megargle, Percy F., 224
Menasco, D.S., 376
Menocal, Jose Estrade, 420
Merz, Charles, 379, 380p
Merz Engineering Company, 391
Metzger, William E., 334p
    background, 334
    exit from EMF, 338
    new partnership with Flanders and Everitt, 340

qualifications, 336
retirement, 346
work at Packard, 134
work at Studebaker, 27
Meyers, Gerald, 459
Miller, George, 205*p*
Miller, Harry, 285*p*
Milton, Tommy, 281, 281*p*, 282, 282*p*, 384*p*
Misch, Herbert L., 192, 193, 196
Miss Americas (speed boats), 155, 155*p*
Mitchell Motors Company, 315
Mitchell, W. Ledyard, 344
Mixter, George W., 116, 117
Mohler, Rebecca, 17, 18*p*
Monahan, Patrick, 192
Mooers, Louis P., 68, 250, 250*p*, 251*p*, 252
Moon Buggy Company, 67–68
Moon, Joseph W., 67*p*
    background, 67
    death, 75
    switch to automobiles, 67–68
Moon Motor Car Company
    car photos (1926), 80*p*, 81*p*
    Diana Motors start, 78
    end of company, 84
    financial impact of Depression, 82
    financial impact of WWI, 74
    incorporation and officers, 68–69
    merchandising plan update, 78, 80
    models and prices (1908-10), 69*p*, 69–70, 70–71*p*
    models and prices (1913-19), 71–75, 72*p*, 73*p*, 74*p*
    models and prices (1920-25), 75–78, 76*p*, 77*p*
    models and prices (1927-29), 81–82
    sales declines (1927), 81
Moore, M.J., 432, 444*p*
Moorehouse, A., 168
Moorehouse, E.A., 148
Mormon Meter by Duesenberg, 294–295, 295*p*
Morton, Wade, 287, 392
Moskovics, Frederick, 386*p*
    background, 386–387
    hiring by Stutz, 387
    resignation from Stutz, 397
Mulford, Ralph, 277
Muller, W.H., 82
Murdock, A. Gordon, 383
Murdock, Samuel T., 383
Murphy, Jimmy, 281, 281*p*, 282, 286
Murray Body Corporation, 316–317

Nance, James J., 51, 191*p*, 191–192, 195
Nash, Charles Warren
    background, 427–428, 447
    management of General Motors, 428
    photos, 427*p*, 444*p*, 446*p*, 447*p*, 453*p*, 454*p*
    purchase of Jeffery company, 429
    start of LaFayette Motors, 432
    work at Buick, 428

Nash-Kelvinator Corporation
    Ambassador cars (1952), 456
    American cars (1956-59), 458–459
    financial status (1930s), 450, 451
    financial status (1940s), 451, 452
    financial status (1950s), 456, 459
    Healy Sports Roadster, 456
    military business, Korean War, 455
    military business, WWII, 452
    partnership with American Motors, 457
    product line and prices (1940-48), 451–453, 452*p*, 453*p*, 454*p*
    product line and prices (1949), 453–454
    product line and prices (1952-54), 456
    product line and prices (450-51), 450–451
    start of company, 449
    Statesman (1952), 456
Nash Motors Company
    Advanced Six car (1926-27), 436*p*, 437, 438*p*
    Ajax Six car (1925), 435–437
    financial impact of Depression, 442
    financial status (1920s), 433, 439, 441
    financial status (1930s), 442, 443, 445, 447
    incorporation, 429
    Light Six car (1926), 437, 437*p*
    merger with Kelvinator, 449
    product line and prices (1918-19), 429–431, 430*p*, 431*p*
    product line and prices (1920-25), 434*p*, 434–435, 436*p*, 437*p*
    product line and prices (1927-29), 438*p*, 438–440, 439*p*, 440*p*
    product line and prices (1930-33), 441, 442*p*, 443*p*, 443–444
    product line and prices (1934-36), 445–449, 448*p*, 449*p*
National Automobile Show, first show, 130
National Bank Holiday, 40
Nehbras, Frederick, 387
Nelson, Emil, 310
New England Westinghouse Company, 13, 15
New Era Motors Incorporated, 82, 83–84, 91, 304–305
Newcomen, Thomas, 1
Noonan, Charles P., 194
Northrup, H.M., 365, 368
Northup, Amos, 238, 318, 326, 329
Norton, Dana, 194

O'Brien, Thomas T., 232
O'Donnell, Eddie, 277, 277*p*, 282
Ohio Automobile Company
    added capitalization from Joy, 133
    car photos, 133*p*, 134*p*
    first model lines (1900-03), 130–132, 131*p*, 132*p*
    name change to Packard, 134
Oldfield, Barney, 252–253, 252*p*, 379
Olds Farm Company, R.E., 245
Olds Motor Vehicle Company Incorporated, 214
Olds Motor Works
    business problems (1907-08), 222

Olds Motor Works *(cont.)*
    company start, 214–215
    curved-dash runabout, 216
    plant fire, 216, 217*p*
    post-fire recovery, 217–218
    return to Lansing, 219
Olds, Pliny and Sarah, 211, 212*p*
Olds, Ransom E., 211*p*, 215*p*, 454*p*
    auto company start, 215
    background, 211, 213
    engine company start, 214
    exit from Olds Motor, 221
    farm company, 235
    post-Reo activities, 245, 245*p*
    Reo Motor start, 222
    retirement from Reo Motor, 232
    return to company, 242–243
O'Madigan, Daniel, 194
Oostdyk, C.A., 368
Orr, Tom, 344
Ostrander, J.P., 193
Owen Motor Car Company, 227–228
Owen, Percy, 403
Owen, Raymond M., 15, 222, 226, 227

Packard Electric Company, 130, 134
Packard, James, 129*p*, 130*p*
    background, 130
    death, 158
    Packard presidency, 134
    retirement, 144
Packard Motor Car Company
    aircraft/marine engines, 154*p*, 154–156, 162*p*, 162–163, 183, 184, 184*p*
    assembly line use, 185–186
    Caribbean sport convertible (1953), 193
    Clipper series (1941-42), 181–183, 181*p*, 182*p*
    Club Sedan, 153
    diesel aircraft engines, 162*p*, 162–163
    eight-in-line engine cars, 151, 151*p*
    eighth series (1930), 164–165
    end of company, 195–196
    fifteenth series (1937), 175–176
    financial status (1913), 143
    financial status (1920s), 149, 152, 153, 154
    financial status (1930s), 169, 170, 172, 174, 176
    financial status (1940s), 186, 187
    first closed model cars, 135
    Gray Wolf, 135*p*, 135
    growth and innovations (1926-28), 156*p*, 156–158, 157*p*
    Liberty aircraft engines, 145*p*, 147–148
    Light Eight (1932), 167–168
    loss of buying public's confidence, 194
    merger with Studebaker, 51–52, 194
    military business, Korean War, 190
    military business, pre-WWII, 180
    military business, WWII, 183–185, 183*p*, 184*p*
    name change from Ohio Automobile, 134
    Nance hiring, 191–192
    nineteenth series (1941), 181
    ninth series (1931), 166–167
    officers (1932), 168
    officers (1945), 185
    120 design (1935), 172–175, 173*p*, 174*p*
    Panther car, 193
    personnel changes (1919), 148
    personnel changes (1952, 1953), 191–192, 193
    personnel changes (1954,1955), 194–195
    post-war recovery, 148–149
    product line and prices (1906-09), 136–139, 137*p*, 138*p*
    product line and prices (1911-14), 140*p*, 140–142, 141*p*
    product line and prices (1922-25), 149–154, 150*p*, 151*p*
    product line and prices (1929), 15*p*, 159
    product line and prices (1930), 160*p*, 160–161, 161*p*
    product line and prices (1932-34), 169–171, 169–172*p*
    product line and prices (1938-40), 176–180, 177*p*, 178*p*, 179*p*
    product line and prices (1947-49), 187*p*, 187–189, 188*p*
    product line and prices (1951-52), 189*p*, 189–191, 190*p*, 191*p*
    product line and prices (1952-53), 192*p*, 192–193
    product line and prices (1956-58), 195, 195*p*, 196*p*
    Safety-Flex suspension use, 173
    Single Six series, 152
    six-cylinder engines, 140–141
    truck production, 136, 136*p*
    Twelve Series (1932), 168, 170
    Twin Six models (1915-19), 144–146
Packard, William, 130*p*, 134, 152
Packer, William M., 175, 178
Page, Robert P., 103
Paige-Detroit Motor Car Company, 411
Palmer, Herman, 297
Panther by Packard, 193
Papin, Dionysius, 1
Parkhurst, E.H., 256
Paton, Clyde R., 163, 168, 183
Pearlescent shade finish, 240
Peerless Manufacturing Company, 249–250
Peerless Motor Car Company
    early businesses, 249–250
    financial status (1909), 254
    financial status (1912), 256
    first autos (1903-04), 250–252, 251*p*, 252*p*
    military business, WWI, 256
    name change, 257
    personnel changes (1912), 256
    product line and prices (1905-12), 253–256, 254*p*, 255*p*
    racing success, 251*p*, 252–253
Peerless Truck and Motor Corporation
    financial impact of Depression, 268–269
    financial status (1919-21), 259
    financial status (1926-27), 266
    financial status (1930), 270
    General Parts deal, 270
    hiring of VerLinden, 264
    litigation against Collins, 261
    officers, 257

    product line and prices (1916-24), 258–259
    product line and prices (1923-25), 262–263, 263p
    product line and prices (1926-27), 264–266, 265p, 266p
    product line and prices (1928-29), 267p, 267–268, 268p
    product line and prices (1930), 269
    reorganization under Collins, 260, 261
    switch to ale, 272
    V-16-cylinder engine use, 271, 271p
    V-8 engines, 257–259, 258p
Peerless Wringer Company, 249
Pelletier, E. LeRoy, 335, 336, 340
Pelton, Carl H., 344
Penberthy Injector Company, 259
Penrose Trophy, 363
Petit, W.S., 28
Pfeffer, C.A., 405
Pierce Arrow Motor Car Company
    car photos, 112p
    control by Studebaker, 37
    dealer representation problems (1932), 123
    end of company, 127
    financial impact of Depression, 120–121
    financial status (1918), 116
    financial status (1932), 40–41
    financial status (1933), 124
    headlamp innovation, 115
    management overhaul (1919), 116–117
    merger with Studebaker, 119
    military business, WWI, 115–116
    moderately priced car line (1924-28), 117–119, 118p
    multi-cylinder engine use, 122
    name change from George N. Pierce Company, 112
    personnel changes (1933), 125
    product line and prices (1910-11), 112–113
    product line and prices (1911-14), 113–115, 114p, 116p
    product line and prices (1921), 117
    product line and prices (1929), 119–120
    product line and prices (1930-31), 120p, 120–121, 121p
    product line and prices (1932), 122–123
    product line and prices (1935-37), 126p, 126–127, 127p
    publicity stunts, 125, 126
    Silver Arrow model (1933-34), 123p, 123–126
    truck business, 113
    truck production (1931), 121–122
Pierce Company, George N.. *See* George N. Pierce Company
Pierce Cycle Company, 110
Pierce, George N., 107, 113
Pierce Great Arrow models, 109–111, 111p
Pierce motorette, 108–109
Pittsburgh Motor Vehicle Company, 92–94
Polychromatic finishes, 240
Porter, John F., 261
Powers, Roy P., 194
Poxson, Elijah G., 237, 243
Pratt, Frederick B., 53–54
Pratt, William and George, 54, 60
President Eight by Studebaker, 36, 36p, 37p
Prince and Whitely, 101–102
Pullen, Eddie, 277p

Race Around the World, 204–206, 205–207p
Raisbeck, John, 193
Ramsey, J.J., 209
Rasmussen, William T., 285, 286
Rathman, Dick, 193
Rawson, R.A., 387
R.C.H. Corporation, 311–312
R.E. Olds Farm Company, 245
Rebadow, Adolph, 209
Redden, Charles F., 343
Rees, Russell R., 193
Reifel, George, 193
Rengers, F.H., 91
Reo Ideal Lawn Mower Company, 245
Reo Motor Car Company
    end of company, 245–246
    endurance runs, 224, 227
    financial impact of Depression, 237
    financial status (1914-17), 229
    financial status (1926), 233
    financial status (1928), 236
    financial status (1931), 240
    financial status, post-WWI, 232
    Flying Cloud (1927), 233–234, 234p
    management changes (1934), 242–243
    military business, WWI, 230–231
    Olds' retirement, 232
    product line and prices (1905-08), 223p, 223–224, 224p, 225p
    product line and prices (1909-10), 226, 226p, 227, 227p
    product line and prices (1911-16), 228–229, 230
    product line and prices (1925-26), 232p, 232–233
    product line and prices (1928-29), 236p, 236–237, 237p
    product line and prices (1931-32), 238p, 238–242, 240p
    product line and prices (1935), 243–244, 244p
    Royale (1931), 238, 238p
    Self Shifter transmission, 242
    Speed Wagon, 230
    start of company, 222
    truck production, 229, 232, 243, 244–245
    Wolverine (1927), 234–235, 235p
Reo Motor Truck Company, 227
Reo Motors Incorporated, 246–247
Resta, Dario, 150
Rickenbaker, Edward V., 344
Ricker, Chester, 144p, 285
Rinehart, John, 192
Robb, J.W., 245
Roberts, Montague, 205p
Robson, Frank, 227
Rockne, Knute E., 38
Rockstell, Glover, 277p
Rodgers, O.E., 193
Roehm, G.E., 327
Roland, Harold A., 237
Rolls-Royce aircraft engines, 180, 184
Romney, George, 373, 453, 457, 457p, 459
Rooney, Tom, 392
Roos, D.G., 35–36, 36p

Roosevelt, F.D., Jr., 172*p*
Rose, John A., 429, 437
Royale by Reo, 238, 238*p*
Rueschaw, Robert C., 230, 232, 233
Russell, Henry, 215, 221
Ruthrauff and Ryan Incorporated, 192
Ruxton Front-Drive by Moon, 82, 304, 305*p*
Ruxton, William V.C., 82

Safety-Flex suspension, 173
Sakhnoffsky, Alexis de, 270
Salzman, George, 203
Saxon Motor Car Company
    early autos and expansion, 402–403
    incorporation and officers, 402
    name change, 403
Saxon Motor Car Corporation
    business problems (1917-19), 404–405
    car photos, 404*p*
    end of company, 405
    incorporation and officers, 403
    reorganization, 405
Scafe, Lincoln R., 403, 404
Schmidlapp, Charles J., 383
Schmidt, Charles, 134–135, 135*p*, 257
Schmidt, William, 194
Schmunk, Robert J., 261
Schribner, Charles D., 194
Schroeder, Eric, 277*p*
Schultz, Glenn, 391*p*
Schuster, George, 205*p*
Schwab, Charles M., 383
Scott, Richard H., 222, 232, 237, 240, 242, 243
Sears, Fay B., 60
Self-propelled land vehicles, 1
Self Shifter transmission, 242
Selfridge Field, 147
Sergardi, Fabio, 232, 233, 239
Shafer, Phil, 287
Shaw, Wilbur, 366
Sheppy, Charles L., 108, 117, 118
Silver Bug by Kissel, 300
Single Six series by Packard, 152
Skinner Oil Rectifying system, 153
Skylark by Hupp, 330–331
Slack, Frederick W., 134, 257, 259, 261, 262, 269
Slack, Lyman W., 172, 175, 187
Smith, Angus, 221, 227
Smith, Don P., 268
Smith, Earl H., 172
Smith, Fred, 215, 221
Smith, Paul, 340
Smith, Samuel L., 214, 221
Snowberger, Russell, 324
Souders, George, 288, 288*p*
Sparrow, Edward, 215
Speed Wagon by Reo, 230
Speedabout by Hudson, 359

Spencer, LeRoy, 191
Spirit of Motion by Graham, 422, 423, 424
Spring, Frank S., 361
Standard Motor Company, 341
Stapp, Egbert, 291*p*, 366
Startix automatic starter, 360
Steam tractors, 1
Steeldraulic brakes, 318
Stevens-Duryea Incorporated, 15*p*, 15–16
Stevens-Duryea vehicles
    business status during WWI, 12–13
    car photos, 12*p*, 13*p*, 14*p*
    product line and prices, 9*p*, 10–12
Stevens, Jay, 9
Storrow, James, 428, 429, 432, 438
Strickland, William R., 257
Studebaker brothers, 17*p*
    background, 17–19
    see also H. and C. Studebaker
Studebaker, Clement, 23
Studebaker, Clement Jr., 28, 29
Studebaker Corporation
    Big Six, Special Six, Light Six (1920-25), 32–34, 32–34*p*
    business problems during Korean War, 50
    business success (1929), 37
    Champion model, 45
    E-M-F and, 27, 27*p*, 338
    eight-cylinder engine design, 35–36
    end of company, 52
    financial loss (1932), 39
    incorporation and officers, 27
    merger with Packard, 51–52, 194
    merger with Pierce, 119
    military business, WWI, 31, 31*p*
    military business, WWII, 47*p*, 48
    personnel changes (1911-12), 28
    Pierce Arrow acquisition, 37
    product line and prices (1913), 28*p*, 28–29
    product line and prices (1914-16), 29, 29*p*
    product line and prices (1917-19), 29, 29*p*
    product line and prices (1930-32), 37–39, 38*p*, 39*p*
    product line and prices (1934-39), 41*p*, 41–44, 42*p*, 44*p*
    product line and prices (1939-41), 45–47, 46*p*
    product line and prices (1946-51), 48–49
    receivership (1933), 40
    recovery and expansion (1934-36), 41–42
    Rockne model, 38
Studebaker, John, 17, 18*p*
Studebaker, John Mohler II, 21
Studebaker, John Sr., 19, 21*p*, 28, 29
Studebaker Manufacturing Company
    car photos, 24–26*p*
    carriage photos, 22*p*, 23*p*
    electric vehicle manufacturing, 23
    incorporation and officers, 21
    move into auto manufacturing, 22–23, 27
    organization into Studebaker Corporation, 27
Stutz Auto Parts Company
    Bearcat product line, 381
    car photos, 378*p*, 379*p*, 380*p*

early sales success (1911-12), 378–379
Indianapolis 500 debut, 377–378
name change to Stutz Motor, 381
racing successes, 379–380
start of company, 376–377
Stutz-Bellanca Airplane Corporation, 385
Stutz, Harry C., 375p, 384p
auto parts company, 376–377
background, 375–376
death, 385
H.C.S. start, 383
resignation, 382
work at Marion, 376
Stutz Motor Car Company
business problems after buyout, 385–386
car photos, 395p
decline in sales (1927), 390
end of company, 400
financial impact of Depression, 398
hiring of Moskovics, 387
management failures (1928-29), 396
Moskovics' resignation, 397
product line and prices (1912-16), 381
product line and prices (1917-20), 382
product line and prices (1929), 396, 397p, 398
product line and prices (1931-33), 398–400, 399p
racing losses, 395–396
racing successes (1928), 392
Ryan's gain of control, 382
Ryan's stock manipulation attempt, 382–383
Schwab takeover, 383
speed record attempt, 393–395
Splendid Stutz (1928), 391–392
stock car racing, 391, 391p
Stutz's resignation, 382
Vertical "8" (1926), 388–390, 389p
Swegles, I.B., 364, 368

Taft, William, 153
Talcott, John C., 118
Taylor, Seneca G., 84
Teague, Richard A., 192, 194, 196
Teel, Harry, 243
Terraplane by Hudson, 362, 364
Terry, F.S., 257, 259
Tetzlaff, Ted, 342p, 343
Thayer, Eugene R., 383
Thomas B. Jeffery Company, 429
Thomas-Detroit Company, 200, 202
Thomas, E.R., 197p
association with Thomas-Detroit Company, 200
exit from company, 207
Thomas, Horace T.
chief engineer position, 243
move to Reo, 222
proxy battle at Reo, 245
self-shifter development, 242
work at Olds, 232

Thomas, Joe, 277p
Thomas Motor Company. *See* E.R. Thomas Motor Company
Thompson, Otis B., 54
Thompson, William N., 385
Times-Herald race, 5–7
T'jarda, John, 330, 423
Tone, Frederick I., 376
Toner, Thomas J., 344
Tower, Jack, 275
Tracey, Joseph, 252
Tracy, W.R., 368
Travelodge by Pierce Arrow, 127
Trego, Frank H., 143, 144p
Tremaine, B.J., 257, 259
Trester, F.A., 261, 267
Trevithick, Richard, 1
Triphagen, Clarence, 232, 233, 243
Trippenzee Body Company, 341
Trucker, Charles A., 267
Trucks, Packard, 136p, 136
Twin Six by Packard, 144–146

United Automobile Workers, 43, 176
United States Motor Company, 341

V-16-cylinder engines, 271
Vail, J.A., 343
Vance, Harold S., 38p, 38, 51
VerLinden, Edward, 264p, 264–265, 267
Vincent, Jesse G., 143, 143p, 147, 183, 191
Vivian, André, 1
Volvo-White Motor Truck Corporation, 106
Von Bachelle, Otto, 310
VonRosen, E.E., 404
Voorhis, Charles B., 429

Wadsworth, George, 257
Wahlberg, Nils Erik, 429, 444p
Walburn, Thomas, 335, 337
Waldon, Sidney D., 132, 134, 139, 144p
Walker, Helm, 82
Ware, Marsden, 183
Warner, B.H., 125
Warner, John A.C., 36
Watt, James, 1
Watts, Frank E., 310
Weber, Orland F., 344
Weinhardt, R.A., 243
Weiss, Erwin A., 172, 195
Weiss, George, 130, 134
Wells, Vic, 277p
Werner, Freidrich, 297
White, D. McCall, 432
White Motor Company, 39, 106, 246–247

Whitman, L.L., 227, 228p
Widman, Edward, 286
Wilcox, Howdy, 379, 380, 384p
Wilkie, Tom, 200p
Williams, Harrison, 257
Willys, John N., 108
Wilson, Charles B. and David, 218
Wilson, E.B., 261
Wilson, James, 429, 444p
Wilson, William R., 237, 240
Windsor White Prince by Moon, 82
Wise, K.M., 125
Wollering, Max F., 28, 35, 360, 360p, 364
Wood, Gar, 155p, 155
Wood, Walter O., 246
Woolson, Lionel M., 163, 164p
Wray, Faye, 157p
Wridgeway, Charles, 252
Wrye, Walter C., 116

Young and Rubicam Incorporated, 191
Young, DuBois, 309p
    background, 308–309
    hiring by Hupp, 312
    leadership of Hupp, 317, 318, 319, 324, 327

Zengel, Len, 378p, 379, 379p
Zero Milestone, 143
Zweiner, Wallace, 328, 329

About the Author

Mr. Michael J. Kollins was born on March 20, 1912, in St. Clairsville, Ohio, the sixth of seven children of Michael A. and Marian (Peck) Kollins. His father held franchises in that rural area of southeast Ohio to sell and service Buick automobiles in 1910, and later several other makes over several decades. Michael Kollins learned of automobile manufacturers early, literally from reading hubcap designations in his father's shop. Chores included cleaning automotive parts, scraping and installing bearings and grind valves as necessary. He also ordered, stocked, and sold parts. His father promoted wide reading of industry magazines of the early period. Michael gained "on the job" training in the automotive industry, intensely aware of many service capabilities given design and manufacturing concerns over quality matters.

Michael and Julia Kollins.

Mr. Kollins has had a unique and long employment history with the industry. As economic times were difficult in his local area, he moved to bustling Detroit in his late teens to join an older brother. He soon became a "co-op" engineering student in the late 1920s at the College of the City of Detroit, now Wayne State University. He began work as a service technician at the Dodge Brothers, then, still a "co-op" student, moved to the Packard Motor Car Company in 1930 as a test driver and engineering analyst of Packard and competitive models, a position that gave him knowledge of most of the cars manufactured in America. He continued at Packard after graduation in 1932, rising to be Chief of Section, Technical Data and Service Engineering from 1945 to 1955. With the demise of Packard, he joined Chrysler in similar activity: He was Service Technical Manager (1955), Manager of Warranty Administration (1964), and Manager of the Highland Park Service center (1968 to retirement in 1975).

Soon after undertaking formal engineering studies, Mr. Kollins became a local car racer on weekends. That experience, in brief, evolved into a long association with the Indianapolis Motor Speedway. Of note, he became Technical Inspector, Vice Chairman of the Technical Committee, Chairman of the Product Certification Committee, and later, Honorary Life Member of the Indianapolis 500 Oldtimers Club in 1952.

In World War II, Michael Kollins served in the U.S. Navy in the Motor Torpedo Boats Squadron 4. He was a Chief Machinists Mate, and later Ensign with engineering assignments. After the war, he continued in the Naval Reserve, retiring in 1972 as Lieutenant Commander.

For almost a half-century, Mr. Kollins has been a member of the Society of Automotive Engineers and the Engineering Society of Detroit. In the early 1980s he became a trustee with the Detroit Public Library National Automotive Collection, writing articles for its quarterly newsletter, *Wheels*. He wrote technical publications for the Navy, Packard, and Chrysler. In particular, he was co-author of a centennial publication of the Engineering Society of Detroit, *ESD Technology Century* (1998). He also is a historian of many automotive developments in the Detroit area.

Mr. Kollins married Julia Advent in 1934 and has two sons and a daughter and three grandchildren. Noticing his intense research in the late 1980s, his older son encouraged him to write this book on the engineering history of the automotive industry, pointing out that he was the logical person to record such happenings through his personal contacts and assocations over the years with several pioneers, executives, and suppliers.

DATE DUE